Cultural Landscapes in Transition: Balancing Heritage, Development, and Sustainability

Cultural Landscapes in Transition: Balancing Heritage, Development, and Sustainability

Guest Editors

Isabel Vaz De Freitas

Fátima Matos Silva

MDPI

Basel • Beijing • Wuhan • Barcelona • Belgrade • Novi Sad • Cluj • Manchester

Guest Editors
Isabel Vaz De Freitas
Department of Tourism,
Heritage and Culture
Portucalense University
Porto
Portugal

Fátima Matos Silva
Tourism, Heritage and
Culture Department
Portucalense University
Porto
Portugal

Editorial Office
MDPI AG
Grosspeteranlage 5
4052 Basel, Switzerland

This is a reprint of the Special Issue, published open access by the journal *Land* (ISSN 2073-445X), freely accessible at: https://www.mdpi.com/journal/land/special_issues/G11CC339T0.

For citation purposes, cite each article independently as indicated on the article page online and as indicated below:

Lastname, A.A.; Lastname, B.B. Article Title. *Journal Name* **Year**, *Volume Number*, Page Range.

ISBN 978-3-7258-7366-1 (Hbk)
ISBN 978-3-7258-7367-8 (PDF)
https://doi.org/10.3390/books978-3-7258-7367-8

Cover image courtesy of Sofia Matos Silva

Contents

About the Editors

Isabel Vaz De Freitas

Isabel Vaz de Freitas is a Full Professor at Universidade Portucalense. She completed her Aggregation in Medieval History in 2016 and earned her PhD in History in 2001, both from the Faculty of Arts and Humanities of the University of Porto. She is a senior researcher affiliated with REMIT (Research on Economics, Management, and Information Technologies) at Portucalense University. She serves as a Corresponding Member of the Portuguese Academy of History, and she is the Director of the Department of Tourism, Heritage, and Culture, as well as the Director of the Conservation and Restoration Clinic at Portucalense University. She represents the Ministry of National Defense on the National Council of Culture (SEPPAAI) and is a member of the Military Higher Education Council (CESM), appointed by the Minister of National Defense within the Government of the Portuguese Republic. She coordinates several study cycles in the areas of Heritage and Tourism at Universidade Portucalense, and she has supervised numerous undergraduate theses, internships, and master's dissertations. She has also served as a member of dozens of master's and doctoral dissertation juries in History, Tourism, and Heritage. As an integrated researcher in research centers, she has organized multiple national and international conferences. She is an editor and serves on the editorial and scientific boards of international indexed journals specializing in History, Heritage, and Tourism. She has coordinated numerous research projects in Heritage, Culture, and Tourism in collaboration with several European countries, the United States, and Brazil. She has contributed to territorial development projects with municipal authorities and local organizations, and she has participated in several funded research projects in Portugal and Spain. Her publications include books, book chapters, and articles, with a focus on territorial conservation and enhancement, water management, border studies within the Peninsular context, and cultural landscapes.

Fátima Matos Silva

Fátima Matos Silva holds a PhD from the University of Granada since 2008, with an associate title of European Doctorate, having been a fellow of the doctoral program PRAXIS XXI of FCT. Since 1988, she has worked at Portucalense University as a professor, researcher and coordinator of study cycles, integrating other bodies, such as the Pedagogical and Scientific Councils. She has extensive experience coordinating national and international research projects and teams. She is a senior researcher affiliated with REMIT (UPT) and CITCEM (FLUP) and has directed archaeological excavations, impact studies, heritage enhancement projects, publications, exhibitions and museum programs, collaborating with various local and national entities related to culture and cultural heritage. She is the author of books, chapters and scientific articles (mostly indexed), focused on archaeological and cultural heritage; has participated in scientific committees, organized international congresses, and reviewed articles for scientific journals. She is also an editor of publications and Special Issues in indexed international journals. She coordinates the Cultura@Portucalense project, under the tutelage of the Rectory of UPT, promoting cultural dissemination in line with the United Nations SDGs.

Preface

The Special Issue *"Cultural Landscapes in Transition: Balancing Heritage, Development, and Sustainability"* was designed to deepen scientific reflection on cultural landscapes as dynamic systems arising from the interactions among natural processes, social practices, and territorial transformations. In a context marked by rapid environmental, social and economic changes, it is becoming increasingly relevant to understand how these landscapes can be preserved, valued and adapted to contemporary needs without compromising their historical and cultural integrity.

The scope of this Special Issue focuses on the analysis of the transformations of cultural landscapes in different geographical and cultural contexts, addressing topics such as heritage management, land use change, the impacts of climate change, urban regeneration, the relationship between cultural identity and ecosystem services, as well as the role of communities in the construction and transmission of heritage values. Among the published contributions, the article "Identifying Coastal Heritage Vulnerabilities: The Case of Historical Fortified Structures in Northern Portugal" stands out, which analyzes the vulnerabilities of the coastal heritage of northern Portugal in the face of climate change and anthropogenic pressure, highlighting the need to integrate risk assessment and adaptation strategies into heritage management policies. Equally relevant is the study "Linking Ecosystem Services, Cultural Identity, and Subjective Wellbeing in an Emergent Cultural Landscape of the Galápagos Islands", which explores the relationship between ecosystem services, cultural identity, and the well-being of local communities, demonstrating how emerging cultural landscapes can play a central role in the construction of social and environmental values.

The motivation for this Special Issue arose from the growing need to promote dialogue among different areas of knowledge—including territorial planning, landscape architecture, geography, heritage studies, and environmental sciences—that can contribute to a more integrated understanding of changing cultural landscapes. The diversity of studies presented from Europe, Asia, Latin America, the Middle East, and Oceania highlights the multiplicity of challenges and opportunities associated with the sustainable management of these landscapes.

This set of articles offers relevant contributions for researchers, practitioners, policymakers, and students interested in the preservation and management of cultural landscapes, as well as in developing strategies that integrate heritage, social, and environmental values. By bringing together different perspectives and methodologies, this Special Issue aims to stimulate new approaches and strengthen the international debate on the future of cultural landscapes in an ever-changing world.

Finally, the guest editors express their sincere gratitude to all the authors who contributed to this Special Issue, as well as to the scientific reviewers for their rigor, dedication, and valuable suggestions that helped raise the quality of the publications. A special thanks is also extended to the Section Managing Editor and all the editorial staff for the editorial follow-up and for the continuous support and professionalism shown throughout the editing and publication process.

Isabel Vaz De Freitas and Fátima Matos Silva
Guest Editors

land

MDPI

Article

Approaches to Collective Cognition in the Historic Centre of Madrid: An Erasmus Interdisciplinary Experience

Mónica Alcindor [1,2,*], Waltraud Müllauer-Seichter [3], Sonia De Gregorio Hurtado [4], Leonor Medeiros [5], Mirella Loda [6] and Delton Jackson [7]

1 CIAUD-UPT—Branch of CIAUD Research Center, Departamento Arquitetura e Multimédia Gallaecia, Universidade Portucalense Infante D. Henrique, Rua Dr. António Bernardino de Almeida, 541, 4200-072 Porto, Portugal
2 CIAUD, Research Centre for Architecture, Urbanism and Design, Lisbon School of Architecture, Universidade de Lisboa, Rua Sá Nogueira, Polo Universitário do Alto da Ajuda, 1349-063 Lisboa, Portugal
3 Departamento de Antropología Social y Cultural, Facultad de Filosofía, Universidad Nacional Educación a Distancia, Paseo Senda del Rey 7, 28040 Madrid, Spain; wmullauer@fsof.uned.es
4 Departamento de Urbanística y Ordenación del Territorio, Escuela Técnica Superior de Arquitectura, Universidad Politécnica de Madrid, 28040 Madrid, Spain; sonia.degregorio@upm.es
5 CHAM—Centro de Humanidades, Departamento de História, Faculdade de Ciências Sociais e Humanas, Universidade Nova de Lisboa, 1069-061 Lisbon, Portugal; leonormedeiros@fcsh.unl.pt
6 Dipartimento di Storia, Archeologia, Geografia, Arte e Spettacolo (SAGAS), Università degli Studi di Firenze, 50121 Firenze, Italy; mirella.loda@unifi.it
7 Studio UrbanArea LLP, Newcastle upon Tyne NE1 1EW, UK; delton@urbanarea.co.uk
* Correspondence: monicaalcindor@upt.pt

Abstract: Beyond their direct use, buildings and heritage places are objects and settings which help to guide community actions. Cognitive perception systems interact directly with the built environment through action and generate experiences that will be used for subsequent actions. This requires a reorientation towards phenomenological perspectives that query the conceptual boundary between cognition and action. Five universities from three countries (Portugal, Italy, and Spain) came together in July 2023 through an Erasmus+ BIP (Blended Intense Programme) experience, developed for the La Latina neighbourhood, in the historical centre of Madrid. The intention was to highlight the importance of different disciplines, and interdisciplinary working, for planning an urban future which includes the goals of socio-economic and environmental sustainability, happiness, and the right of residents to maintain longstanding emotional connections with their neighbourhoods. The novelty of this experience compared to existing Master's and PhD programmes in Europe was the early and intense contact of students with the subject through the development of fieldwork over two weeks. This was led by teachers from different disciplines to provide interdisciplinary perspectives for a training programme which included architecture, urbanism, urban anthropology, geography, history, and archaeology. Through this training, the intended outcomes were twofold: to equip students with the necessary knowledge and criteria to critically address these issues and to raise awareness among local stakeholders about the negative transformations affecting historic centres and their impact on residents' quality of life.

Keywords: Erasmus+; built heritage; heritage communities; La Latina; gentrification; Madrid; collective cognition

1. Introduction

Prior to the advent of late modernity, the preservation and safeguarding of material heritage relied on the integration of heritage into community consciousness [1]. The importance of engaging local communities in this process has also been emphasized by the Charter for the Conservation of Historic Towns and Urban Areas [2]. Furthermore, active community participation has played a crucial role in supporting tourism development and managing heritage conservation efforts [3,4]. This is because a strong sense of belonging and comprehensive local knowledge enables communities to coexist harmoniously with the unique assets of their regions, which are increasingly appreciated for their distinctiveness [5]. The Faro Convention (COE 2005) was instrumental in explicitly introducing the concept of "heritage communities" and advocating for a people-centred approach to heritage [6]. Nowadays, however, the globally dominant economic ideology of neoliberalism opposes the interests of local people and favours those of speculators and bureaucrats, subjugating the lives of residents to impersonal imperatives under the guise of improvement and development [7,8].

This paper presents an experience of teaching (supported by an Erasmus+ programme) that was focused on understanding the importance of an interdisciplinary approach when studying a city. This study aims to present the goals of this Erasmus+ experience, which sought to understand the relationship between the different roles of neighbourhood places over time and how the emotional bonds associated with them were developed by residents. Particular attention is given to the effects of touristification, gentrification, and the intensity of leisure activities on transforming the social tissue and essence of places and the associated impacts on their historical communities (understood as land-, use-, and place-based associations in the traditional sense). This is founded on the understanding that knowledge is polyhedral and enriched by the confluence of different perspectives, which in turn strengthen joint action by communities. This approach enables a consideration of the different perspectives which are present in public spaces, providing a broader view of reality. The experience implied that a re-evaluation of practices which might normally be taken for granted should be conducted, as well as one of norms and management styles.

Five universities from three countries (Portugal, Italy, and Spain) participated in this Erasmus+ BIP, whose lecturers brought an interdisciplinary perspective to the training. The aim was to put into practice the combination of the methodological logics of architecture, urban planning, urban anthropology, and other relevant social and human sciences. To this end, it was proposed to work on the historic centre of Madrid, as this was a suitable area for the application of this methodological logic. This association made it possible to guarantee learning results by investigating questions such as convergences in space and time, the speeds and temporalities of the city, permanence and reinvention, and the narratives and symbolic values that underlie the identities of a neighbourhood. This concept was then tested through a specific case study, an examination of Madrid's La Latina neighbourhood.

Students were first taught to examine without prejudice the rationality behind certain activities that take place in a given community, in order to avoid them becoming discontinuity agents, and to be able to proceed with an awareness of existing dynamics before undertaking action in any place or location.

Although experiences like this have already been undertaken and analysed (see [9,10]), there are still few experiences that combine architecture and social sciences, i.e., the tangible and the intangible when studying cities in an intensive period. Most of the scholars investigating urban matters with the assistance of Erasmus students use them as social actors to understand urban dynamics [9,11] rather than focussing on the educational experience itself. However, there are some studies that have highlighted the interdisciplinary approaches on the study of the city [12] and also the importance of its didactical dimension [13].

This paper is organized into four main sections. In the first, the question is explored of how the connections of the local community mind work, in particular with regard to the cognitive processes of bonding with places. In the following section, the neighbourhood place aspects of the area of study are examined before moving to the next, where the methodology used in the Erasmus+ experience is presented. The fourth section presents an analysis of the experience, provided by the outcomes of the students who participated across their different fields of knowledge together with a critical analysis by the teaching staff involved. The final thoughts are presented at the end.

2. Theoretical Framework

The connection of individuals with the built environment is more than simply a function of experience, but it also represents how people construct their own identities, through a mixture and interaction between their past histories and place or places of residence. The images and forms of heritage environments generate feelings of security and trust known as ontological security; that is to say, they maintain trust (in the form of confidence and belief) in the continuity of the past, present, and future while connecting this trust with routine social practices [14–17]. In this way, the physical environment continues to embody and represent the memories of the community. These collective memories vary in people's individual memories according to gender, age, and education or training [18]. As Debray emphasizes, there is an indissolution of matters considered material and immaterial [19]. The material level of form transmitting is necessary to inform the physical world by manufacturing consultable stores of externalized memory through available technologies, such as conserving, inscribing, inventorying, and distributing recorded traces of cultural expression. In architecture, these are usually conveyed through engravings, reliefs, and paintings on façades and by the buildings themselves [19]. Meanings are created as the direct products of a continuing cycle between humans and the environment [20]. A nodal point appears where materialized organization and organized matter interact [19], and the built environment becomes corporeally absorbed into our behaviour in the pre-reflexive and mechanically repetitive forms of cultural praxis. It is imposed on us in a materially objectivized form by our spaces and buildings, which have become a compendium of meaningful signs and symbols (acting as orientation schemata [21]) and which incorporate routinized networks. In other words, the cultural construction of space comprises not only organized space, but in a real sense, it also represents the spatial relationships between objects and their relative positions. At the same time, the inclusion of subjects implies that there is information about those using and occupying those spaces as well. It is not about simple associations. Instead, they consist of cultural schemata that are both process and structure; that is, they combine the organization and relative disposition of physical elements with the approach, sequence, and succession of behaviours or elements in relation to an end [21].

Cultural practices have the potential to elicit feelings of "being at home" while generating a sense of community [22]. They are symbolic actions that transmit and represent the values and sense of order that keep a community cohesive, generating a community without communication. They transform "being in the world" into "being at home" [23]. Conducting things in unison carries the message that the community exists [24] and maintains control of its environment. In fact, studies have demonstrated that ritual actions can increase the feelings of control when people are faced with dilemmas or uncertain situations [25,26]. According to Yin Keli [27], cultural practices directly enhance the perceptions of control, while symbolic meaning also enhances perceived control through the mediation of positive emotions. In fact, these meanings and practices promote more participatory experiences while enhancing the enjoyment and pleasure of consumption [28], imbuing it with a sense of comfort derived from practices which remind individuals that they belong to something that is larger than themselves [29].

Therefore, through the built environment, materialized organizations (such as institutions) can set out and establish communitarian arrangements, namely the diverse forms of group cohesion that bring together the human agents of particular transmissions [19]. Paradoxically, the explicit identification or classification of "heritage environments" represents a fetishized memory of material forms, presenting a facticity which causes us to overlook the origins, processes, and activities of how and why these environments were built in the first place [19]. This becomes visible in narrative testimonies, when people use a passive construction that indicates an official history [30,31].

In order to understand the range and complexity of the qualitative functions that public spaces should offer in today's society, we must also consider the new functions of urban spaces. These urban spaces are increasingly important as a form of therapy, as a "shock absorber", and as a place of social encounter to channel and mitigate the growing potential for aggressiveness and depression for a series of groups in our society, in relation to hard-to-solve problems such as unemployment, immigration, the increasing proportion of the elderly in European cities, and the increasing problem of unwanted loneliness in urban areas. These groups are not homogeneous, but they are united by some characteristics warranting attention, in particular a surplus of free time coupled with limited economic resources [18]. The observations made show that analysing the treatment of free time in the social sciences is obligatory; it must be analysed both in its traditional form, as recounted in the testimonies of interviews with local residents, and by working at the point where society starts to stigmatize this free time, which must, nevertheless, occupy real space in society.

In this context, growing pressures on heritage communities and neighbourhoods associated with the dynamics of urban gentrification and touristification have become particularly intense in the recent past [32]. Gentrification is understood in multiple ways. However, in debates about gentrification, an increasing emphasis has been placed on the relationship between socio-economic processes (such as the uneven circulation of capital and class dynamics) and aesthetic and physical changes in housing stock, the transformation of national heritage, searches for specific lifestyles and identity creation, and even emotional and affective issues [33]. Thus, the meaning of the term remains fluid and far from univocal, having been generated by variegated geographies of gentrification [32,34]. For this paper, however, it is important to understand the perspectives regarding new forms of political struggle over space, interlaced with the socio-spatial configurations produced within contemporary capitalism. Understanding the way in which this is culturally practiced in communities is key to working in historic environments. To achieve the teaching of this goal, the pedagogical approach of this Erasmus+ BIP was drawn based on two key characteristics: interdisciplinarity and practice-oriented education.

The implementation of interdisciplinarity in a short-term programme was based on the idea of raising awareness and recognition among students of the importance of generalist qualities, because being an effective specialist requires more than just deep knowledge in a narrow area. It involves the ability to link specialized knowledge to broader issues [35]. Students should understand that working in an urban environment involves learning from different viewpoints regarding urban development problems, that every discipline has its own value for understanding and offering definitions and solutions to these problems, and that monodisciplinary perspectives rarely deliver solutions without any negative effects [13]. However, according to Snellen [36], different rationales could be epistemologically impossible when it comes to integrating these perspectives completely. This is why practical experience in working with different disciplines forces the development of the reflective skills needed to effectively understand the difficulties of generating integrated solutions, both in general practice and in addressing touristification in historical centres.

At the same time, case-driven education supports students in situated learning, where the emphasis is on the learning environment and the social aspects within it, an approach which leads to activities that are considered "authentic" [37]. This involves a highly active learning approach, where principles and materials are absorbed, integrated, and then applied to new situations [38].

3. The Historical Background of the La Latina Neighbourhood in Madrid

The La Latina neighbourhood, situated in the heart of the Spanish capital Madrid, traces its origins back to the medieval period. La Latina's story begins with its name, honouring Beatriz Galindo, known as "La Latina", a renowned 15th-century scholar, humanist, and educator [39]. However, the neighbourhood's history predates this era, dating back to the Islamic period when it was part of the Muslim citadel that stood where the Royal Palace now resides. After the Christian Reconquista, the area was gradually resettled by Christians, evolving into a small community.

During the Middle Ages, La Latina thrived as a bustling marketplace and a hub for artisans, merchants, and Jewish and Moorish communities. Its strategic location near the river Manzanares and the presence of the Puente de Toledo (Toledo Bridge) contributed to its commercial significance. The neighbourhood flourished, witnessing the construction of churches, convents, and noble residences, shaping its architectural character.

However, a significant turning point occurred in the 16th and 17th centuries when the area faced a decline due to epidemics and economic downturns. This decline led to the exodus of nobility, and the neighbourhood gradually became a working-class district. The shift in demographics and economic status laid the groundwork for La Latina's transformation and re-emergence as a vibrant, culturally diverse area.

During the 20th century, La Latina experienced urban redevelopment initiatives, particularly during the dictatorship of General Francisco Franco, which aimed to modernize and reshape parts of Madrid. Despite these changes, La Latina retained its unique historical essence.

Beyond characteristics such as its gastronomic allure, La Latina is home to significant landmarks, including the San Francisco el Grande Basilica, a magnificent example of Spanish Baroque architecture, and the Plaza de la Cebada, a historic square that has witnessed various transformations throughout the centuries [40].

In recent years, the area of La Latina has undergone an intense process of transformation and gentrification, attracting a diverse community; reinforcing the presence of leisure activity; and becoming a hotspot for culture, arts, and nightlife [41]. This case study offers an insight into the changing perceptions of the local community over time. Its present-day concerns can be summarized as follows: failure to comply with opening hours by nightclubs and bar and restaurant terraces; the lack of street cleaning and failure to comply with the garbage pickup schedule; premises used as STRs (short-term rentals) without certificates of habitability; issues with noise (street music with sound amplifiers and percussion); the over-occupation of public spaces by restaurant and bar terraces; the replacement of ground-floor business premises with tourist apartments; a heavy increase in apartment rent; the loss of local populations; and an increased need for Municipal Police (neighbourhood police) patrols. Some of the problems mentioned stem directly from the increasing number of tourists and the conversion of the old city centre of Madrid into an area of restaurants, bars, and leisure uses [42], while others must be seen as the result of the changing occupants of the neighbourhood. All this is resulting in the loss of the neighbourhood character of the La Latina area and the feeling of the loss of identity on the part of its residents.

4. Methods

The Erasmus+ educational plan was structured around three main functions, diagnostic, formative, and summative [43], in its approach to working in the city. The diagnostic phase focused on identifying knowledge about the place through various techniques drawn from diverse disciplines (see Scheme 1). These methods provided a robust set of perspectives set within the same context, enabling the generation of a broader and more comprehensive description of reality. Through this program, we achieved a deeper understanding of the existing connections between heritage places and their inhabitants, as well as the critical importance of this relationship for the quality of life. Qualitative techniques from anthropology (interviews and participant observation), geography (cognitive mapping and frequency charts), in situ archaeological research, freehand drawings, and Jan Gehl's toolkit mapping were combined to obtain these insights.

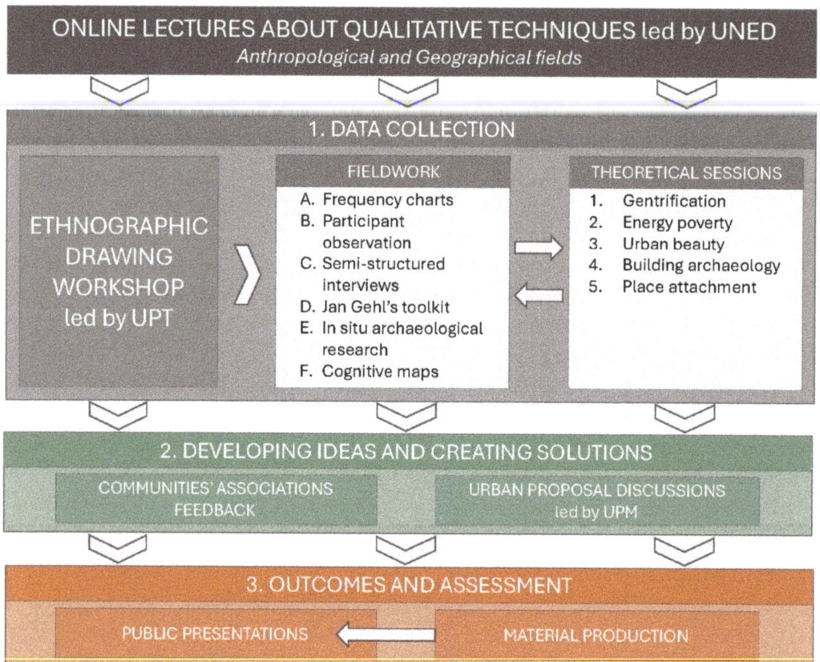

Scheme 1. Educational methodology. Source: self-development.

Ethnography, involving participant observation and interviews, focused on analysing the semiotic keys which would allow students to decipher the varied backgrounds obtained from a place in order to develop a more conscious analytical perspective. Interviews were deemed a critical part of the methodology, aimed at extracting the intersubjectivity of spatial matters. In this way, it was possible to study the meanings behind the public life produced by individuals and their relationship to public spaces [44]. In the collection of testimonies, the perception of change in the time axis and proposals for improvement were investigated from a bottom-up philosophy. This was because it was based on the thesis that local residents had a valuable role in the process of citizen participation with respect to creating a vision for improvement in the study area (which must increasingly coexist with mass tourism and increased levels of tourism accommodation). The aim was to demonstrate the integral value in using the perceptive, the sensorial, and the emotional dimensions as fundamental chains of expression that must be considered before progressing towards the proposal phase of architectural/urban work.

Also, in addition to these useful qualitative tools, we taught the students Erving Goffman's approach [45], which makes a distinction between what he terms "back" and "front" regions. Front regions are those social areas where people feel the need to construct performances and manage impressions in front of an audience. Back regions, by contrast, are those places where individuals can more easily relax, be themselves, and suspend contrived representations [46]. In this sense, it was important to convey that every field research activity involves the presence of the researcher in the actions of the people being observed.

Freehand drawing helps us to observe and see more deeply. In fact, all kinds of drawings enhance the understanding of the physical world as more than a composition of completed objects but rather as part of an unfolding cultural process interwoven with actions and articulating behaviours (Ingold cited in [47]). Kuschnir [48] in particular highlighted the importance of incorporating ethnographic drawings into anthropological research, through a review of the most significant authors of traditional anthropology.

However, drawing as a tool is hardly discussed in contemporary ethnographic manuals, with the exception, as mentioned earlier, of the cognitive maps created, in this case, by the interviewees, not by the researchers. Within the context of this project, drawings were created by the student researchers during the research phase, rather than simply at the proposal phase which is much more common amongst architects.

Last but not least, Jan Gehl's Mapping toolkit emphasizes the observation and documentation of how people interact with their surroundings, including the patterns of movement, social interactions, and stationary activities. This toolkit incorporates quantitative and qualitative metrics to assess factors such as accessibility, comfort, safety, and sociability.

However, in order to carry out the fieldwork, the creation of a field guide was necessary, enabling students to establish a filter for understanding the fundamentals which shaped the worldview of this location. Therefore, different theoretical sessions were set up dealing with concepts such as gentrification, energy poverty, urban beauty [49], building archaeology, and place attachment.

The formative phase was carried out during the two weeks of on-site interaction, where corrections were made, questions clarified, and observations refined in real time. This phase involved contributions from professors across multiple disciplines, compelling students to continuously think from different perspectives. This dynamic approach effectively regulated the learning process and fostered the development of the educational project [50]. Finally, the summative phase culminated in the production of written and graphic documentation, which was presented publicly. This final output served as a synthesis of the entire process, showcasing the integration of disciplinary insights and the educational outcomes of the project.

This training was aimed at students interested in reflecting on the new challenges arising from recent socio-cultural and economic phenomena that materialise in the public realm and spaces of the city, particularly in the historic centres. For this, it was paramount to explore the urban conditions and physical context which underpin and form part of socio-cultural structures. In particular, an attempt was made to overcome the difficulties of giving a voice to local people and residents by providing the future professionals who will work in these areas with a broader perspective supported by the interdisciplinarity of different fields of knowledge and allowing factors which are significant for the local communities (who are being served) to be captured. The difference between "space" and "place" was a key concern that guided this experience.

Students learned to use diagnostic tools and techniques that augment and complement the results of the ethnographic method, and they used them to make proposals for action in the public and symbolic space of the historic centre.

The objective of the discipline of architecture is to produce or provide solutions to problems of space, but Omahna [51] argues that anthropology focuses on understanding the causes in specific problems. Considering the dynamics of both disciplines, the issue is to seek different kinds of a common form of expression in order to successfully work with the subjective perception of the dialectic between people's habitual capabilities and that of constructed space [51]. Regarding the link between the two fields of architecture and anthropology, Omahna [52] lists a series of synergic issues which include the following: research on historical buildings and urban landscapes that indicates the process of socio-cultural change; constructive typologies and the organization of space; the culture of building in itself; and, finally, the use of space by social agents and institutions which, according to "the logics of their specific practices in relation to the space provoke changes in it" [52] (p. 41).

Promoting the understanding of the synergy between architecture and anthropology establishes the use of reflection as a process of knowing a place as a means of better understanding the worldview of the communities who live there [53]. This is why the development of a diagnostic methodology based on ethnography was proposed as a method of microcosm research of particular value to architects entering into the proactive field of architecture. The particular relevance of the ethnographic method is that its use provides a tool for identifying and capturing the various issues which must be addressed with urban regeneration projects. These range from specific issues related to the direct orders or requirements of individual clients to other variables related to the community such as touristification. Importantly, it should not be forgotten that individual processes continue to be social, since they always originate from a social reference [54]. In this way, we aimed to transmit to our students that environmental knowledge is "constructed" in the sense that individuals invent models which enable them to both understand and deal with reality. The nature of the structures or models that are developed by individuals determines what that individual "sees" in the real world. It was very important for students to realise the existence of several realities, which in a certain way are layers that overlap each other. It was also important to transmit to the group that knowledge results from interaction between factors internal to the individual (needs, personality, motivation) and by the social situations in which that person is operating. This means that the implicit knowledge of local residents built up through goal-oriented activities (e.g., shopping, going to work) constituted important dynamics which students needed to be aware of in their practice of data collection [45].

However, a common problem which emerges in the use of qualitative techniques is "researcher bias", and trying to neutralize distorting biases was one of the essential aims of this experience. Objectivity in ethnography, if it exists, focuses on the choice of strategy employed by the researcher. From an interactive and cognitive perspective, there is not a single cultural reality, and it must be recognized and understood that there may instead be different cultural realities [53]. The key is not to lose sight of the intrinsic logics, rhythms, and/or tendencies toward disassociation and division which often occur as dominant patterns in cities.

Starting from this principle, one of the objectives pursued was to reduce the production of data informed or influenced by the principles, beliefs, and values in which students have been endo-cultured as social agents, in order to recognize and then overcome cultural relativism. In seeking a position of relative objectivity, the proposed ethnography worked with a series of related techniques and different sources, as well as borrowing from human geography (frequency charts, cognitive maps, stress levels, the social eye, etc.) [55]. Together with techniques from the field of urbanism and archaeology, this enabled the triangulation of data, providing internal validity to this method while acknowledging that external validation is not possible, given that this is not allowed for by qualitative work [56].

In addition to the aforementioned quality mechanism, objectivity was also ensured by having several researchers working in the same place so that this study was carried out by more than one student. This ensured that there were re-studies of the same place, enabling the accumulation of individual biases to annul one another and ensuring a better and closer understanding of what is conceived as objectivity.

After developing an integrated diagnosis [57], the students made the leap to the propositional sphere, where they proposed actions aimed at the public interest and the collective quality of life of residents, as well as to the maintenance of their traditions, identity, and memory, seeking to make them compatible with economic activity (tourism, hospitality, etc.). All this was developed from a logic where sustainability, equitable resilience, and progress towards recognition and response to ecological imperatives were at the heart of the proposals. Moreover, this training was developed in full alignment with the Urban Agenda for the European Union, the United Nations 2030 Agenda, and the Spanish Urban Agenda.

5. The Erasmus+ Experience

The group involved with this project comprised 22 students, including 10 Portuguese, 5 Spanish, and 7 Italian students and those from various disciplines, including 11 geography students, 7 anthropology students, 1 archaeology student, 2 architecture students, and 1 urban planning student.

The experience began with the creation of interdisciplinary groups of different nationalities and fields of knowledge (mainly comprising five students) to carry out the work of analysis and the development of an urban micro-project of soft regeneration that would address the problems of the different public spaces of La Latina in the Palacio neighbourhood. The overall area was divided into sub-areas formed by Plaza de la Paja, Plaza de Puerta de Moros, and Plaza Lina Morgan + Plaza de la Cebada. The particular areas analysed included the streets that interconnected the plazas mentioned: Costanilla de San Andrés, Carrera de San Francisco, Calle Toledo, and Calle del Humilladero (see Figure 1).

The aim of this study, from an anthropological perspective, was to focus on the testimonies of individual and collective memories, working on the feeling of belonging with respect to the different places that made up the study area intersected with an analysis of the current situation of its public spaces in terms of use and dynamics. This study was based on the concept of the "vital place" as a resource of well-being, personal security, and mental health for individuals and the concept of the "city of proximity".

Although the students received instruction on the ethnographic method and techniques via PowerPoints coupled with comments in a shared online space, the first week of activity was used for four online classes to introduce the teaching team and the group of students from the three countries. In these classes, the unit of study and the application of the techniques to the specific terrain where the fieldwork was going to be carried out were presented (see Scheme 1) [55].

Data collection began with a workshop on the use of drawing. These drawings helped students internalize the proportions, materials, and architectural language characteristic of the La Latina neighbourhood (see Figure 2). Through the use of this tool, they identified and explored the neighbourhood's character of narrow, winding streets and historic buildings with façades of brick and stone, wrought iron balconies, and tiled roofs, which collectively create a cohesive urban fabric. They were also able to analyse public space in terms of aesthetics, urban scene, scale, structure, the presence of vegetation and street furniture, the presence of bar terraces, bioclimatic conditions, activities (or lack of crucial activities supportive of daily life) on the ground floor of buildings facing public spaces, etc. The fieldwork also entailed the mapping of the main public facilities in the area (markets, schools, centres for the elderly, etc.) and the presence of cars in the public spaces (circulating or parked).

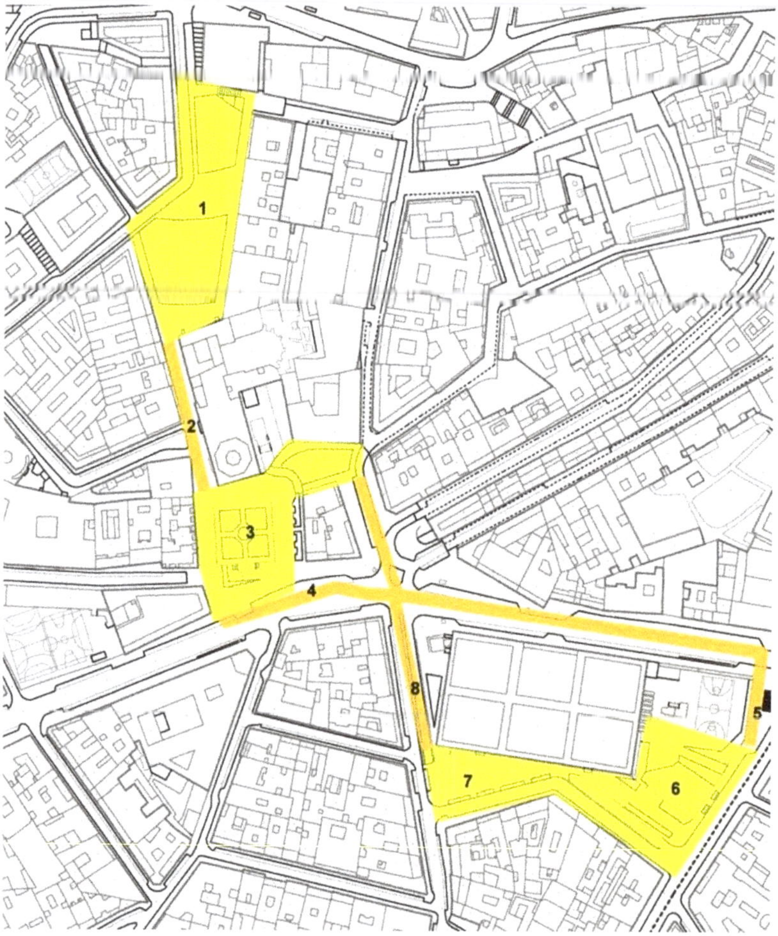

Figure 1. Study areas of La Latina, Madrid. (1) Plaza de la Paja, (2) Costanilla de San Andrés, (3) Plaza de Puerta de Moros, (4) Carrera de San Francisco, (5) Calle Toledo, (6) Plaza Lina Morgan, (7) Plaza de la Cebada, (8) Calle del Humilladero. Source: self-produced course material.

Also, from the outset, the student groups carried out combined tasks which included the creation of frequency charts, participant observation, and writing in field diaries for their assigned study areas. In this way, they were able to integrate information from different sources and create a basis for what would, from then on, be their growing empirical material and at the same time their knowledge base for formulating questions to enable them to carry out semi-structured interviews.

A particular characteristic of the neighbourhood is its dualistic nature—during the day, its public squares, such as Plaza de la Cebada and Plaza de la Paja, provide public space for the daily life of local people, while at night, the area transforms into a lively social scene with tapas bars and terraces (see Figure 3).

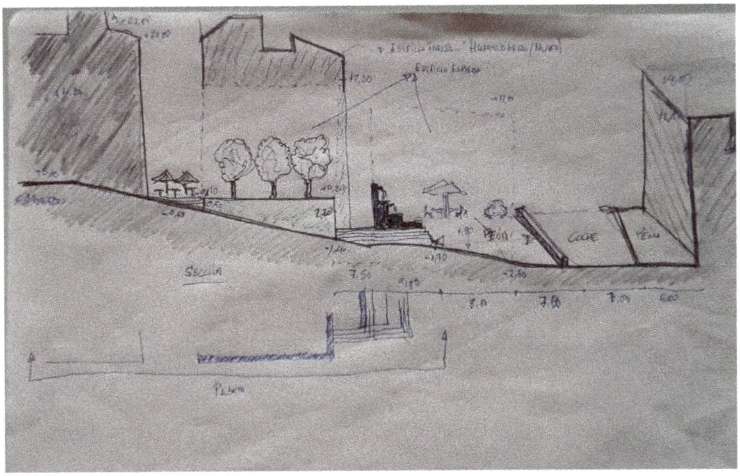

Figure 2. Hand drawing. Source: Valiño, J.

Figure 3. Daytime and evening activities at the Plaza de la Paja. Source: Azevedo, C.; Giach E.; Gomes E.; Niccoli, M.; Bartoli, T.; and Usón Gasca, T.

The toolkit developed by Jan Gehl for the diagnosis of public life was used [58] for gathering data on the age, gender, and flow of people who frequented the study area and the physical characteristics of spaces and how they enrich the experience of those spending time there.

Additionally, this toolkit also revealed that private furniture for bar terraces in the squares, such as tables, chairs, and umbrellas, varies significantly throughout the day. For instance, in the Plaza de la Paja at 10 a.m., it is possible that no furniture might be seen, while by 10:10 a.m., more than 20 tables and 40 chairs from various establishments may have been set up (see Figure 4). While there are also public seating areas, by contrast, these are often unkempt and neglected.

Pedestrian movement predominantly occurs along the edges of the squares, especially during the hottest part of the day, as the centres lack shade. This study also revealed the limited presence of children, likely due to the end of the school term. In fact, the majority of individuals aged 15 to 24 observed in the squares were tourists accompanied by their parents. Adults aged 25 to 64, however, represented the largest group (comprising similar proportions of men and women), either strolling through the square or seated at bar terraces, and they were made up of both residents and tourists. People aged 65 and older were frequently seen sitting on benches or walking their dogs, with this group consisting primarily of local residents (see Figure 5).

Figure 4. Jan Gehl's mapping. Source: Azevedo, C.; Giach E.; Gomes E.; Niccoli, M.; Bartoli, T.; and Usón Gasca, T.

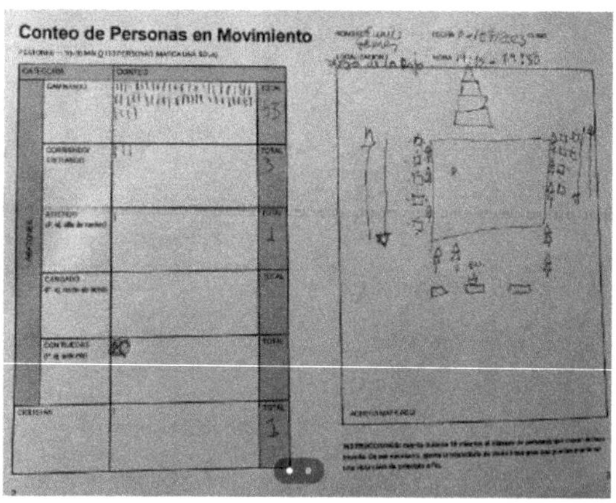

Figure 5. Jan Gehl's mapping. Source: Azevedo, C.; Giach E.; Gomes E.; Niccoli, M.; Bartoli, T.; and Usón Gasca, T.

Simultaneously, the fieldwork incorporated the idea of a lived-in past that remained imbricated in the new uses and contemporary objects of the urban spaces studied. Through exercises of observation and interpretation, the team explored how the past emerged in the palimpsest of uses by residents, tourists, and others. This enabled the students to develop an appreciation of the timespans over which the transformations had taken place, aimed principally at revealing how cultural heritage could support the suggestion of improvements being designed as part of the project [59].

The interviews were conducted with workers, residents, and institutional representatives, and they revealed significant insights into the neighbourhood's transformation due to gentrification and touristification. They reflected how residents occupied, used, and related to their places throughout the year, above all regarding the people who have lived in this area for decades. The narrative recognized and recorded places and landmarks that were part of these people's entire lives, which they shared with parents and grandparents and now with their own children. Among these memories were the public swimming pool that once stood at the intersection of Plaza de la Cebada and Calle de Toledo, which has now been substituted by a modern sports centre, and the vibrant local shops that once characterized Calle de Cava Baja (now completely transformed into a street for restaurants and bars). Recollections also referenced popular festivals and personal routines, such as Sunday walks, social gatherings, and recreational activities, all of which have contributed to the rich and evolving identity of the neighbourhood.

Workers in local restaurants and bars consistently highlighted the dominance of tourists among their clientele, with estimates ranging from 70% to 80%. Nonetheless, some businesses continue to attract local residents, especially on weekends or through offerings tailored to neighbourhood preferences. Despite the economic benefits, many workers noted that the increasing presence of tourists is contributing to the gradual erosion of the neighbourhood's identity. This sentiment was echoed in observations about gentrification, which has significantly altered the demographic and social fabric of La Latina.

> *'Nos afecta mucho el tema de los pisos turísticos ya que no se ponen en funcionamiento de acuerdo a la normativa y han subido mucho el precio de pisos y alquileres.'*
>
> *'We are very affected by the issue of tourist flats, as they are not being put into operation in accordance with the regulations and the prices of flats and rents have risen a lot.'*
>
> (Resident 55 years old)

Also, institutional representatives, such as the director of the Santa Bárbara Institute, pointed out that the local population has declined and aged, with many residents converting their homes into short-term rental apartments for higher profits. This trend has severely impacted local schools, which now rely on students commuting from other districts, such as Lavapiés and Carabanchel, with approximately 80% of their attendees coming from outside La Latina.

On the other hand, residents' perceptions of these changes are mixed. While some older residents feel unaffected, younger and middle-aged respondents acknowledge both positive and negative aspects of the neighbourhood's evolution. They praise La Latina's charm and its ability to retain a sense of community but expressed concerns about issues such as noise, odours, and the rising cost of housing driven by unregulated tourist apartments.

Another local stakeholder, however, the religious institution Hermanitas del Cordero, has experienced distinct and nuanced effects. For instance, it has benefitted significantly from the influx of tourists, as the increased visits to the Capilla have provided funding for its restoration. They regard this as an opportunity to extend their mission and breathe new vitality into their space. This dynamic, often subtly interwoven with contemporary elements, highlighted how touristification not only engages with the rich historical legacy of La Latina but also influences the preservation and reinterpretation of that heritage. Consequently, it demanded thoughtful consideration in the site analysis to balance historical conservation with evolving uses.

In summary, the interviews revealed the complexity involved in intervening in historic centres in order to find integrated solutions. They also highlighted the importance of avoiding monodisciplinary perspectives as they rarely deliver solutions without negative effects.

The sessions with the three neighbourhood associations, followed by a walk in the Parque de la Cornisa with one of these, were strategic. The first took place in the Municipal

Public Library Iván de Vargas, located on the border of the study area (see Figure 6). The representatives of the Neighbourhood Associations Plaza Mayor and Cavas-La Latina provided a historical framework for the area studied, and information on the evolution of neighbourhood participation in struggles against new problems was shared. These included issues arising from malaise and the declining quality of life in the area where we carried out our research, with some of the most important problems mapped being the prevalence of tourist apartments, the touristification of the area, and the intensive use of public spaces to locate bar terraces. The second session took place in the green area nearer to La Latina (Parque de la Cornisa) (see Figure 7). The local collective "Vecinas Corniseras" shared with students the refusal of the municipality to engage in dialogue with the local community about the rehabilitation of the park. From these sessions, students developed an understanding that despite a clear need, the problems, demands, and necessities of the local community were not being responded to by the municipality.

Figure 6. A session with the two neighbourhood associations in the Public Library Iván de Vargas. Source: De Gregorio Hurtado, S.

Figure 7. A walk in the Parque de la Cornisa with one of the neighbourhood associations (Vecinas Corniseras). Source: De Gregorio Hurtado, S.

From an urban perspective, different intervention proposals (micro-projects for soft regeneration) were proposed in the different study units that sought to include specific measures (social, economic, governance of the place, etc.) on the basis of an integrated approach (see Figure 8). These were aimed at tackling current problems, as well as strengthening the contribution of this part of the historic centre of Madrid to a model of a sustainable city that moves towards equitable resilience, always bearing in mind social cohesion, ecological and climatic challenges, quality of life, equality, and the development of "vital places".

Figure 8. The different student groups generating their proposals in a classroom of the Universidad Politécnica de Madrid. Source: De Gregorio Hurtado, S.

A number of key findings stand out, especially with regard to the area of the La Cebada market where there was an opportunity to adapt a new space recently created by building work for a new sports centre, introducing new uses in a place that would normally be associated with intense collective memories. Dynamizing this "virgin" ground, which as yet had no meaning for "long-term" residents, offered a chance to create a new image which could work to better reconcile the new needs generated by a high degree of touristification in recent years. We were also able to highlight new perspectives regarding "cultural assets", in particular wine cellars on the edges of the Plaza de la Paja which are currently in a neglected state but could be restored.

Also, tangible proposals arose from the understanding of the meaning of "being at home", reinforcing the existing symbolic meaning which enhances perceived control through the full mediation of positive emotions for "long-term" residents. In other words, the strategy was to present proposals for better incorporating and reducing the flow of tourists usings shared spaces, in order to recover the notion of belonging and identity connected to the place. This involved highlighting traditional uses in different parts of the study area, removing the invasion of the public spaces by bars and restaurants, and improving the environmental quality and urban scene of public spaces (see Figures 9 and 10).

Figure 9. Plaza de la Paja proposal. Source: Azevedo, C.; Giach E.; Gomes E.; Niccoli, M.; Bartoli, T.; and Usón Gasca, T.

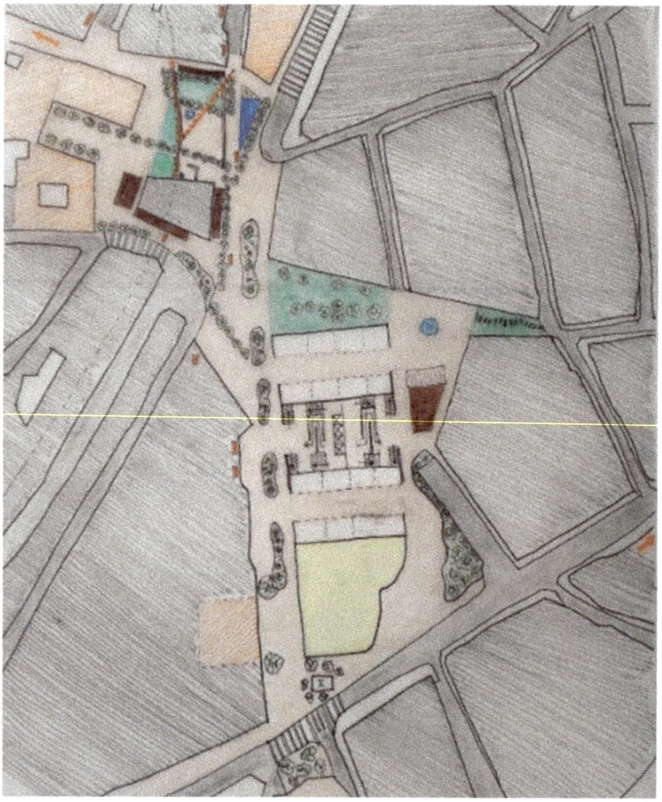

Figure 10. Plaza de Puerta de Moros proposal. Source: Mendes, A.B.; Miranda, A.L.; Fittipaldi, A.; Valiño, J.; Masili, M.; and Corrêa, R.

Understanding and addressing the preservation of cultural practices associated with a neighbourhood's public space, i.e., the contemporary rituals of the old residents who gather there, enriched the different rationalities of students from different disciplines. At the same time, the attempt to develop comprehensive proposals that considered both the needs of residents and tourists highlighted the value of applying specialized knowledge to address larger, more general challenges.

Finally, through an anonymous survey, the perceptions of the experiences of 19 of the 22 students who participated in the program were collected. The interdisciplinary work between different cultures and fields of knowledge turned out to be one of the most valued aspects, since the learning came not only from theoretical classes but also from the discussions students had amongst themselves. It was also apparent that the work of the teachers in relating all of these to each other was of great importance in regard to the formative dimension of the educational experience.

When it comes to teaching, and therefore learning, two activities occur at the same time. On the one hand, there is the academic task environment, comprising the curriculum which researchers and cognitive psychologists focus on. On the other, there is the structure of social participation, which is the way in which this governs the sequence and articulation of interactions [60]. The Erasmus+ BIP experience allowed equal attention to be paid to both tasks, as a means of overcoming the difficulties involved in understanding the complexity of the city. It was crucial to activate the different "expertise" provided by each of the students (coming from different courses and different disciplines, as mentioned above). This involved a highly active learning approach, where principles and materials were absorbed, integrated, and then applied to an empirical case in the form of a real location. It forced them to develop reflective skills to understand and act to find integrated solutions for touristification in historical centres.

It was particularly important to establish the structure of social participation by students [60] in order to identify any gaps which needed to be addressed, in order to ensure the development of required and desirable skills. In the accompanying analysis of the city, cooperative and interdependent actions dominated. This was derived from what Vygotsky [61] referred to as the zone of proximal development, in which the teacher's reaction depended upon the degree of internalization of the contents by the students.

This experience was an essential precursor to developing and conveying the meaning of "context", holistically understood as constituting the enactment of activity involving people and spaces, since it involves specific objects and goals together with specific settings—the conception of context as a transformative relationship between people, spaces, and artefacts (c.f. [62]). Through the Erasmus+ BIP experience, students were encouraged to actively question the rationale behind various community activities, rather than focusing solely on the physical environment. This approach prevented them from becoming agents of disruption, as they developed an understanding of the existing social dynamics before taking action in any location. By fostering interdisciplinarity and involvement, this experience allowed students to consider the intangible dimensions present in public spaces, offering a more comprehensive understanding of reality. It also prompted a re-evaluation and re-examination of assumed norms, practices, and management styles, encouraging a more thoughtful and context-sensitive approach.

6. Final Thoughts

Public spaces not only provide conditions for being human but also for responding to the human condition. There is a need to pay attention to the philosophy underlying material manifestations that, roughly, are induced in projects by architects on behalf of communities yet still act as detached and independent agents with very different worldviews. Therefore, architects and other "experts" are unsurprisingly viewed with suspicion [63], and to overcome this, it is paramount to explore the public space conditions that support and are part of socio-cultural structures.

This methodology is based on taking a broad perspective and allowing factors which are significant to the communities studied to be captured. In this way, students were supported in their search for hidden rules and patterns, for developing a sense of future orientation, and for making strategic judgements about the meaning and relevance of design [64] in facing the transformation of the city. This allowed for a critique of the spirit of the place that characterizes La Latina and that which has emerged from its long history, its adaption to the social and cultural changes in urban design, and the uses that are now taking place in its streets and plazas.

Briefly, the learning outcomes of the Erasmus+ BIP experience are the following:

Improved communication across disciplines and diverse backgrounds which will help close the gap between fields such as anthropology, archaeology, geography, architecture, and urban planning.

A heightened awareness and sensitivity towards the preservation of heritage communities was successfully cultivated among the participants with regard to the threat of touristification.

All participants were exposed to and engaged with tools from a variety of knowledge areas.

However, to understand the extent to which students were able to construct an understanding of daily life in the territory which formed the study area, a series of issues must be mentioned that could potentially result in an incorrect view of the spaces investigated. As was described, the fieldwork was carried out in only 15 days. From the perspective of qualitative research, this is a very short time to carry out the rigorous fieldwork needed to generate sufficient empirical material required to produce serious and weighty conclusions. In fact, the fieldwork was undertaken in the month of July when the school year had finished, an important point to consider as the study area is a vital space for after-school activities during term time and a meeting point for parents and children in the hours before class is dismissed (see Figures 11 and 12). It was important to present this reality to our students, enabling them to realistically understand and assess the cyclic use of the study areas throughout the year, rather than relying on observations shaped during the most touristed months of July and August when the area was simultaneously almost free of children.

Figure 11. Carros plaza in July 2023. Source: Müllauer-Seichter, W.

Figure 12. Carros plaza in February 2024. Source: Müllauer-Seichter, W.

Another aspect that can lead to the incorrect interpretation of the empirical material gathered in this space of time is the fact that very few zones have vegetation and trees. Historical city centres, due to their nature, present narrow streets and alleys, and the oldest city centres do not offer many options for incorporating parks, unless there are empty lots that make it possible. Because of this, both in general and in our specific case, the long-time residents of these historical spaces have their strategies for using the plazas and the "urban green" [18], depending on the time that they have available (plazas for a short amount of time and for social encounters; if they have an hour or more, they go to the gardens that are relatively close). This was also the case in our study area, which at first sight might seem to be a "hard space" characterized by plazas, in general paved with old stone and accompanied by narrow sidewalks. What is less apparent, particularly when visiting as a stranger with limited local knowledge, is the enormous greenbelt presented by the nearby Las Vistillas; the Parque de la Cornisa (which we visited with the local collective "Vecinas Corniseras"); the Jardines de la Cuesta de la Vega; the Parque de Atenas; the Campo del Moro; and, since 2009, the Madrid Río linear park that merges all of these interconnected green areas with the Casa de Campo. In total, these collectively offer an enormous belt of urban nature for the historical centre and particularly for the Palacio neighbourhood in which the area of La Latina is located.

The Erasmus+ BIP experience was built upon the fundamental recognition of the vital connections between heritage places and their heritage communities. However, the neoliberal context of recent decades has steered conservation efforts towards a predominantly economic focus, often at odds with the interests and well-being of historical communities [65]. Through this training, our goal was twofold: to equip students with the necessary knowledge and criteria to critically address the current challenges facing European traditional city centres from a people-based approach and to raise awareness among local stakeholders about the negative transformations affecting historic centres and their impact on residents' quality of life. In this sense, the training aligned with our ongoing commitment to fostering meaningful reflections on social and urban challenges through our roles as educators and researchers.

Author Contributions: Conceptualization, M.A., W.M.-S. and S.D.G.H.; methodology, M.A., W.M.-S., S.D.G.H. and L.M.; formal analysis, M.A., W.M.-S., S.D.G.H. and L.M.; investigation, M.A., W.M.-S., S.D.G.H. and L.M.; writing—original draft preparation, M.A., W.M.-S., S.D.G.H. and L.M.; writing—review and editing, D.J.; project administration, M.A., W.M.-S., S.D.G.H., L.M. and

M.L.; funding acquisition, M.A. All authors have read and agreed to the published version of the manuscript.

Funding: This work is financed by national funds through FCT—Fundação para a Ciência e a Tecnologia, I.P., under the Strategic Project with the references UIDB/04008/2020 and UIDP/04008/2020.

Data Availability Statement: The original contributions presented in this study are included in the article. Further inquiries can be directed to the corresponding author.

Conflicts of Interest: Author Delton Jackson is a director of the UK company Studio UrbanArea LLP. The remaining authors declare that the research was conducted in the absence of any commercial or financial relationships that could be construed as a potential conflict of interest.

References

1. Martinón-Torres, M. El megalito ha muerto.¡ Larga vida al megalito! *PH Boletín Inst. Andal. Patrim. Histórico* **2008**, *16*, 67. [CrossRef]
2. ICOMOS. *The ICOMOS Charter for the Conservation of Historic Towns and Urban Areas;* ICOMOS: Washington, DC, USA, 1987.
3. Nicholas, L.N.; Thapa, B.; Ko, Y.J. Residents' perspectives of a world heritage site: The Pitons Management Area, St. Lucia. *Ann. Tour. Res.* **2009**, *36*, 390–412. [CrossRef]
4. Stephens, J.; Tiwari, R. Symbolic estates: Community identity and empowerment through heritage. *Int. J. Heritage Stud.* **2015**, *21*, 99–114. [CrossRef]
5. Jaafar, M.; Noor, S.M.; Rasoolimanesh, S.M. Perception of young local residents toward sustainable conservation programmes: A case study of the Lenggong World Cultural Heritage Site. *Tour. Manag.* **2015**, *48*, 154–163. [CrossRef]
6. Colomer, L. Heritage on the move. Cross-cultural heritage as a response to globalisation, mobilities and multiple migrations. *Int. J. Heritage Stud.* **2017**, *23*, 913–927. [CrossRef]
7. Herzfeld, M. Engagement, Gentrification, and the Neoliberal Hijacking of History. *Curr. Anthr.* **2010**, *51*, S259–S267. [CrossRef]
8. Harrison, R. Heritage as social action. In *Understanding Heritage in Practice;* Manchester University Press: Manchester, UK, 2010; pp. 240–276.
9. Marin-Pantelescu, A.; Tăchiciu, L.; Oncioiu, I.; Ștefan-Hint, M. Erasmus Students' Experiences as Cultural Visitors: Lessons in Destination Management. *Sustainability* **2022**, *14*, 2553. [CrossRef]
10. Acevedo Pardo, C.; Farjas Abadía, M.; Zazo Moratalla, A.; Georgopoulos, A.; Mielczarek, M.; Parenti, R.; Parseliunas, E.; Schramm, T.; Stefanakis, E.; Tapinaki, S. *Experiences Gained from the ERASMUS Intensive Programme HERICT;* IATED Digital Library: Madrid, Spain, 2013; pp. 4424–4431.
11. Calvo, D.M. Procesos de revalorización patrimonial en el barrio de Alfama: El papel de los estudiantes Erasmus en la tematización de la ciudad. Etnográfica. *Rev. Cent. Rede Investig. Antropol.* **2013**, *17*, 31–50. [CrossRef]
12. Prisco, M. *Seeing the City: Interdisciplinary Perspectives on the Study of the Urban;* Verloo, N., Bertolini, L., Eds.; Amsterdam University Press: Amsterdam, The Netherlands, 2020; 291p.
13. Chen, Y.; Daamen, T.A.; Heurkens, E.W.T.M.; Verheul, W.J. Interdisciplinary and experiential learning in urban development management education. *Int. J. Technol. Des. Educ.* **2020**, *30*, 919–936. [CrossRef]
14. Giddens, A. *The Consequences of Modernity;* John Wiley & Sons: Hoboken, NJ, USA, 2013.
15. Altman, I.; Low, S.M. *Place Attachment;* Springer Science & Business Media: Berlin, Germany, 2012.
16. Nine, C. The Wrong of Displacement: The Home as Extended Mind. *J. Politi-Philos.* **2018**, *26*, 240–257. [CrossRef]
17. Mydland, L.; Grahn, W. Identifying heritage values in local communities. *Int. J. Heritage Stud.* **2012**, *18*, 564–587. [CrossRef]
18. Müllauer-Seichter, W. *Mujeres y Tiempo Libre: Pasear por el Espacio Verde Urbano Como Práctica Cultural;* Instituto Universitario de Estudios de la Mujer: Madrid, Spain, 2004; pp. 151–160.
19. Debray, R. *Transmitting Culture;* Columbia University Press: New York, NY, USA, 2000.
20. Määttänen, P. *Mind in Action: Experience and Embodied Cognition in Pragmatism;* Springer: Berlin/Heidelberg, Germany, 2015.
21. Velasco, H. Hablar y pensar, tareas culturales. In *Temas de Antropología Lingüística y Antropología Cognitiva;* Editorial universitaria Ramón Areces: Madrid, Spain, 2009.
22. Antonsich, M. Searching for Belonging—An Analytical Framework. *Geogr. Compass* **2010**, *4*, 644–659. [CrossRef]
23. Han, B.-C. *The Disappearance of Rituals: A Topology of the Present;* John Wiley & Sons: Hoboken, NJ, USA, 2020.
24. Cruces Villalobos, F. Símbolos en la ciudad. In *Lecturas de Antropología Urbana;* UNED: Madrid, Spain, 2010.
25. Lang, M.; Krátký, J.; Shaver John, H.; Jerotijevic', D.; Xygalatas, D. Effects of Anxiety on Spontaneous Ritualized Behavior. *Curr. Biol.* **2015**, *25*, 1892–1897. [CrossRef]
26. Norton, M.I.; Gino, F. Rituals alleviate grieving for loved ones, lovers, and lotteries. *J. Exp. Psychol. Gen.* **2014**, *143*, 266–272. [CrossRef] [PubMed]

27. Yin, K.; Lan, M.; Li, H.; Zhao, Z. How ritual actions, symbolic meanings, and positive emotions enhance perceived control: A dual path way mechanism. *Acta Psychol. Sin.* **2021**, *54*, 54–65. [CrossRef]
28. Vohs, K.D.; Wang, Y.; Gino, F.; Norton, M.I. Rituals Enhance Consumption. *Psychol. Sci.* **2013**, *24*, 1714–1721. [CrossRef]
29. Hobson, N.M.; Schroeder, J.; Risen, J.L.; Xygalatas, D.; Inzlicht, M. The Psychology of Rituals: An Integrative Review and Process-Based Framework. *Pers. Soc. Psychol. Rev.* **2017**, *22*, 260–284. [CrossRef]
30. Rotenberg, R. *Landscape and Power in Vienna, London*; John Hopkins University Press: Baltimore, MD, USA, 1996.
31. Müllauer-Seichter, W. La intervención ciudadana en la transformación de los espacios públicos. Disparidades. *Rev. De Antropol.* **2007**, *62*, 167–185.
32. Lees, L. The geography of gentrification: Thinking through comparative urbanism. *Prog. Hum. Geogr.* **2011**, *36*, 155–171. [CrossRef]
33. Phillips, M. Baroque rurality in an English village. *J. Rural. Stud.* **2014**, *33*, 56–70. [CrossRef]
34. Phillips, M. Other geographies of gentrification. *Prog. Hum. Geogr.* **2004**, *28*, 5–30. [CrossRef]
35. Campbell, H. *Specialists and Generalists: Are There Too Many Hedgehogs and Not Enough Foxes?* Taylor & Francis: Abingdon, UK, 2014.
36. Snellen, I. Conciliation of Rationalities: The Essence of Public Administration. *Adm. Theory Prax.* **2002**, *24*, 323–346. [CrossRef]
37. Johnston, A.S. CitySection: A Pedagogy for Interdisciplinary Research and Collaboration in Planning and Environmental Design. *J. Plan. Educ. Res.* **2014**, *35*, 86–92. [CrossRef]
38. Hale Feinstein, A.; Mann, S.; Corsun, D.L. Charting the experiential territory: Clarifying definitions and uses of computer simulation, games, and role play. *J. Manag. Dev.* **2002**, *21*, 732–744. [CrossRef]
39. Fernández Villalba, V. *Beatriz Galindo La Latina, La Semilla de un Barrio Madrileño*; Nivola: Madrid, Spain, 2023.
40. Paises Monfort, A. *Barrio de La Latina: Su Historia, Sus Calles y Sus Gentes*; Temporae: Madrid, Spain, 2017.
41. Aramayona, B.; García-Sánchez, R. Decoding middle-class protest against low-cost nocturnal tourism in Madrid. In *Tourism and the Night*; Routledge: England, UK, 2021; pp. 10–23.
42. De Gregorio Hurtado, S. *Resistir en el Centro*; Eldiario.es: Madrid, Spain, 2022.
43. Bloom, B.; Hastings, T.; Madaus, G.F. *Manual de Avaliação Formativa e Somativa do Aprendizado Escolar*; Livraria Pioneira Editora: Rio de Janeiro, Brazil, 1971.
44. Díaz De Rada, Á. El Taller del Etnógrafo. In *Materiales y Herramientas de Investigación en Etnografía*; UNED Universidad Nacional de Educación a Distancia: Madrid, Spain, 2011.
45. Goffman, E. The presentation of self in everyday life: Selections. In *The Production of Reality: Essays and Readings on Social Interaction*; SAGE Publications: Thousand Oaks, CA, USA, 2010; p. 262.
46. Karp, D.A.; Stone, G.P.; Yoels, W.C.; Dempsey, N.P. *Being Urban: A Sociology of City Life*; Bloomsbury Publishing USA: New York, NY, USA, 2015.
47. Causey, A. *Drawn to See: Drawing as an Ethnographic Method*; University of Toronto Press: Toronto, ON, Canada, 2017.
48. Kuschnir, K. Ethnographic Drawing: Eleven benefits of using a sketchbook for fieldwork. *Vis. Ethnogr.* **2016**, *5*, 103–134.
49. De Gregorio Hurtado, S. La Altana del Palacio Falconeri y la "Belleza Urbana". Locuslab. 2022. Available online: https://www.locuslab.eu/2020/09/10/la-altana-del-palacio-falconieri-y-la-belleza-urbana/ (accessed on 6 September 2024).
50. Perrenoud, P. *Avaliação: Da Excelência à Regulação das Aprendizagens—Entre Duas Lógicas*; Artmed: Porto Alegre, Brazil, 1999.
51. Rolshoven, J.; Omahna, M. *Reziproke Räume. Texte zu Kulturanthropologie und Architektur*; Jonas Verlag: Marburg, Germany, 2013.
52. Omahna, M. Kulturanthropologie und Architektur. Episteme temporärer Begegnungen. In *Reziproke Räume. Texte zu Kulturanthropologie und Architektur*; Rolshoven, J., Omahna, M., Eds.; Jonas Verlag: Marburg, Germany, 2013.
53. Alcindor, M. Ethnography applied to the architectural designs studio. In Proceedings of the International Ankara Multidisciplinary Studies Congress, Ankara, Turkey, 13–14 October 2023; Sahin, M., Ed.; IKSAD Global: Ankara, Turkey, 2023; pp. 878–885.
54. Cuadrado-Guirado, M.I.; Gaviria-Stewart, E.; López-Sáez, M. *Introducción a la Psicología Social. Teoría y Cuaderno de Investigación*; Sanz y Torres: Madrid, Spain, 2013.
55. Müllauer-Seichter, T. Rendimiento y utilidad de técnicas prestadas de la geografía humana: Trabajando la percepción individual. Disparidades. *Rev. De Antropol.* **2003**, *58*, 47–70. [CrossRef]
56. Corbetta, P. *Social Research: Theory, Methods and Techniques*; SAGE Publications Ltd.: London, UK, 2003.
57. De Gregorio Hurtado, S. La regeneración urbana como política clave frente a los retos del presente. In *Políticas Urbanas y Localización de los Objetivos de Desarrollo Sostenible: Teoría y Práctica*; Alonso Ibañez, R., González Medina, M., De Gregorio Hurtado, S., Eds.; Tirant lo Blanch: Valencia, Spain, 2021.
58. Gehl, J.; Svarre, B. *How to Study Public Life*, 2nd ed.; Island Press: Washington, DC, USA, 2013.
59. Medeiros, L. Heritage-led Development: A Systemic Approach. In *Other Landscapes of Cultural Heritage(s): History and Politics*; Lopes, N., Rosa, W., Bandeira Jerónimo, M., Eds.; Imprensa da Universidade de Coimbra: Coimbra, Portugal, 2023.
60. Erickson, F. El discurso en el aula como improvisación: Las relaciones entre la estructura de la tarea académica y la estructura de la participación social en clase. In *Lecturas de Antropología Para Educadores. El Ámbito de la Antropología de la Educación y de la Etnografía Escolar*; Velasco Maillo, H.M., García Castaño, F.J., Díaz de Rada, A., Eds.; Trotta: Madrid, Spain, 1993.

61. Vygotsky, L.S. *Mind in Society: Development of Higher Psychological Processes*; Harvard University Press: Cambridge, MA, USA; London, UK, 1978.
62. Nardi, B.A. Studying context: A comparison of activity theory, situated action models, and distributed cognition. In *Context and Consciousness: Activity Theory and Human-Computer Interaction*; The MIT Press: Cambridge, MA, USA, 1996; Volume 69102, pp. 35–52.
63. Pallasmaa, J. The existential image: Lived space in cinema and architecture. *Phainomenon* **2012**, *25*, 157–174. [CrossRef]
64. Suri, J.F. Poetic observation: What designers make of what they see. In *Design Anthropology*; Springer: Berlin/Heidelberg, Germany, 2011.
65. Alcindor, M.; Jackson, D.; Alcindor-Huelva, P. Heritage places and the place attachment of adolescents: The case of the Castelo of Vila Nova de Cerveira (Portugal). *J. Rural Stud.* **2021**, *88*, 410–421. [CrossRef]

land

MDPI

Article

The Ephemeral Cultural Landscape of an Australian Federal Election

Dirk H. R. Spennemann [1,*] and Deanna Duffy [2]

1 Gulbali Institute, Charles Sturt University, P.O. Box 789, Albury, NSW 2640, Australia
2 Spatial Analysis Network, Charles Sturt University, P.O. Box 789, Albury, NSW 2640, Australia; dduffy@csu.edu.au
* Correspondence: dspennemann@csu.edu.au

Abstract: This paper explores the concept of ephemeral cultural landscapes through the lens of public election advertising during the 2025 Australian Federal election in the regional city of Albury, New South Wales. Framing election signage as a transient cultural landscape, the study assesses the distribution of election signage (corflutes) disseminated by political candidates against demographic and socio-economic criteria of the electorate. The paper examines how corflutes and symbolic signage reflect personal agency, spatial contestation, and community engagement within urban and suburban environments. A detailed windscreen survey was conducted across Albury over three days immediately prior to and on election day, recording 193 instances of campaign signage and mapping their spatial distribution in relation to polling booth catchments, population density, generational cohorts, and socio-economic status. The data reveal stark differences between traditional party (Greens, Labor, Liberal) strategies and that of the independent candidate whose campaign was marked by grassroots support and creative symbolism, notably the use of orange corflutes shaped like emus. The independent's campaign relied on personal property displays, signaling civic engagement and a bottom-up assertion of political identity. While signage for major parties largely disappeared within days of the election, many of the independent's symbolic emus persisted, blurring the temporal boundaries of the ephemeral landscape and extending its visual presence well beyond the formal campaign period. The study argues that these ephemeral landscapes, though transitory, are powerful cultural expressions of political identity, visibility, and territoriality shaping public and private spaces both materially and symbolically. Ultimately, the election signage in Albury serves as a case study for understanding how ephemeral landscapes can materially and symbolically shape public space during moments of civic expression.

Keywords: cultural landscapes; ephemerality; cultural heritage; demographics; elections; political advertising

1. Introduction

Cultural landscapes represent past and present human interactions with their environment, thereby commonly leaving observable and tangible traces of events and activities on a larger micro- or macro-landscape scale [1]. Commonly, such cultural landscapes are ascribed a value by the community and are regarded as cultural heritage. On other occasions, cultural landscapes are more ephemeral or transient and do not persist long enough for heritage values to emerge.

Broadly speaking, cultural heritage emerges from the interactions between people and their environment, as well as the interactions between people themselves. Intangible cultural heritage is derived from human relationships and is primarily reflected in elements such as language, technical skills, folklore, customs, and traditions [2]. Intangible heritage also exists in the sensorial space, primarily visual and auditory [3] but also olfactory [4]. In contrast, the interaction between people and their environment generates aspects of tangible heritage as manifested in the form of built structures (e.g., buildings, roads) or modifications to the natural environment (e.g., mining, agriculture). Tangible heritage also encompasses a variety of movable objects and artifacts of different forms and materials (e.g., tools, objects of art). In addition, we need to consider an emergent third domain, that of virtual heritage, which considers an intangible digital dimension of human–machine interaction [5]. What defines these interactions as "heritage" is the cultural significance that a community assigns to them [6,7]. In addition to cultural heritage, we need to consider natural heritage, an environmental space that is valued by the current generations for its seemingly unspoiled and unmodified characteristics [1]. These assigned values, which are mutable qualities [8], motivate community efforts to preserve and manage this natural, tangible and intangible heritage for the benefit of the current population [9–11] and, purportedly, also for future generations [12]. While most heritage assets are managed or curated as single units, some may be aggregated in the form of heritage conservation areas [13]. A different kind of spatially circumscribed heritage asset is cultural landscapes.

In its traditional/orthodox framing, an interpretation and analysis of cultural landscapes is focused on the tangible manifestations, such as landscape modification change, archeological evidence, structural evidence, or cultural plantings. Even though exceptions exist [14], cultural landscapes do not tend to represent an unadulterated ensemble of past land use. Rather, cultural landscapes are a palimpsest of one or more phases of past cultural interactions that are overlain by modern contemporary land use [15,16]. Viewed from a First Nations, as well as an Indigenist, standpoint [17], this landscape space extends, and always has extended, to the cultural attachment and the valuing of seemingly 'natural', unmodified landscape features that have cultural and spiritual meaning and significance to First Nations communities [18–20]. An intangible manifestation of cultural landscapes are toponymic landscapes, where settler colonialist societies supplanted the First Nations appellations of localities, natural and spiritual features with colonial imports or phonetic bastardizations of First Nations names [21–23]. In addition to these permanent cultural landscapes, we need to consider the phenomenon of ephemeral landscapes and place these into the cultural heritage context.

1.1. Ephemeral Landscapes

By definition, ephemeral landscapes are impermanent transient manifestations of natural or cultural phenomena within a spatially circumscribed dimension. The attributes of these landscapes, in common parlance, are that they are tangible and visually interpretable. Excluded from this paper are additional intangible, sensory dimensions that may add to or enhance the perceived value and significance of these ephemeral landscapes, such as soundscapes [24–26] and smellscapes [4,27,28].

Ephemeral landscapes are commonly seen through a geomorphological lens [29], associated with natural phenomena. Examples are the flooding of Kati Thanda-Lake Eyre, an endorheic lake in the far north of South Australia [30], the annual inundation of river floodplains [31] or ephemeral wetlands [32]. Beyond this, ephemeral landscapes can be regularly reoccurring, such as seasonal landscapes; transitory, such as peri-urban zones; or as temporary transformations of persistent landscapes.

Seen through another environmental lens, seasonal landscapes not only reflect the visual appearance of seasonally variable vegetation (natural or planted as crops), its flowering and coloring [33–36], but also find manifestation in harvesting features (large haybales) [36], as well as in economic and communications systems [37]. Transhumance, for example, has been a long established way of life in the alpine areas of Europe [38–40], South America [41,42], and Asia [43,44] but also North America [45,46], and Australia [47,48]. Likewise, some tracks and roads may be closed due to snow and ice in winter (e.g., some mountain passes) or due to thawed and boggy ground in summer, while frozen lakes and estuaries may allow for direct communication across the ice that is impossible in summer and requires longer, circum-estuary trips or ferry rides [49].

In the urban planning space, the concept of ephemeral landscapes has been applied to peri-urban areas within the urban growth corridor. Traditionally, an expanding urban space saw the peri-urban belt of orchards being replaced by dwellings, while new orchards were developed on former pastures and arable land. More recently, these orchard belts are no longer replanted, with the urban fringe dominated by pastures and arable land that see a lower level of care, especially when already purchased by developers waiting for the right economic conditions to commence construction [50]. Here, the ephemerality is represented by transitory existence between two stable states.

A third 'class' of ephemeral landscapes are temporarily circumscribed cultural landscapes. Commonly, these are transformations planned by authorities, such as garden festivals that transform a peri-urban space into a complex, primarily flower-based tourist landscape replete with structures (e.g., the German Bundesgartenschau) [51,52], or International Exhibitions (Expos) that create temporary communities with usually ephemeral architecture [53]. Likewise, theme parks are imagined ephemeral landscapes [54–56] that will persist as long as they remain commercially viable. Other examples, on a smaller scale, are transformations of roads into urban parklets [57,58], roads along river fronts into temporary beaches [51], town squares converted into Christmas market villages [59] and streets illuminated with Christmas decorations [60,61]. While intended as ephemeral in their 'original' presentations, some elements of these ephemeral transformations may not revert back to the prior status quo once the festival ends, but may be retained, thereby generating a new permanent cultural landscape layer [62,63]. Examples are Balboa Park in San Diego as a remnant of the 1915 Panama–California Exposition [64] or elaborate film sets that have been retained as tourist attractions, such as "Hobbiton' (Lord of the Rings) in New Zealand [65] or 'Tatooine' (Star Wars) in Tunesia [66].

Finally, there are event-based ephemeral landscapes that are reactive to the event and manifest themselves for its duration. An example are the ad hoc testing stations that were established during the COVID-19 pandemic [67,68].

1.2. Personal Agency in Ephemeral Landscapes

While the majority of ephemeral landscapes are generated by natural phenomena or through communal or governmental control, some ephemeral landscapes can be created by individual or personal agency. These landscapes can be created by individual or collective expressions of custom, tradition or political opinion. These can range from sanctioned [69] and unsanctioned ('guerilla') privately created gardens in public spaces [70] to ritualized expressions of nationalism, such as homeowners flying the US flag each year on Independence Day (4 July) [71] or the yellow ribbon campaign during the first Iraq war (1990) [72,73]. Collective, socially acceptable personal agency can generate public landscapes by decorating the exterior of private residences for Halloween [74,75] or Christmas [76]. Communal art projects reclaiming urban spaces, such as yarn bombing [77,78], create ephemeral landscapes that are welcomed or tolerated by the wider community. Similar personal ephemeral art projects, reclaiming urban spaces in the form of graffiti [79–82], are widely perceived as anti-social behavior. While yarn bombing and graffiti are expressions of territoriality in an ephemeral landscape, Halloween and Christmas decorations are individual, yet collective expressions of community-wide customs. The display of national flags as well as political stickers and posters are expressions of political opinion that create a visually recognizable ephemeral landscape.

The temporality of these ephemeral landscapes differs. While some are defined by cyclical events at fixed (Halloween, Christmas) or semi-regular intervals (elections) with a well-defined duration, others can be event-based and manifest themselves for the duration of that event (yellow ribbon campaign).

Another type of ephemeral landscapes are political landscapes created in the run-up to elections, which find their visual expression on election posters on power poles, fences and houses. Election posters are not merely visual artifacts of political messaging but reflect deep-seated power dynamics related to political dominance, socioeconomic stratification, and territorial 'control' as manifested where posters appear, how densely they are distributed, and on whose property they are placed.

1.3. Political Landscapes

Political landscapes are commonly considered in their intangible form, conceptualizing the relationships of political parties with their voters in the social and geographical [83,84] or the digital space [85]. In these contexts, the term 'landscape' is used metaphorically rather than referring to a physical entity [86]. Yet political landscapes can also have a permanent, or at least semi-permanent, tangible form where political power manifests itself in structures of dominance such as the now largely demolished socialist architecture of the former German Democratic Republic [87], or structures of exclusion and inclusion, such as the Berlin Wall, and the border wall between Mexico and the USA [88]. Political landscapes can also be tangible, yet transitory, such as temporary urban destruction following civic riots [89,90], or they can be fully ephemeral, such as election signage of protest signage. There is a limited body of work that considers election signage in the form of yard signs as expressions of political cultural landscapes in the USA [91–94] and papers that contextualize the nature and spatial distribution of protest signage, responding to a natural disaster event in Florida (USA) [95] and responding proposed development in an Australian setting [96]. Some studies have suggested that the socio-economic makeup of a suburb or neighborhood will find expression in the nature and messaging on formal election signage as well as individualized yard signage [91,92,94,97]. A very small number of studies looked at the spatial distribution and density of campaign signage in suburban spaces correlating them with socio-economic and ethnic data [91,92].

In this paper we will consider the ephemeral cultural landscape of an Australian Federal election as it is represented by public election advertising, using a regional community as an example. We will demonstrate that Australian ephemeral landscapes are culturally influential, if not significant, and that they can shape public and private spaces both materially and symbolically. To do so, we will first provide a contextual background on the Australian electoral system as well as the study community. Following an explanation of the methodology espoused, we will discuss the spatial distribution of electoral signage in relation to demographic and socio-economic parameters.

2. Background

2.1. Background to Federal Elections in Australia

The House of Representatives of the Commonwealth of Australia is re-elected every three years as is half of the Senate, unless a double dissolution has been called (when the entire Senate is reelected). The House of Representatives is made up of directly elected representatives chosen by popular vote from all citizens residing and being registered in the 150 electoral divisions (henceforth 'electorates') which are single member constituencies [98].

The electorates are broadly of the same population size (~120,000 people) but because of the uneven spatial distribution of Australia's population, these electorates differ widely in geographical size [98]. Thus they can be spatially very small, such as the electorate of Wentworth which comprises parts of northern Sydney (NSW) with an area of 31.2 km^2 (127,588 voters), or very large, such as the electorate of Durack in Western Australia with an area of 1.41 million km^2 (118,845 voters) [99]. The boundaries of the electorates are defined by an independent, apolitical body, the Australian Electoral Commission. Gerrymandering, a common feature of U.S. politics [100,101], does not exist in Australia. The candidate with the highest number of votes in each electorate after the distribution of preferences (see below) takes the seat in parliament. The Senate is made up of party-nominated candidates based on the proportion of votes that party receives in a given state, with the number of seats per state defined in the constitution of 1901 [98]. Voting in an Australian Federal or State election is compulsory for every citizen over the age of 18 residing in Australia and registered to vote [64].

Voting occurs via in-person voting at any of the voting centers ('polling booths') within the electorate, in-person voting at polling booths outside the electorate (absentee voting), in person voting prior to election day at specific pre-poll voting centers within the electorate (pre-poll voting), or mailed-in ballot papers (postal ballot) and phone votes for people with severe physical disabilities [64]. The distribution and location of polling booths is driven entirely by spatial and access considerations. School halls tend to be the preferred venue as they are held in public hand (land?), are vacant on election days (which are always held on a Saturday) and, being in primary and high schools, tend to be spatially well distributed.

Voting is based on the preferential system, where voters are given printed ballot papers (green for the House of Representatives and white for the Senate), enter a cardboard booth and are required to indicate their preferences by entering a number in delineated boxes next to the candidates' names. For ballots for the House of Representatives, voters are required to number all boxes 1 (for their first preference), 2 (for their second preference), 3 and so on until all boxes are numbered. Failing to number boxes, omitting or doubling up numbers, or annotating the ballot paper in any way results in the ballot paper being declared invalid ('informal') and excluded from the count of the valid ('formal') votes [64]. In the case of the 2025 election, the ballot paper for the electorate of Federal (Farrah?), which contains Albury, carried nine names. The order in which these names are listed are drawn at random [102]. Being on top of the list provides a small vote windfall as some people, simply in order to fulfill their voting obligations, number the candidates sequentially from top to bottom ('donkey vote') [64]. The ballot paper for the Senate comprises the parties arranged (again by random allocation) from left to right. Voters are given the opportunity to vote for a party above a horizontal line on the ballot paper and which case they are obliged to allocate their preference to at least six parties, or they may choose to allocate their preferences to individual candidates, arranged by party, below that line. In that case, 25 candidates need to be allocated sequential preference numbers (Figure 1) [64].

After close of voting, usually 6 p.m. on voting day, all ballots are counted by sorting them according to the first preferences (i.e., boxes with the number 1). The candidate with the lowest number of primary votes is excluded and the ballot papers with first preferences for that candidate redistributed to the other candidates based on the second preference marked on the paper. That exclusion and redistribution process continues until only two candidates remain, with the candidate obtaining the highest number of votes being declared the winner of the parliamentary seat [64].

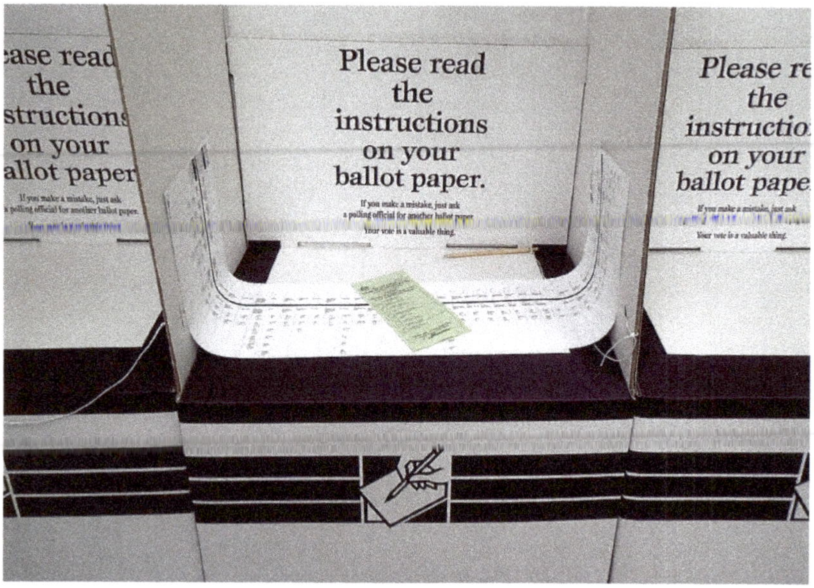

Figure 1. Example of a cardboard voting booth with pencil and the ballot papers for the House of Representatives (green) and the Senate (white). Note how the long Senate ballot paper curls up on both sides. Federal election 2013. (Photograph DHRS 2013).

2.2. Background to Election Advertising

Campaigning in the 2025 Federal election involved a variety of media ranging from 'traditional' physical mailouts, letter box drops and advertising posters, 'corflutes' (weather-resistant 3–5 mm thick sheets of corrugated polypropylene plastic), as well as television advertisements, and social media campaigns via Instagram, and Facebook (Figure 2). One party made 'a name for itself' by sending out unsolicited text messages to people's phones [103]. On election day, party volunteers positioned near the entrances to the polling booths were handing out how to vote cards (Figure 3) that are aimed at undecided voters and that indicate a party's preferred sequence of preferences (Figure 4).

While the public display of election material is permitted from the time an election is called until the end of election day, and while there are rules related to the nature and conduct of election campaigning and advertising [104,105], there are no restrictions from the perspective of electoral law other than a requirement they not be placed within 6 m of a polling place [105,106]. The public posting of election posters and corflutes is governed by laws and regulations relating to political signage as promulgated by state government (for example with regard to placing placards on electricity poles) and local council (in regard to placing placards on private fences and properties that can be seen from the street). In consequence, there is considerable variation across Australia [106–108].

Commonly permitted are advertising posters mounted on private properties as long as they do not exceed the standard ISO A1 size (841 × 594 mm). Advertising material larger than this may require council approval, which may not be granted, requiring candidates to take down the offending posters (Figure S6) [109].

While this is not the venue to discuss in detail the relative merits of advertising and campaign approaches and the target audiences that these are reaching, or the amount of waste that they are generating [110,111], some comments are warranted to contextualize the data set of corflutes that will be drawn on for this paper. Broadly, social media advertising has become a major plank of political messaging prior to and during the election period, largely displacing static media such as election posters, as well as reducing the relevance of TV advertising (Figure 2) [112–114]. Yet, because of the low cost and relative permanence, election posters persist in the tool kit of campaign managers.

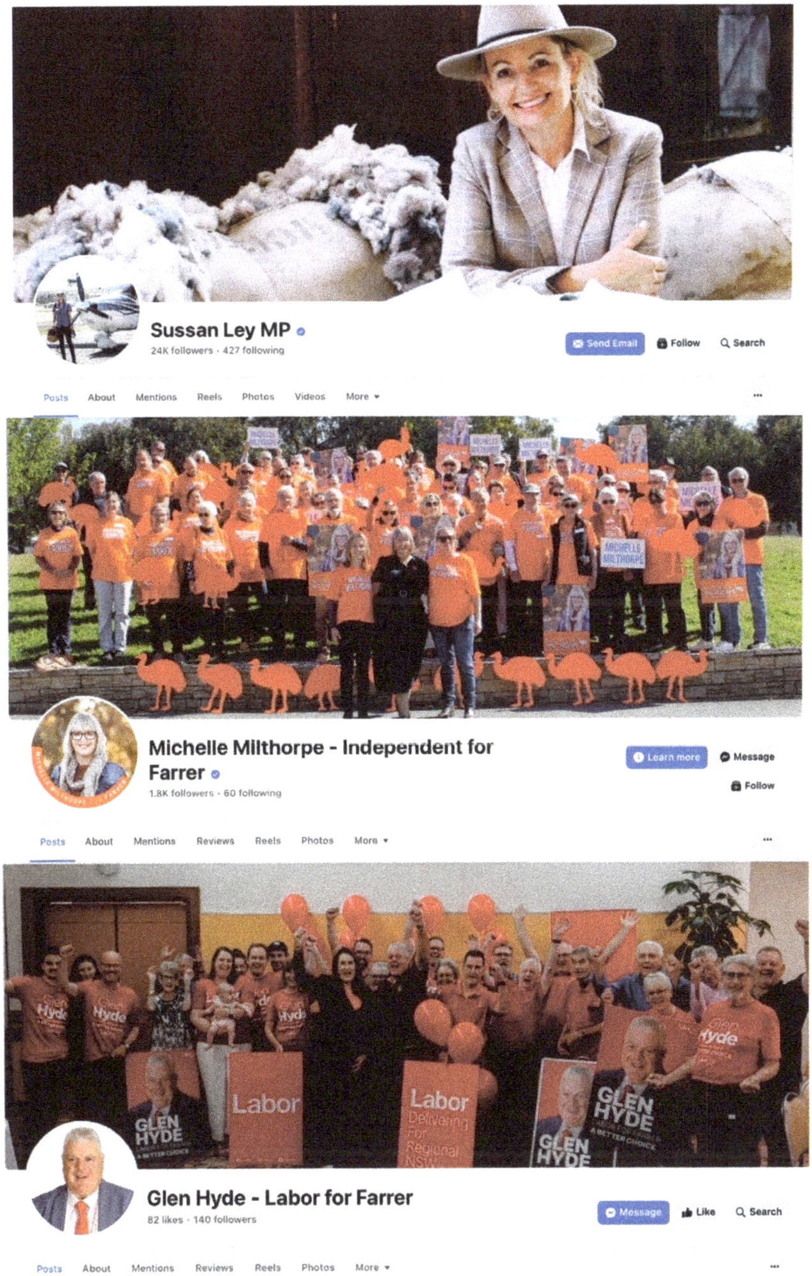

Figure 2. Screen shots of Facebook banners of three candidates polling the most votes in the electorate of Farrer in the 2025 Federal election.

Figure 3. Volunteers handing out how-to-vote cards for the 2025 Federal election at the Thurgoona East polling booth, Albury. Note the corflutes. (Photo DHRS 2025).

Figure 4. A selection of how-to-vote cards produced for the Farrer electorate for the 2025 Federal election (Photo DHRS 2025).

Corflutes, which because of their durability have long since replaced paper and carboard posters, are a well-established means of political communication [115] that provides visual exposure and recognition for candidates, in particular those that lack established recognition (such as incumbency) or exposure in other media [107,115–117]. The placement of corflutes has also been theorized as a means of projecting strength [117,118] and visually claim territory by deliberate and strategic placement in prominent positions, thereby generating free news media coverage [107] through repeated recognition of a familiar face in order to generate a personal vote [119].

2.3. Background to Albury

Albury is a regional service center in the southern Riverina area of New South Wales (Australia) located at the northern banks the Millewa/Murray River [120]. Based on the 2021 census, Albury's population in 2025 was 56,084 people, approximately 77.5% of which were of election age [121]. The local government area (LGA) of Albury comprises the town's historic nineteenth century core in the south (Central and South Albury), early twentieth century subdivisions surrounding that core, mid-twentieth century subdivisions in the north (merging the village of Lavington), late twentieth century subdivisions in the west and north together with the emerging satellite community of Thurgoona in the north east, as well as twenty-first century subdivisions in the north and in the Thurgoona area (Figure 5) [120,122].

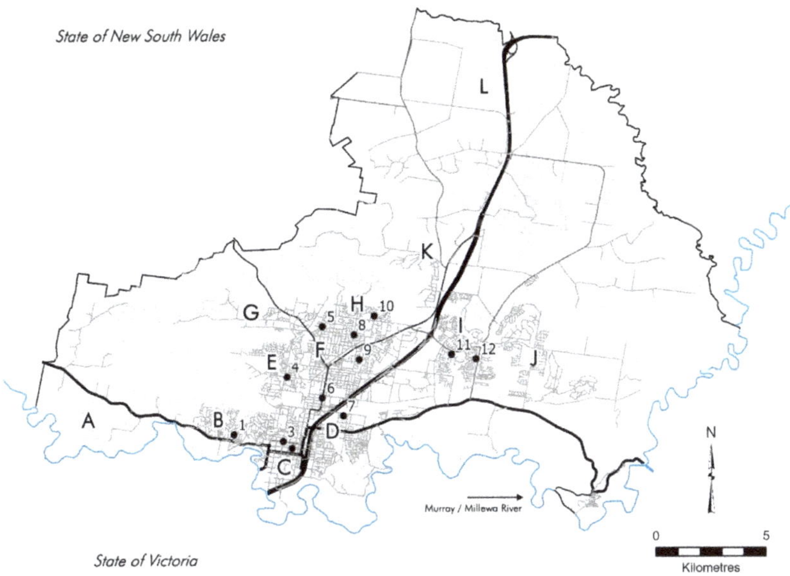

Figure 5. Map showing the Albury local government area with suburb terms used as well as the locations of the polling booths. *Suburbs*: A—Splitters Creek; B—West Albury; C—South Albury; D—East Albury; E—Glenroy; F—Lavington; G—Hamilton Valley; H—Springdale Heights; I—Thurgoona; J—Wirlinga; K—Ettamogah; L—Table Top. *Polling Booths*: 1—Albury West Public School; 2—Albury Public School (Hall); 3—Albury High School; 4—Glenroy Public School; 5—Lavington Public School; 6—Albury North Public School; 7—Albury East; 8—Lavington East Public School; 9—Hume Public School; 10—Springdale Heights Public School; 11—Thurgoona Public School; 12—Thurgoona Community Centre.

2.4. Background to the Federal Electorate of Farrer

Albury is the largest community in the predominantly rural electorate of Farrer, which, with an area of 126,563 km^2, forms the geographically largest electorate in NSW, extending from the South Australian border in the west to Jingellic in the east. The boundaries of the electorate for the 2025 election were circumscribed by the boundaries of local government areas, encompassing the cities of Albury and Griffith (population 27,086) as well as the rural shires of Albury City, Balranald, Berrigan, Carrathool, Edward River, Federation Council, Greater Hume, Hay, Leeton, Murray River, Murrumbidgee, Narrandera, and Wentworth (Figure S1) [123].

Given the largely rural population, the electorate of Farrer is traditionally a politically conservative area that has been held by conservative politicians since its creation in 1949 [124]. In the past seven elections it has been held by the Liberal Party by a considerable margin, ranging from 61.2% in 2007 to 70.5% in 2016 (two-party preferred vote) [125,126]. When considering the two-party preferred votes over time, it should be noted that the boundaries of the electorate had been readjusted to the current boundaries in 2016 [123,127]. This entailed excluding the City of Broken Hill that had been included in the electorate since 2007 [128,129] and that had a more left leaning population.

2.5. Election Results 2025

At the 2025 election, nine candidates nominated for the electorate of Farrer, the positions on the ballot paper were determined on 11 April 2025 [130] (Table S2). The total eligible (enrolled) voting population in the electorate of Farrer for the 2025 election was 123,827 persons [131] with 112,522 votes cast [132]. Of these, 8.4% were informal [133]. The election was called on 28 March 2025 [134]. Pre-polling opened on 21 April, two working weeks before the election on 3 May 2025. Albury was serviced by two pre-polling centers; one located in the Albury CBD and one in Lavington. Pre-poll votes made up 58.5% of the total of 32,133 votes cast in combined Albury polling booths (Table 1).

Table 1. First preferences (in % of formal votes) received at the Albury booths [133]. Abbreviations: G'roy—Glenroy; HS—High School; Lav—Lavington; PS—Public school; SHts—Springdale Heights; Tb Top—Table Top.

		Albury							Lavington			Thurgoona		Tb Top	Pre-Poll	
n°	Name MapCode	PS 2	HS 3	East 7	North 6	West 1	G'roy 4	SHts 10	Hume 9	PS 5	East 8	PS 12	East 11	—	Alb —	Lav. —
1	Ley	27.4	28.8	29.2	26.2	28.3	30.3	27.0	26.3	28.9	29.1	27.4	32.1	38.0	38.7	37.6
2	Hyde	13.3	12.6	10.2	16.1	10.4	17.5	18.1	14.8	15.7	18.2	17.9	13.7	8.1	13.2	15.5
3	Milthorpe	34.0	35.5	36.1	29.5	41.8	27.2	25.8	29.5	25.4	21.4	26.7	29.0	33.9	31.7	26.9
4	Hendrie	10.3	10.9	10.4	9.3	8.1	8.6	8.7	11.8	9.4	10.5	9.4	6.6	4.3	5.4	4.9
5	Hicks	4.5	3.9	5.6	7.2	4.1	6.8	7.9	6.8	7.7	8.5	5.0	5.7	4.6	3.9	5.3
6	Sinclair	3.1	1.8	2.6	2.9	2.0	3.1	3.3	4.7	4.5	3.4	3.7	2.5	4.4	2.0	2.9
7	O'Reilly	2.6	2.4	2.6	2.5	2.1	1.3	2.2	1.8	2.0	2.5	3.3	3.5	2.4	1.7	1.8
8	Scriven	2.8	2.7	1.2	3.8	1.4	2.1	3.5	2.1	4.0	4.1	3.8	4.0	3.5	1.7	2.6
9	Hargraves	2.1	1.4	2.0	2.6	1.7	3.1	3.5	2.3	2.4	2.4	2.8	2.7	0.7	1.7	2.6
	total formal	1443	940	801	1100	801	1265	473	708	1382	v	1513	1091	540	12,372	6600
	% informal	9.9	4.8	7.7	10.6	8.5	9.6	10.4	9.9	8.8	11.3	8.7	6.7	6.1	4.9	9.6
	overall total	1602	987	868	1231	875	1407	549	811	1549	1129	1675	1178	575	13,011	7302

After full distribution of preferences, Sussan Ley won the election with 56.19% of all eligible votes cast [132]. The previously referred to conservative character of the rural spaces in Farrer becomes clear when one considers the percentages of the top five vote winners (Table 2).

Table 2. First preferences (in % of formal votes) cast in Albury compared with votes cast in the rest of the Farrer electorate.

n°	Name	Albury	Rest of Farrer	Formal Votes
1	Ley	34.5	48.2	41,694
2	Hyde	14.3	15.4	14,367
3	Milthorpe	30.0	15.7	19,613
4	Hendrie	6.9	3.5	4439
5	Hicks	5.1	7.3	6247

3. Methodology

3.1. Survey

The survey area focused on the key populated areas of Albury as well the arterial roads in and out of the Albury CBD (Figure 6). The spatial distribution of the public advertising material, comprised on posters or corflutes, cut out party symbols (emus, see below) as well as other election messaging, was mapped via a windscreen survey. The windscreen survey, which was carried out on 1 and 2 May 2025 (both DHRS and DD) as well as the morning of 3 May (election day) (DHRS only), entailed systematically driving all streets in the survey area and recording the presence of public advertising material via a field map data collection form hosted on ArcGIS pro 3.5 [135,136]. The survey area was defined as the core of the Albury LGA, omitting the low-density areas of Splitters Creek in the west and Ettamogah/Table Top in the north (see Figure 5 for locations). Figure 6 shows the streets traversed on each of the three days. Not recorded were the corflutes outside the polling stations on election day, as well as the corflutes outside the two pre polling stations in Albury and Lavington [137]. The data recorded were the location of the election advertising, the party it represented, its type (corflute, emu) and whether it was placed on a public or a private space.

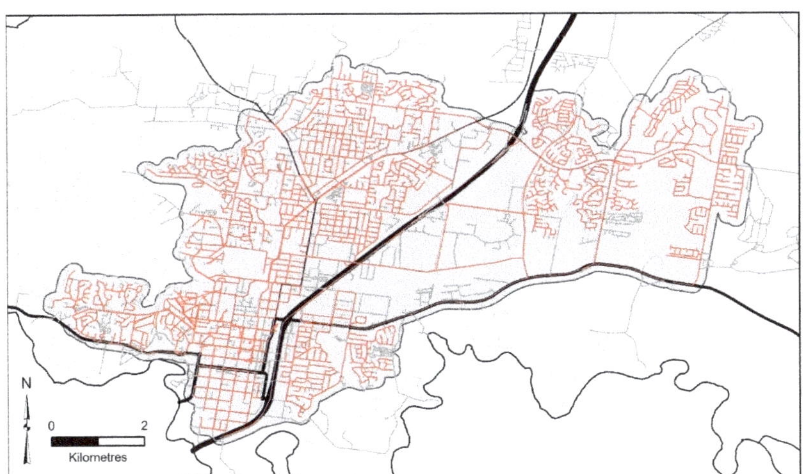

Figure 6. Map showing the area of Albury Local government area surveyed for this study (shaded in grey) with the roads covered in the windscreen survey.

3.2. Analysis

It was posited that most people who had not cast their vote at a pre-poll station would attend the polling booth closest to them. The study area was segmented into twelve polling booth 'catchments' based on a detailed road network including footpaths and consisting of an attribute of length in meters [138]. A Closest Facility analysis established routes from the centroid of each property to the closest polling booth, followed by an aggregation into booth 'catchments' via a Service Area analysis (Figure S2).

The distribution of election posters is plotted against Australian Bureau of Statistics census data (census 2021) at the census mesh block level where specific data sets are available (e.g., population density) [139] or otherwise against Statistical Area Level 1 (cell) data (e.g., Index of Relative Socio-economic Advantage and Disadvantage score) [140].

While census data are presented in 5 year age brackets starting with 0–4, it made more sense to aggregate the ages into generational cohorts (e.g., Generation Y, Baby Boomers) as they share common socio-cultural characteristics (for correlation of cohorts, see Table S4).

3.3. Limitations

The election signage comprises co standard ISO A1 sized (841 × 594 mm) corflutes as well as signs in the shape of emus (large emus ~440 × 550 mm, small emus ~220 × 250 mm) (Figure S3d,e). While corflutes were usually mounted at the property fence or boundary facing the street, some emus were mounted on walls or Colorbond fences that were aligned perpendicular to the street. As these emus could be missed depending on their size and on the direction of vehicle movement, it is likely that the number of orange emus has been undercounted. At least on two occasions such emu signage was noted when the survey sequence required the authors to backtrack on a street that had already been surveyed.

3.4. Statistics

Summary data and frequencies as well as *t*-test probabilities were calculated using MS Excel, while the statistical comparisons of proportions were established with an on-line comparison of proportions calculator [141].

3.5. Auxillary Data

To improve the readability of the paper, a number of additional tables and figures have been related to a supportive data document. These can be identified by the prefix 'S' (e.g., Table S1) and can be accessed from the journal's web page.

4. Results

The survey identified locations with 193 corflutes and other stationary election signage. Signage for Michelle Mithorpe dominated with 77.2% of all locations, followed by the Liberals (10.0%) and the Greens (7.3%) (Table 3). Setting aside the displays that were displayed at the polling booths on election day, all election signage by the Labor Party, the Greens, the Independent and the Shooters, Fishers & Farmers party was mounted on private properties. On the other hand, one third of the Liberal party's election signage and the sole National party corflute were placed on public spaces.

Unlike the traditional parties, which all relied on standard A1-sized corflutes, the campaign of the independent also used orange corflutes in the shape of an emu of varied sizes) either blank or with the stenciled text 'Milthorpe for Farrer') (Figure S3d,e). These orange emus as a symbol of the independent 'Voices of Farrer' campaign, started to appear in some areas well before the candidate had been decided. The symbolism of the orange emu is three-fold: orange as a color for an independent political voice [142] which was successfully used in the success of Independent candidates in the adjacent electorate of Indi [143,144]; the choice of an emu as a typical bird in the Riverina; and the choice of an emu as a bird that symbolizes progress as it cannot easily moved backwards, and that for that reason (like the Kangaroo) is used in the Commonwealth Coat of Arms [145].

Considering only the corflutes promoting the independent candidate, 40.9% of locations displayed just an emu and 35.6% of locations displayed a corflute. Just under a quarter of locations (23.5%) displayed both a corflutes and one or more emus.

Spatially, the majority of corflutes are concentrated in the election booth catchments of central, north central and west Albury, while the lowest numbers were noted in East Albury, Springdale Heights, Lavington and Thurgoona (Table 3). A correlation of the number of a party's corflutes in a polling booth catchment with the percentage of votes received for that party show no statistical significance for any of the four key parties: Liberals (r = 0.262530807), Milthorpe (r = 0.360746872), Labor (r = −0.374007451) or the Greens (r = 0.454955675).

Broadly speaking, it is not surprising that the presence of corflutes appears, at first sight, to be correlated with the density of population (Figure 7). Yet, at a booth catchment level, the number of corflutes is not correlated with the average population density, nor is it correlated at that level with specific generations (baby boomers, Gen X, Gen Y, Gen Z).

Figure 7. Distribution of election advertising mapped against population density (based on census mesh blocks [139]) in the area surveyed for this study (see Figure 6 for details). Map faces north.

A finer-grained analysis at the census mesh block level (Figure 7) showed that the proportion of corflutes per population density quintile is significantly correlated with the frequency of mesh block quintile (r = 0.9864) (Table S5), indicating that the corflute placement follows the general distribution of Albury's population. When aggregating the corflute data into left-leaning (Labor, Greens) and right leaning (Liberals, National, Farmers, Fishers & Shooters) 'traditional' parties, as well the independent (Milthorpe) (Table S6), differences emerge. The corflutes associated with right leaning parties are concentrated in the lower two quintiles of population density. There are significantly more corflutes associated with right leaning parties in the lowest quintile than the frequency of mesh blocks in the quintiles (p = 0.0061), while corflutes associated with the independent are significantly fewer in that quintile (p = 0.0334).

Table 3. Absolute number of corflutes per booth catchment observed during the survey. For map codes see Figure 8.

	Albury							Lavington			Thurgoona			
Booth Name	**PS**	**HS**	**East**	**North**	**West**	**G'roy**	**SHts**	**Hume**	**PS**	**East**	**PS**	**East**	**n**	**%**
Map Code	2	3	7	6	1	4	10	9	5	8	12	11		
Milthorpe (emu)	14	11	1	4	12	10		1	2	1	2	3	61	31.6
Milthorpe (Corflute)	10	13	3	8	9	5	1	2			2		53	27.5
Millthorpe (emu & CF)	6	9	1	3	10	1			1		1	3	35	18.1
Liberal	2	5	1	2	4	3		1	2			1	21	10.9
Greens	2	2		1	3	1		2		2	1		14	7.3
Labor			1		2	1			1				5	2.6
Nationals					1								1	0.5
Farmers, Fishers & Shooters									1				1	0.5
Other (non-party)		1				1							2	1.0
total	34	41	7	18	41	22	1	6	7	3	6	7	193	

The average age of the entire population in the SA1 cells is 40.17 ± 24.41 years (n = 29,584) with the average age of the population eligible to vote 51.36 ± 18.33 years (n = 21,676). The overall population of voting age in the surveyed area (n = 22,283, Statistical Area Level 1 data) comprises 7.8% Generation Z, 32.0% Generation Y, 33.4% Generation X and 26.8% Baby Boomers (and older). A finer-grained analysis of generational data (only possible at Statistical Area Level 1) (Figures S16–S22) showed that the generations are not evenly distributed in the surveyed area with both Generation Z (Figure S19) and Baby Boomers (Figure S17) showing some clustering. The distribution of population density quintiles of the generation of Baby Boomers and older is skewed as they are clustered in those SA1 cells that contain retirement homes.

Looking at the frequency distribution of quintiles of total the population in the 69 SA1 cells, the majority of the cells belong to the fourth (44.8%) and third quintile (31.3%) (Table S7). Not surprisingly, using this frequency distribution as a base, the overall distribution of corflutes mirrors the distribution of the population. When looking at the generational breakdown, however, differences emerge (Table S7). Corflutes tend to be concentrated in areas with a low density of Baby Boomers r = −0.6018), Generation Y ($r = -0.0220$) and Generation Z ($r = -0.2763$), which suggests it is not these generations that reflect the posting of the corflutes. The frequency distribution of Generation X (SA1 ages 46–64) (Table S7), however, correlates very well with the distribution total the population ($r = 0.9601$)

With six exceptions (of a total of 21), the Liberal Party corflutes were all displayed tied to the poles of traffic signs along the arterial roads of Albury (Figure S20) reflecting party-organizational placement of these with little personal engagement. In contrast, the corflutes of all other candidates were displayed on private property.

Albury has seen period expansion of its urban envelope from the core in the south to the north (Lavington, Springdale Heights, Hamilton Valley), west (West Albury), east (East Albury), and north-east (Thurgoona, Wirlinga) (Figure 5). Correlating the distribution of corflutes with the growth periods of Albury showed that greatest number of corflutes was placed in the oldest (pre-1930s, 13.3%) and older suburbs (1930s, 35.7%), with a further peak in the 1970s (14.8%). There is no significant correlation with the distribution of the overall mesh blocks per period with the actual distribution of the corflutes (Tables S9 and S10).

Figure 8 shows the distribution of corflutes against the Index of Relative Socio-economic Advantage and Disadvantage score (ISRAD) expressed in quintiles. There is no significant correlation between the number of corflutes and the average ISRAD when assessed at the booth catchment at the overall level ($r = 0.2346$) nor at candidate levels

(Ley $r = -0.3097$; Milthorpe $r = 0.3003$) (Table 4). A finer-grained analysis at Statistical Area Level 1, however, shows that Statistical Areas with a higher ISRAD score also show a higher number of corflutes. When considering the distribution of corflutes displayed by supporters of the independent candidate, then residents in areas with the highest ISRAD score displayed very significantly more corflutes than residents in areas with the second highest ISRAD score ($p = 0.0051$), which in turn displayed very significantly more corflutes than residents in areas with the third highest ISRAD score ($p = 0.0056$). The distribution of corflutes by the other parties showed no significant differences which may be due to the small sample size for each of these. Worth noting is that there are no differences in corflute distribution between the 'traditional' parties when the data are concatenated into left-leaning (Labor, Greens) and right leaning (Liberals, National, Farmers, Fishers & Shooters) and the ISRAD score categories are aggregated into 'low' (two bottom quintiles) and high (two upper quintiles).

Table 4. Frequency of corflutes (in %) per Index of Relative Socio-economic Advantage and Disadvantage (ISRAD) quintile based on Statistical Area Level 1 boundaries. FFS—Farmers, Fishers & Shooters.

ISRAD Score	Milthorpe	Labor	Green	Liberals	FFF	Nationals	All
646–796	—	20.0	7.1	9.5	100.0	—	2.6
797–895	18.8	—	35.7	19.0	—	—	19.4
896–972	13.4	40.0	7.1	28.6	—	—	15.2
973–1037	26.2	20.0	35.7	23.8	—	100.0	26.7
1038–1129	41.6	20.0	14.3	19.0	—	—	36.1
n	149	5	14	21	1	1	191

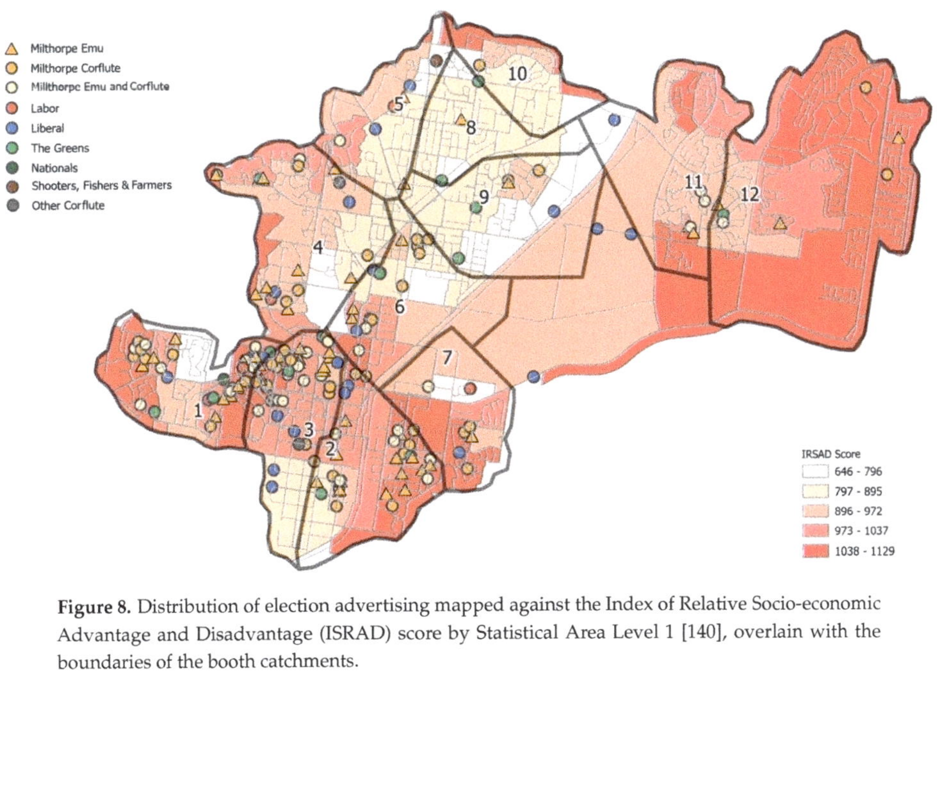

Figure 8. Distribution of election advertising mapped against the Index of Relative Socio-economic Advantage and Disadvantage (ISRAD) score by Statistical Area Level 1 [140], overlain with the boundaries of the booth catchments.

5. Discussion

While election posters are commonly interpreted as spatial visual communications [115,146–148], they exist within the context of the urban space in which they are placed. This space is not neutral, but shaped by power, practice, and ideology [149]. The spatial distribution of election and protest posters mirrors power relations, access to visibility, and contested territories that can be interpreted as a cultural landscape of political power [97,150], race relations [91,92] and community identity [95,96]. From a cultural landscape perspective, the deliberate and strategic placement of election corflutes in prominent positions represents, in visual form, the claiming of territory by the candidate [107]. In a theoretical framing, prime poster locations at busy intersections, commuter corridors, or public squares, are more than neutral display surfaces. Given that visibility is a form of power, such locations represent high-value political real estate, establishing a perception of territoriality [96].

In marginal and thus hotly contested electorates, this territoriality manifests itself in 'poster wars,' where mundane surfaces are turned into arenas of contestation and control. These 'wars' are characterized by a monopolization and saturation of prime spaces, as well as unauthorized removal (Figure 9 and Figure S13), vandalism and defacement [150,151]. The practice of layering, that is pasting posters on top of those by opposing parties, which had been common in earlier campaigns, has largely ceased with the introduction of the stiff corflutes.

Figure 9. Election posters at the Albury Public School polling booth. Election day, 3 May 2025, 7 a.m. Note the vandalized poster of the Australian Labor party on the ground.

These traditional poster or corflute wars were absent in Albury during the 2025 Federal election, with the exception of the fences at some polling booths (Figure 9) where proximity to the entrance gate was 'premium real estate' (Figure S12). While Sussan Ley as the long-term incumbent could rely on established recognition and a personal vote, [119] the Independent candidate Michelle Milthorpe, without a personal electoral track record and without the backing of the brand recognition afforded by a political party, had established public recognition. Thus, the corflutes of the Liberal party were primarily placed, at random intervals and locations, on the poles of traffic signs on Albury's arterial roads, signaling a top-down placement approach with limited citizen involvement.

As corflutes provide considerable visual exposure and recognition for candidates that lack established recognition or longer-term exposure in other media [107,115–117,146], the Milthorpe campaign relied on sympathizers to display the corflutes on their properties. This strongly signaled grassroots support through individual endorsement and personal engagement, suggesting that residents actively participated in shaping the visual political landscape. The creative and symbolic use of orange corflutes shaped like emus, used even before candidacy was confirmed, illustrates a grassroots and identity-driven strategy distinct from traditional campaign methods. The resonance of the orange emus as both a campaign emblem and a protest symbol suggests a deeper connection of political messaging with local values and dissatisfaction with traditional representation. Much like yarn bombing or Christmas lights, they represent and reflect personal and collective agency, and, collectively, demonstrate community-driven expressions of identity [96]. They not only map a transient landscape of political expression and social affiliation but are also situated within a lineage of bottom-up production of ephemeral cultural landscapes [94].

While the visual ephemeral landscape of an election is legally circumscribed as the time period between the day the election was called (in this case, 28 March 2025) and election day itself (in this case, 3 May 2025) [134], this was not necessarily true in the case discussed here. Under 'normal' circumstances, it takes a few days after an election is called to ramp up the production, distribution and placement of corflutes. In consequence, the density and spatial patterning of electoral signs will increase in the first two weeks and plateau towards election day. Normally, election posters that had been placed in public spaces are taken down on election day as soon as the polls have closed or on the day after, because leaving posters hanging for a longer time tends to generate public backlash. Election posters placed on private property are also usually removed on the day of, or the day after the election. Thus, the visual urban landscape undergoes a temporally constrained transformation.

While the established parties (Liberal, Labor, Greens, etc.) followed that pattern, the independent campaign differed in that regard. Established in June 2023 as a grass roots campaign to emulate the success of an Independent candidate in an adjacent electorate [152], the groups selected their candidate in September 2024 [153]. Well before the candidate was selected, the orange emu had been adopted as campaign logo and had begun to appear on people's fences and walls from April 2024 onwards, as a sign of protest and need for political change. Once the election was called, these were augmented by standard A1 corflutes. This multi-phase approach reflects staged messaging tactics and movement-building prior to electioneering. While all standard A1-sized corflutes promoting the independent candidate have long since been removed, several of the orange emus still persist at the time of writing, 60 days after the polls closed (Figure 10B). The fact that several of these emus still persist well past the election period attests to their dual function as protest symbols as well as election posters. This echoes observations in Queensland where individually designed and executed protect signage persisted for a prolonged period [96]. It is unclear how long these residual emus will persist. Ultimately it depends on the will of their owners continuing to display them and the physical preservation of the emu signage. Given that they are made from corflute material (polypropylene plastic) covered with orange paint, they are prone to an, albeit slow, decay due to exposure to UV radiation once the orange paint starts to break down.

While everybody has their own cultural positioning in terms of spirituality/religion and political leanings, it is a commonly accepted notion for most, but not all members of society, that such polarizing topics best be avoided in 'polite conversation' as well as at the workplace [154]. Setting aside issues based protest movements, such as the anti-nuclear debate of the 1970s and 1980s [155,156], only a minority will carry their values 'on their sleeve' and express them through buttons, car stickers or postcards [157–161]. On occasion, controversial development proposals can result in protest signage that crosses standard political boundaries [96]. With the exception of a usually uncontestable public expression of national pride in the form of the national flag being flown on private flag poles, people's political positioning, however, is rarely expressed in the urban landscape. This changes during election times.

Figure 10. A corflute and multiple emus promoting the independent candidate, Michelle Milthorpe. Seen in Central Albury. (**A**) prior to election day, 2 May 2025; (**B**) persistent presence of the emus, nine weeks after the election, 5 July 2025.

Depending on what is (perceived to be) at stake at the election, supporters of political parties will engage in a public display of political protest and allegiance, which shapes public and private spaces materially and symbolically [91,94]. Personal observations in the author's neighborhood suggest that this public display is not instant/overnight, but grows incrementally during the declared election campaign period, culminating in the last days before election day itself. Gradually the visual appearance of a streetscape is reshaped from politically neutral to engaged by property owners taking a stance and signaling this to all passers-by. While it can be surmised that personal agency and engagement in the political process are the motivating factors, it appears that people are initially hesitant to project their own political opinions in such a public fashion but follow once others have taken the step. Future research into the motivations and timing could prove a rich filed of enquiry at the next state or Federal election.

The interpretation of contemporary urban landscapes is commonly limited to an understanding of broad power relationships that position places of authority (government buildings, religious structures) against places of commerce (shops) and private residences. Visual elements of streetscapes, such as established leafy suburbs and heritage precincts/conservation areas, or modern ostentatious housing contrasted with social housing complexes, signal socio-economic strata within a community. But they do not illuminate the cultural and political positioning of the community let alone in its spatial patterns.

While ephemeral, these temporally constrained political landscapes project individual cultural and political positioning beyond the confines of family and personal acquaintances. At a time when general disengagement with established political parties is on the rise [162] in particular among the youth [163], Independent candidates have gained traction [143,164]. The cultural landscape of election signage reflects this well. The cultural landscape elements generated by established parties are both top-down in nature and overall limited—which may well be a tangible expression of a sentiment that takes the electorate of Farrer for granted. The cultural landscape of independent campaigns is generally characterized by bottom-up active construction and engagement across the community. The abundance of election signage for the Independent candidate, all of which was displayed on private property, reflects this well. When seen in juxtaposition to the election signage displayed by, and for, the established parties, the public expression of support for the Independent candidate visually acts a sign of the public's political voice, of protest and community power [94,96]. Although ephemeral and transient, election signage generates a culturally influential landscape.

Finally, a major unknown is the future of such ephemeral, political cultural landscapes. Whereas in the past media advertising was limited to free-to-air and commercial television as well as print media, the advent of social media has reshaped the communications landscape [165]. This entails passive media (such as campaign Facebook pages) but also active media in the form of advertisements inserted into people's personal social media feeds. Facebook advertising has been commonplace since the late 2010s [166–168], followed by Instagram [169,170] and most recently TikTok [171,172]. Social media campaigning allows candidates to project both a more detailed, a more nuanced and, significantly, also a more immediate response to election issues as they emerge, providing greater exposure to lesser-known candidates. Location and demographic data of Facebook users allows political parties to target specific demographics in geographically circumscribed locations (state, region) with focused messaging [165,173]. Anecdotally, based on personal observations of the authors, the overall number of election signage posted during Federal elections at arterial roads and public spaces has continually declined over the past decade. Similar decreases have been observed elsewhere [93]. While the overall trend may point downwards, with ongoing supplantation by digital media, the case study discussed in this paper has shown that grassroots mobilization still can generate a tangible, albeit ephemeral, political cultural landscape.

While these digital political cultural landscapes are likewise ephemeral, the individual elements (posts), lack the ontic dimensions of materiality and thus will be difficult to accurately document and collect [5]. Future research should explore the dissonance between traditional tangible cultural political landscapes that can be witnessed by every person in the community and the emerging digital political landscapes that can be witnessed only by the viewer and that may well be highly personalized.

6. Conclusions

This paper has demonstrated that ephemeral cultural landscapes, such as election signage, are not merely passive backdrops to political events but actively shape and reflect political engagement, territorial expression, and community identity. The 2025 Federal election in Albury provided a valuable case study in which the distribution, form, and persistence of campaign materials reflected divergent political strategies, levels of community involvement, and socio-spatial dynamics. The marked contrast between the top-down placement of Liberal Party corflutes on public infrastructure and the grassroots, individually endorsed deployment of independent candidate Michelle Milthorpe's signage, particularly the symbolic orange emus, highlights differing models of political engagement and visibility.

Significantly, the study reveals that ephemeral landscapes can transcend their designed temporality. The continued presence of the orange emus beyond the legislated campaign period illustrates how visual elements of political communication can acquire extended symbolic lives, becoming enduring features in the socio-political landscape. This challenges conventional understandings of election advertising as temporally bounded and instead frames it as part of a broader continuum of community expression, resistance, and identity formation.

This study expands the concept of cultural landscapes by highlighting the significance of ephemeral political signage as a transient yet meaningful layer of spatial expression. It demonstrates that cultural landscapes are not solely shaped by enduring structures or historic use, but also by short-lived, community-driven interventions that reflect identity, agency, and political sentiment. The research reveals how such ephemeral elements, especially when grassroots in origin, can challenge dominant spatial narratives and contribute to evolving cultural memory. By foregrounding temporality and personal engagement, the study offers a nuanced perspective on how cultural landscapes are continuously produced, negotiated, and contested.

Supplementary Materials: The following supporting information can be downloaded at: https://www.mdpi.com/article/10.3390/land14081610/s1. References [109,133,139,140,174] are cited in the Supplementary Materials.

Author Contributions: Conceptualization: D.H.R.S.; Methodology: D.H.R.S.; Data collection: D.H.R.S. and D.D.: Formal analysis: D.H.R.S. and D.D.; Writing—original draft preparation: D.H.R.S. and D.D.; Writing—review and editing: D.H.R.S. and D.D.; Visualization: D.H.R.S. and D.D. All authors have read and agreed to the published version of the manuscript.

Funding: This research received no external funding.

Institutional Review Board Statement: Not applicable.

Data Availability Statement: The data file can be accessed at https://doi.org/10.26189/767de167-7457-4e53-829a-0d59dff4cf9f.

Conflicts of Interest: The authors declare no conflicts of interest.

References

1. Papayannis, T.; Howard, P. Nature as heritage. *Int. J. Herit. Stud.* **2007**, *13*, 298–307. [CrossRef]
2. Smith, L.; Akagawa, N. *Intangible Heritage*; Routledge: London, UK, 2009.
3. Parker, M.; Spennemann, D.H.R. Classifying sound: A tool to enrich intangible heritage management. *Acoust. Aust.* **2021**, *50*, 23–39. [CrossRef]
4. Parker, M.; Spennemann, D.H.R.; Bond, J. Sensory perception in cultural studies—A review of sensorial and multisensorial heritage. *Senses Soc.* **2024**, *19*, 231–261. [CrossRef]
5. Spennemann, D.H.R.; Spennemann, R.J. The New Frontier of Heritage Management: Conceptualizing, Defining and Managing the Heritage of the Digital Age. *Heritage*, 2025; *submitted*.
6. de la Torre, M. Values and heritage conservation. *Herit. Soc.* **2013**, *6*, 155–166. [CrossRef]
7. Fredheim, L.H.; Khalaf, M. The significance of values: Heritage value typologies re-examined. *Int. J. Herit. Stud.* **2016**, *22*, 466–481. [CrossRef]
8. Spennemann, D.H.R. The Shifting Baseline Syndrome and Generational Amnesia in Heritage Studies. *Heritage* **2022**, *5*, 2007–2027. [CrossRef]
9. Spennemann, D.H.R. The Nexus between Cultural Heritage Management and the Mental Health of Urban Communities. *Land* **2022**, *11*, 304. [CrossRef]
10. Smith, G.S.; Messenger, P.M.; Soderland, H.A. *Heritage Values in Contemporary Society*; Routledge: Abingdon, UK, 2017.
11. Mydland, L.; Grahn, W. Identifying heritage values in local communities. *Int. J. Herit. Stud.* **2012**, *18*, 564–587. [CrossRef]
12. Spennemann, D.H.R. Futurist rhetoric in U.S. historic preservation: A review of current practice. *Int. Rev. Public Nonprofit Mark.* **2007**, *4*, 91–99. [CrossRef]
13. Spennemann, D.H.R. What actually is a Heritage Conservation Area? A Management Critique based on a Systematic Review of NSW Planning Documents. *Heritage* **2023**, *6*, 5270–5304. [CrossRef]
14. Spennemann, D.H.R. *The Cultural Landscape of the World War II Battlefield of Kiska, Aleutian Islands. Findings of a Cultural Heritage Survey, Carried Out in June 2009*; Institute for Land, Water and Society, Charles Sturt University: Albury, Australia, 2011; p. 552.
15. Mitin, I. Constructing urban cultural landscapes & living in the palimpsests: A case of Moscow city (Russia) distant residential areas. *Belgeo Rev. Belg. Géogr.* **2018**, *4*. [CrossRef]
16. Vâlceanu, D.-G.; Kosa, R.-A.; Tamîrjan, D.-G. Urban landscape as palimpsest. *Urban. Arhit. Constr.* **2014**, *5*, 17.
17. Phillips, J. Indigenous Australian Studies, indigenist standpoint pedagogy, and student resistance. In *The Oxford Encyclopedia of Global Perspectives on Teacher Education*; Oxford University Press: Oxford, UK, 2019; Volume 257.
18. Baird, M.F. 'The breath of the mountain is my heart': Indigenous cultural landscapes and the politics of heritage. *Int. J. Herit. Stud.* **2013**, *19*, 327–340. [CrossRef]
19. Carter, J. Displacing indigenous cultural landscapes: The naturalistic gaze at Fraser Island World Heritage Area. *Geogr. Res.* **2010**, *48*, 398–410. [CrossRef]
20. Campolo, D.; Bombino, G.; Meduri, T. Cultural landscape and cultural routes: Infrastructure role and indigenous knowledge for a sustainable development of inland areas. *Procedia-Soc. Behav. Sci.* **2016**, *223*, 576–582. [CrossRef]
21. Carter, P. *Naminng No Man's Land: Postcolonial Toponymies*; Palgrave Macmillan Cham: Cham, Switzerland, 2024.
22. Carter, P. Proper Names: Differences Between Aboriginal and Colonial Toponymy. In *Naming No Man's Land: Postcolonial Toponymies*; Springer: Berlin/Heidelberg, Germany, 2024; pp. 59–94.
23. Tentand, J.; Blair, D. Motivations for naming: The development of a toponymic typology for Australian placenames. *Names* **2011**, *59*, 67–89. [CrossRef]
24. Parker, M.; Spennemann, D.H.R.; Bond, J. The Effect of the COVID-19 Pandemic on Heritage Festival Soundscapes—A Critical Review of Literature. *Curr. Pollut. Rep.* **2024**, *10*, 277–285. [CrossRef]
25. Parker, M.; Spennemann, D.H.R. To ring or not to ring: What COVID-19 taught us about religious soundscapes in the community. *Heritage* **2022**, *5*, 1676–1686. [CrossRef]
26. Parker, M.; Spennemann, D.H.R. Conceptualising sound making and sound loss in the urban heritage environment. *Int. J. Urban Sustain. Dev.* **2022**, *14*, 264–286. [CrossRef]
27. Henshaw, V. *Urban Smellscapes: Understanding and Designing City Smell Environments*; Routledge: Abingdon, UK, 2013.
28. Parker, M.; Spennemann, D.H.R.; Bond, J. Using destination reviews to explore tourists' sensory experiences at Christmas markets in Germany and Austria. *J. Herit. Tour.* **2023**, *19*, 172–200. [CrossRef]
29. Bedoya, J.L.; Mikkan, R.A.; Díaz, M.V. Ephemeral geomorphodiversity: Conceptual debate, valuation, and heritage management. *Geomorphology* **2025**, *471*, 109569. [CrossRef]
30. Kingsford, R. *Lake Eyre Basin Rivers: Environmental, Social and Economic Importance*; CSIRO Publishing: Canberra, Australia, 2017.

31. Chen, Y.; Syvitski, J.P.; Gao, S.; Overeem, I.; Kettner, A.J. Socio-economic impacts on flooding: A 4000-year history of the Yellow River, China. *Ambio* **2012**, *41*, 682–698. [CrossRef]
32. Liu, X.; Watts, R.J.; Allan, C. Local ecological knowledge and wise use of ephemeral wetlands: The case of the Cowal system, Australia. *Wetl. Ecol. Manag.* **2023**, *31*, 791–804. [CrossRef]
33. Atha, M. Ephemeral landscapes. In *The Routledge Companion to Landscape Studies*; Routledge: Abingdon, UK, 2018; pp. 113–126.
34. Wright, K. Armidale's Imported Autumn. *Transform. J. Media Cult.* **2012**, *21*, 1–14.
35. McClellan, A. *The Cherry Blossom Festival: Sakura Celebration*; Bunker Hill Publishing, Inc.: Chicago, IL, USA, 2005.
36. Brassley, P. On the unrecognized significance of the ephemeral landscape. *Landsc. Res.* **1998**, *23*, 119–132. [CrossRef]
37. Palang, H.; Fry, G.; Jauhiainen, J.S.; Jones, M.; Sooväli, H. Landscape and seasonality—Seasonal landscapes. *Landsc. Res.* **2005**, *30*, 165–172. [CrossRef]
38. Luick, R. Transhumance in Germany. In Proceedings of the European Forum on Nature Conservation and Pastoralism, Bunschoten, The Netherlands, 22–24 October 2008; pp. 137–154.
39. Davies, E. The patterns of transhumance in Europe. *Geography* **1941**, *26*, 155–168.
40. Jurt, C.; Häberli, I.; Rossier, R. Transhumance farming in Swiss mountains: Adaptation to a changing environment. *Mt. Res. Dev.* **2015**, *35*, 57–65. [CrossRef]
41. Stewart, N.R.; Belote, J.; Belote, L. Transhumance in the central Andes. *Ann. Assoc. Am. Geogr.* **1976**, *66*, 377–397.
42. Baied, C.A. Transhumance and land use in the Northern Patagonian Andes. *Mt. Res. Dev.* **1989**, *9*, 365–380. [CrossRef]
43. Duncan, A.J.; Rahman, A.; Miller, D.; Frutos, P.; Gordon, I.J.; Rehman, A.-U.; Baig, A.; Ali, F.; Wright, I.A. Transhumance livestock production in the Northern Areas of Pakistan: Nutritional inputs and productive outputs. *Agric. Ecosyst. Environ.* **2006**, *117*, 195–204. [CrossRef]
44. Shirasaka, S.; Watanabe, T.; Song, F.; Liu, J.; Miyahara, I. Transhumance in the Kyrgyz Pamir, Central Asia. *Geogr. Stud.* **2014**, *88*, 80–101. [CrossRef]
45. Huntsinger, L.; Forero, L.C.; Sulak, A. Transhumance and pastoralist resilience in the western United States. *Pastoralism* **2010**, *1*, 9–36.
46. White, L. Transhumance in the sheep industry of the Salt Lake Region. *Econ. Geogr.* **1926**, *2*, 414–425. [CrossRef]
47. Scott, P. Transhumance in Tasmania. *N. Z. Geogr.* **1955**, *11*, 155–172. [CrossRef]
48. O'Connor, M. Mapping Australia's transhumance: Snow lease and stock route maps of NSW. *Globe* **2004**, *56*, 13–24.
49. Palang, H.; Sooväli, H.; Printsmann, A. Seasonality and landscapes. In *Seasonal Landscapes*; Palang, H., Printsmann, A., Sooväli, H., Eds.; Springer: Dordrecht, The Netherlands, 2007; pp. 1–16.
50. Qviström, M.; Saltzman, K. Exploring landscape dynamics at the edge of the city: Spatial plans and everyday places at the inner urban fringe of Malmö, Sweden. *Landsc. Res.* **2006**, *31*, 21–41. [CrossRef]
51. Madanipour, A. Ephemeral landscape and urban shrinkage. *Landsc. Res.* **2017**, *42*, 795–805. [CrossRef]
52. Berressem, H.; Berressem, K.-G. *Das Niddatal: Der Umbau einer Landschaft für die Bundesgartenschau*; Die Grünen Frankfurt: Frankfurt am Main, Germany, 2007; p. 22.
53. Monclús, J. International exhibitions and urban design paradigms. In *Exhibitions and the Development of Modern Planning Culture*; Freestone, R., Amati, M., Eds.; Routledge: Oxfordshire, UK, 2016; pp. 241–258.
54. Baker, C. The prevalence of storyworlds and thematic landscapes in global theme parks. *Ann. Tour. Res. Empir. Insights* **2023**, *4*, 100080. [CrossRef]
55. Young, T.G.; Riley, R.B. *Theme Park Landscapes: Antecedents and Variations*; Dumbarton Oaks, Trustees for Harvard University: Washington, DC, USA, 2002; Volume 20.
56. Borrie, W.T. Disneyland and Disney World: Constructing the environment, designing the visitor experience. *Loisir Soc.* **1999**, *22*, 71–82. [CrossRef]
57. Jarman, N.; Stratford, E. Whose rights to the city? Parklets, parking, and university engagement in urban placemaking. *Aust. Geogr.* **2024**, *55*, 115–136. [CrossRef]
58. Stevens, Q.; Morley, M. The contested value of parklets. *J. Urban Aff.* **2024**, 1–19. [CrossRef]
59. Spennemann, D.H.R.; Parker, M. "Endlich wieder Strietzeln": The restart of German Christmas Markets after the COVID-19 pandemic. *Int. J. Event Festiv. Manag.* **2025**. [CrossRef]
60. Edensor, T.; Millington, S. Illuminations, class identities and the contested landscapes of Christmas. *Sociology* **2009**, *43*, 103–121. [CrossRef]
61. Eldridge, A.; Pappalepore, I. Festive space and dream worlds: Christmas in London. In *Destination London: The Expansion of the Visitor Economy*; University of Westminster Press London: London, UK, 2019; pp. 183–203.
62. Loschwitz, G. Neue Landschaften-Garten nach dem Uranerzabbau Bundesgartenschau 2007 in Gera und Ronneburg. *Gart. Landsch.* **2007**, *117*, 4.
63. Minner, J.; Abbott, M. How urban spaces remember: Memory and transformation at two Expo sites. *Int. Plan. Hist. Soc. Proc.* **2018**, *18*, 1063–1074.

64. Amero, R.W. The Making of the Panama-California Exposition, 1909–1915. *J. San. Diego Hist.* **1990**, *36*, 1–47.
65. Singh, K.; Best, G. Film-induced tourism: Motivations of visitors to the Hobbiton movie set as featured in the Lord of the Rings. In Proceedings of the International Tourism and Media Conference Proceedings, Melbourne, Australia, 24–26 November 2004; pp. 98–111.
66. Escher, A.; Riempp, E.; Wust, M. Auf den Spuren von Sternenkriegern und Seepiraten. Auswirkungenv on Hollywoodfilmenin Tunesien. *Geogr. Rundsch.* **2008**, *60*, 42–48.
67. Spennemann, D.H.R. COVID-19 on the ground: Heritage sites of a pandemic. *Heritage* **2021**, *3*, 2140–2162. [CrossRef]
68. Spennemann, D.H.R. *Ephemeral COVID-19 Testing Sites in Central Manhattan, New York City (USA)*; School of Agricultural, Environmental and Veterinary Sciences; Charles Sturt University: Albury, Australia, 2025. [CrossRef]
69. Hunt, S.; Maher, J.; Swapan, M.S.H.; Zaman, A. Street Verge in Transition: A Study of Community Drivers and Local Policy Setting for Urban Greening in Perth, Western Australia. *Urban Sci.* **2022**, *6*, 15. [CrossRef]
70. Mikadze, V. Ephemeral urban landscapes of guerrilla gardeners: A phenomenological approach. *Landsc. Res.* **2015**, *40*, 519–529. [CrossRef]
71. Hansen, D.G.; Ryan, M.P. Public ceremony in a private culture: Orange county celebrates the fourth of July. In *Postsuburban California. The Transformation of Orange County since World War II*; Kling, R., Olin, S.C., Poster, M., Eds.; University of California Press: Los Angeles, CA, USA, 1991; pp. 165–189.
72. Santino, J. Yellow ribbons and seasonal flags: The folk assemblage of war. *J. Am. Folk.* **1992**, *105*, 19–33. [CrossRef]
73. Pershing, L.; Yocom, M.R. The yellow ribboning of the USA: Contested meanings in the construction of a political symbol. *West. Folk.* **1996**, *55*, 41–85. [CrossRef]
74. Santino, J. Halloween in America: Contemporary customs and performances. *West. Folk.* **1983**, *42*, 1–20. [CrossRef]
75. Rogers, N. *Halloween: From Pagan Ritual to Party Night*; Oxford University Press: Oxford, UK, 2002.
76. Werner, C.M.; Peterson-Lewis, S.; Brown, B.B. Inferences about homeowners' sociability: Impact of Christmas decorations and other cues. *J. Environ. Psychol.* **1989**, *9*, 279–296. [CrossRef]
77. Mann, J. Towards a politics of whimsy: Yarn bombing the city. *Area* **2015**, *47*, 65–72. [CrossRef]
78. Hahner, L.A.; Varda, S.J. Yarn bombing and the aesthetics of exceptionalism. *Commun. Crit./Cult. Stud.* **2014**, *11*, 301–321. [CrossRef]
79. Murphy, K.D.; O'Driscoll, S. The art/history of resistance: Visual ephemera in public space. *Space Cult.* **2015**, *18*, 328–357. [CrossRef]
80. Cercleux, A.-L. Graffiti and street art between ephemerality and making visible the culture and heritage in cities: Insight at international level and in Bucharest. *Societies* **2022**, *12*, 129. [CrossRef]
81. Edwards-Vandenhoek, S. You aren't here: Reimagining the place of graffiti production in heritage studies. *Convergence* **2015**, *21*, 78–99. [CrossRef]
82. Dovey, K.; Wollan, S.; Woodcock, I. Placing graffiti: Creating and contesting character in inner-city Melbourne. *J. Urban Des.* **2012**, *17*, 21–41. [CrossRef]
83. Marks, G.; Attewell, D.; Rovny, J.; Hooghe, L. The changing political landscape in Europe. In *The EU Through Multiple Crises*; Routledge: Abingdon, UK, 2020; pp. 20–44.
84. Bjarnadóttir, J.Ý. How parties respond to protests in a changing political landscape. *J. Eur. Public Policy* **2025**, *32*, 2068–2097. [CrossRef]
85. Khazraee, E. Mapping the political landscape of Persian Twitter: The case of 2013 presidential election. *Big Data Soc.* **2019**, *6*, 2053951719835232. [CrossRef]
86. Förster, F.; Grossmann, R.; Iwe, K.; Kinkel, H.; Larsen, A.; Lungershausen, U.; Matarese, C.; Meurer, P.; Nelle, O.; Robin, V.; et al. What is Landscape? Towards a Common Concept within an Interdisciplinary Research Environment. *eTOPOI* **2012**, *3*, 169–179.
87. Dellenbaugh-Losse, M. The Cultural Landscape of the Berliner Republic: Undoing the Socialist Past. In *Inventing Berlin: Architecture, Politics and Cultural Memory in the New/Old German Capital Post-1989*; Springer: Berlin/Heidelberg, Germany, 2019; pp. 87–133.
88. Till, K.E. Political landscapes. In *A Companion to Cultural Geography*; Duncan, J., Johnson, N.C., Schein, R.H., Eds.; Wiley: New York, NY, USA, 2004; pp. 347–364.
89. Zilberg, E. A Troubled Corner: The ruined and rebuilt environment of a Central American barrio In post-Rodney-King-riot Los Angeles. *City Soc.* **2002**, *14*, 31–55. [CrossRef]
90. Hirsch, A.B. Restoring Los Angeles's Landscapes of Resistance. *J. Archit. Educ.* **2018**, *72*, 248–272. [CrossRef]
91. Bass, J.J. Posted: The campaign sign landscape, race, and political participation in Mississippi. *J. Cult. Geogr.* **2009**, *26*, 223–241. [CrossRef]
92. Bass, J.J. Posted redux: Campaign signs, race, and political participation in Mississippi, 2008. *Southeast. Geogr.* **2011**, *51*, 473–494. [CrossRef]

93. Makse, T.; Sokhey, A.E. Yard Sign Displays and the Enthusiasm Gap in the 2008 and 2010Elections. *PS Political Sci. Politics* **2012**, *45*, 694–699. [CrossRef]
94. Makse, T.; Sokhey, A.E. The displaying of yard signs as a form of political participation. *Political Behav.* **2014**, *36*, 189–213. [CrossRef]
95. Alderman, D.H.; Ward, H. Writing on the plywood: Toward an analysis of hurricane graffiti. *Coast. Manag.* **2007**, *36*, 1–18. [CrossRef]
96. Hamerlinck, J.D.; McKinlay, A.; Baldwin, C. Signs in the landscape: Material culture, protest, and place identity in coastal South East Queensland. *J. Cult. Geogr.* **2025**, 1–26. [CrossRef]
97. Makse, T.; Minkoff, S.; Sokhey, A. *Politics on Display: Yard Signs and the Politicization of Social Spaces*; Oxford University Press: Oxford, UK, 2019.
98. Parliament of Australia. Electoral Divisions. Determination of Divisions. Available online: https://www.aph.gov.au/About_Parliament/House_of_Representatives/Powers_practice_and_procedure/Practice7/HTML/Chapter3/Electoral_divisions (accessed on 20 July 2025).
99. Commonwealth of Australia. Digital Atlas Australia. Federal Electoral Divisions (March 2025). 2025. Available online: https://digital.atlas.gov.au/maps/digitalatlas::federal-electoral-divisions-march-2025 (accessed on 20 July 2025).
100. Bickerstaff, S. *Election Systems and Gerrymandering Worldwide*; Springer: Basel, Switzerland, 2020.
101. Durst, N.J. Racial gerrymandering of municipal borders: Direct democracy, participatory democracy, and voting rights in the United States. *Ann. Am. Assoc. Geogr.* **2018**, *108*, 938–954. [CrossRef]
102. Dick, S. AEC Draws Order of Candidates on Federal Election Ballot Papers Using Blindfolds, Bingo-Like System. 11 April 2025. Available online: https://www.abc.net.au/news/2025-04-11/aec-draws-ballots-federal-election-2025/105165884 (accessed on 20 July 2025).
103. Bahr, J. Trumpet of Patriots Spam Texts: Why Are You Receiving Them, and Are They Legal? 2025. Available online: https://www.sbs.com.au/news/article/trumpet-of-patriots-texts/vsqj4l75x (accessed on 5 May 2025).
104. Australian Electoral Commission. Authorising Electoral Communications. 15 April 2025. Available online: https://aec.gov.au/about_aec/Publications/Backgrounders/authorisation.htm (accessed on 5 May 2025).
105. Australian Electoral Commission. Voter's Guide: Communication Channels Catalogue. 31 January 2025. Available online: https://www.aec.gov.au/About_AEC/voters-guide-campaign-channels.htm (accessed on 5 May 2025).
106. Hughes, A. Democracy on Display or a Public Eyesore? The Case for Cracking Down on Election Corflutes. The Conversation. 29 April 2025. Available online: https://theconversation.com/democracy-on-display-or-a-public-eyesore-the-case-for-cracking-down-on-election-corflutes-255219 (accessed on 5 May 2025).
107. Smith, R. Australian Election Posters. In *Election Posters Around the Globe: Political Campaigning in the Public Space*; Holtz-Bacha, C., Johansson, B., Eds.; Springer: Berlin/Heidelberg, Germany, 2017; pp. 53–76.
108. Kelsall, T. No Corflutes, No Election? Some Voters Left in the Dark by SA Poster Ban. *ABCNews*. 26 May 2025. Available online: https://www.abc.net.au/news/2025-05-26/some-voters-left-in-dark-by-sa-poster-ban/105329218 (accessed on 2 June 2025).
109. Greenway, B. Size Matters: Border Election Candidate Asked to Remove Signs After Complaint. *The Border Mail*. 11 April 2025. Available online: https://www.bordermail.com.au/story/8939846/ (accessed on 5 May 2025).
110. Jakimow, T. Discarded Candidates: Waste as Metaphor in Local Government Elections in Australia (and Elsewhere). *Cult. Anthropol.* **2025**, *40*, 276–300. [CrossRef]
111. Valentine, J. Once the Election Campaign Is Over, can You Recycle Corflutes? *ABC News*. 23 May 2022. Available online: https://www.abc.net.au/news/2022-05-23/corflute-election-placards-recycle-reuse/101077148 (accessed on 2 June 2025).
112. McSwiney, J. Social networks and digital organisation: Far right parties at the 2019 Australian federal election. In *Understanding Movement Parties Through Their Communication*; Routledge: Abingdon, UK, 2023; pp. 75–92.
113. Decker, H.; Angus, D.; Bruns, A.; Dehghan, E.; Matich, P.; Tan, J.; Vodden, L. Topic diversity in social media campaigning: A study of the 2022 Australian federal election. *Politics Gov.* **2024**, *12*, 8155. [CrossRef]
114. Grantham, S. The rise of TikTok elections: The Australian Labor Party's use of TikTok in the 2022 federal election campaigning. *Commun. Res. Pract.* **2024**, *10*, 181–199. [CrossRef]
115. Geise, S. Theoretical perspectives on visual political communication through election posters. In *Election Posters Around the Globe: Political Campaigning in the Public Space*; Holtz-Bacha, C., Johansson, B., Eds.; Springer: Berlin/Heidelberg, Germany, 2017; pp. 13–31.
116. Breux, S.; Couture, J.; Koop, R. Influences on the Number and Gender of Candidates in Canadian Local Elections. *Can. J. Political Sci.* **2019**, *52*, 163–181. [CrossRef]
117. Dumitrescu, D. The importance of being present: Election posters as signals of electoral strength, evidence from France and Belgium. *Party Politics* **2011**, *18*, 941–960. [CrossRef]
118. Qadir, E.M. Visual Rhetoric in Election Posters: A Multimodal Critical Discourse Analysis Approach. *Koya Univ. J. Humanit. Soc. Sci.* **2023**, *6*, 136–159. [CrossRef]

119. Bean, C. The Personal Vote in Australian Federal Elections. *Political Stud.* **2006**, *38*, 253–268. [CrossRef]
120. Spennemann, D.H.R. "No Entry into New South Wales": COVID-19 and the Historic and Contemporary Trajectories of the Effects of Border Closures on an Australian Cross-Border Community. *Land* **2021**, *10*, 610. [CrossRef]
121. Id Profile. Albury City. Service Age Groups. 2025. Available online: https://profile.id.com.au/albury/service-age-groups (accessed on 5 May 2025).
122. Spennemann, D.H.R. *Albury Heritage Review 2022*; Historic Themes for Albury (NSW); Albury as a Regional Service Centre and the Principal Heritage Sites associated with this theme; Report to Albury City; School of Agricultural, Environmental and Veterinary Sciences; Charles Sturt University: Albury, Australia, 2024.
123. Australian Electoral Commission. Profile of the Electoral Division of Farrer (NSW). 10 October 2024. Available online: https://www.aec.gov.au/profiles/nsw/farrer.htm (accessed on 5 May 2025).
124. Green, A. Farrer Federal Election 2025 Results. 2025. Available online: https://www.abc.net.au/news/elections/federal/2025/guide/farr (accessed on 6 May 2025).
125. Australian Electoral Commission. NSW Division—Farrer. First Preferences and Two Candidate Preferred. 12 December 2007. Available online: https://results.aec.gov.au/13745/website/HouseDivisionFirstPrefs-13745-118.htm (accessed on 5 May 2025).
126. Australian Electoral Commission. 2016 Federal Election. NSW Division—Farrer. 2 August 2016. Available online: https://results.aec.gov.au/20499/website/HouseDivisionPage-20499-118.htm (accessed on 5 May 2025).
127. Australian Electoral Commission. 2016 Federal Election: Profile of the Electoral Division of Farrer (NSW). 5 July 2016. Available online: https://www.aec.gov.au/Elections/federal_elections/2016/profiles/nsw/farrer.htm (accessed on 5 May 2025).
128. Australian Electoral Commission. 2013 Federal Election: Profile of the Electoral Division of Farrer (NSW). 4 September 2013. Available online: https://www.aec.gov.au/Elections/federal_elections/2013/profiles/nsw/farrer.htm (accessed on 5 May 2025).
129. Australian Electoral Commission. 2007 Federal Election: Profile of the Electoral Division of Farrer. 16 September 2011. Available online: https://www.aec.gov.au/elections/federal_elections/2007/profiles/farrer.htm (accessed on 5 May 2025).
130. McKinley, M. Sussan Ley Takes Top Spot for Ballot Paper, Order doesn't Worry Candidates. *The Border Mail*. 11 April 2025. Available online: https://www.bordermail.com.au/story/8940691/election-2025-farrer-ballot-positions-revealed-incumbent-leads (accessed on 5 May 2025).
131. Australian Electoral Commission. 2025 Federal Election Enrolment Statistics. 10 April 2025. Available online: https://www.aec.gov.au/enrolling_to_vote/enrolment_stats/national/2025-fe.htm (accessed on 5 May 2025).
132. Australian Electoral Commission. House of Representatives Downloads. Two Candidate Preferred by Candidate by Vote Type. 5 June 2025. Available online: https://tallyroom.aec.gov.au/Downloads/HouseTcpByCandidateByVoteTypeDownload-31496.csv (accessed on 6 June 2025).
133. Australian Electoral Commission. House of Representatives Downloads. First Preferences by Candidate by Polling Place—NSW. 5 June 2025. Available online: https://tallyroom.aec.gov.au/Downloads/HouseStateFirstPrefsByPollingPlaceDownload-31496-NSW.csv (accessed on 6 June 2025).
134. Australian Electoral Commission. Australia's Largest ever Federal Election Kicks Off. 28 March 2025. Available online: https://www.aec.gov.au/media/2025/federal-election.htm (accessed on 5 May 2025).
135. ESRI. *ArcGIS Field Maps Designer*; Environmental Systems Research Institute, Inc.: Redlands, CA, USA, 2025.
136. ESRI. *ArcGIS Pro*, version 3.5; Environmental Systems Research Institute, Inc.: Redlands, CA, USA, 2025.
137. Spennemann, D.H.R.; Duffy, D. *The Ephemeral Cultural Landscape of an Australian Federal Election—Data Set*; Charles Sturt University: Albury, Australia, 2021. [CrossRef]
138. Whitsed, R.; Black, R.; Harvey, R.; Knight, A.; Watkins, S.; Armstrong, D. Better parks for people. *Australas. Parks Leis.* **2016**, *19*, 38–39.
139. Australian Bureau of Statistics. Digital Boundaries for Mesh Blocks, July 2021. 2021. Available online: https://www.abs.gov.au/statistics/standards/australian-statistical-geography-standard-asgs-edition-3/jul2021-jun2026/access-and-downloads/digital-boundary-files (accessed on 13 May 2025).
140. Australian Bureau of Statistics. Socio-Economic Indexes for Areas (SEIFA), 2021 Statistical Area Level 1 (SA1). 2023. Available online: https://www.abs.gov.au/statistics/people/people-and-communities/socio-economic-indexes-areas-seifa-australia/latest-release#data-downloads, (accessed on 13 May 2025).
141. *MedCalc Software*, Comparison of Proportions Calculator Version 22.032; MedCalc Software Ltd.: Oostende, Belgium, 2018. Available online: https://www.medcalc.org/calc/comparison_of_proportions.php (accessed on 20 June 2025).
142. Sawer, M. Wearing your politics on your sleeve: The role of political colours in social movements. *Soc. Mov. Stud.* **2007**, *6*, 39–56. [CrossRef]
143. Curtin, J.; Costar, B. The Contest for Rural Representation: The celebrated contest over Indi and the fate of the independents. In *Abbott's Gambit: The 2013 Australian Federal Election*; Johnson, C., John, W., Lee, H.-A., Eds.; ANU Press: Canberra, Australia, 2015; p. 275.
144. Hendriks, C.M. Citizen-led democratic reform: Innovations in Indi. *Aust. J. Political Sci.* **2017**, *52*, 481–499. [CrossRef]

145. Baskerville, B. Coat of arms. In *Symbols of Australia*; Harper, M., White, R., Eds.; National Museum of Australia Press and UNSW Press: Canberra, Australia, 2010; pp. 107–113.
146. Dumitrescu, D. *Spatial Visual Communications in Election Campaigns: Political Posters Strategies in two Democracies*; Doctor of Philosophy, Ohio State University: Columbus, OH, USA, 2009.
147. Seidman, S.A. The visual design of election campaign posters. In *The Graphic Communication Handbook*; Routledge: Oxford, UK, 2013; pp. 92–111.
148. Schott, M.; Wolf, J. Election poster persuasion. *Soc. Psychol.* **2018**, *49*, 3–15. [CrossRef]
149. Lefebvre, H. From the production of space. In *Theatre and Performance Design*; Collins, J., Nisbet, A., Eds.; Routledge: Abingdon, UK, 2012; pp. 81–84.
150. Fox, C.A. Visualizing politics in Indonesia: The design and distribution of election posters. *Int. J. Commun.* **2022**, *16*, 4187.
151. Philipps, A. Defacing Elect. Posters: A Form Political Cult. Jamming? *Pop. Commun.* **2015**, *13*, 183–201. [CrossRef]
152. Mardling, X. Voices of Farrer out to Uncover Independent Challenger to MP Sussan Ley. *Border Mail*. 29 June 2023. Available online: https://www.bordermail.com.au/story/8250890/farrer-raise-your-voice-how-these-women-hope-to-uncover-sussan-leys-next-challenger/ (accessed on 20 June 2025).
153. Ikenasio, L. Michelle Milthorpe Endorsed as Independent Candidate for Farrer. *Riverine Herald*. 20 September 2024. Available online: https://www.riverineherald.com.au/news/michelle-milthorpe-endorsed-as-independent-candidate-for-farrer/ (accessed on 20 June 2025).
154. Winters, M.-F.; Reese, M.; Reese, M.N. *We Can't Talk About That at Work!: How to Talk About Race, Religion, Politics, and Other Polarizing Topics*; Berrett-Koehler Publishers: Oakland, CA, USA, 2024.
155. Vigsø, O. Extremist stickers: Epideictic rhetoric, political marketing, and tribal demarcation. *J. Vis. Lit.* **2010**, *29*, 28–46. [CrossRef]
156. Ebel, H. Wearing of anti-nuclear-buttons during working hours. *Oeffentliche Verwalt.* **1980**, *12*, 437–439.
157. Case, C.E. Bumper Stickers and Car Signs: Ideology and Identity. *J. Pop. Cult.* **1992**, *26*, 107–119. [CrossRef]
158. Spennemann, D.H.R. QSL: Subliminal messaging by the Nuclear Industry in Germany during the 1980s. *Heritage* **2021**, *4*, 2054–2080. [CrossRef]
159. Endersby, J.W.; Towle, M.J. Tailgate partisanship: Political and social expression through bumper stickers. *Soc. Sci. J.* **1996**, *33*, 307–319. [CrossRef]
160. Bodden, S.; Awcock, H. Stuck up, peeled off, covered up, shared and scribbled out: Doing ordinary politics with political stickers. *GeoHumanities* **2024**, *10*, 131–149. [CrossRef]
161. Doyle, K.; Tranter, K. F#ck Your Family!: The Visual Jurisprudence of Automobility. *Int. J. Semiot. Law—Rev. Int. Sémiotique Jurid.* **2017**, *30*, 1–22. [CrossRef]
162. Flinders, M. The general rejection? Political disengagement, disaffected democrats and 'doing politics' differently. *Parliam. Aff.* **2015**, *68*, 241–254. [CrossRef]
163. Zhang, W. Political disengagement among youth: A comparison between 2011 and 2020. *Front. Psychol.* **2022**, *13*, 809432. [CrossRef] [PubMed]
164. Reid, R.; Hendriks, C.M.; Gauja, A. An alternative to the party? Australia's movement for community independents. *Party Politics* **2025**, 13540688251356900. [CrossRef]
165. Fowler, E.F.; Franz, M.M.; Martin, G.J.; Peskowitz, Z.; Ridout, T.N. Political advertising online and offline. *Am. Political Sci. Rev.* **2021**, *115*, 130–149. [CrossRef]
166. Baviera, T.; Sánchez-Junqueras, J.; Rosso, P. Political advertising on social media: Issues sponsored on Facebook ads during the 2019 General Elections in Spain. *Commun. Soc.* **2022**, *35*, 39–49. [CrossRef]
167. Beraldo, D.; Milan, S.; de Vos, J.; Agosti, C.; Sotic, B.N.; Vliegenthart, R.; Kruikemeier, S.; Otto, L.P.; Vermeer, S.A.; Chu, X. Political advertising exposed: Tracking Facebook ads in the 2021 Dutch elections. *Internet Policy Rev.* **2021**, *11*.
168. Bene, M.; Kruschinski, S. Political advertising on Facebook. In *Campaigning on Facebook in the 2019 European Parliament Election: Informing, Interacting with, and Mobilising Voters*; Springer: Berlin/Heidelberg, Germany, 2021; pp. 283–299.
169. Pierri, F. Political advertisement on Facebook and Instagram in the run up to 2022 Italian general election. In Proceedings of the Proceedings of the 15th ACM Web Science Conference 2023, Austin, TX, USA, 30 April 2023; pp. 13–22.
170. Russmann, U.; Svensson, J.; Larsson, A.O. Political parties and their pictures: Visual communication on Instagram in Swedish and Norwegian election campaigns. In *Visual Political Communication*; Springer: Berlin/Heidelberg, Germany, 2019; pp. 119–144.
171. Cervi, L.; Marín-Lladó, C. What are political parties doing on TikTok? The Spanish case. *Prof. Inf.* **2021**, *30*, 1–17. [CrossRef]
172. Medina Serrano, J.C.; Papakyriakopoulos, O.; Hegelich, S. Dancing to the partisan beat: A first analysis of political communication on TikTok. In Proceedings of the Proceedings of the 12th ACM Conference on Web Science, Online, 6–10 July 2020; pp. 257–266.

173. Papakyriakopoulos, O.; Tessono, C.; Narayanan, A.; Kshirsagar, M. How algorithms shape the distribution of political advertising: Case studies of Facebook, Google, and TikTok. In Proceedings of the 2022 AAAI/ACM Conference on AI, Ethics, and Society, Oxford, UK, 1–3 August 2022; pp. 532–546.
174. Australian Electoral Commission. House of Representatives Downloads. Two Party Preferred by Polling Place. 5 June 2025. Available online: https://tallyroom.aec.gov.au/Downloads/HouseTppByPollingPlaceDownload-31496.csv (accessed on 6 June 2025).

land

MDPI

Article

Land Use Change and Biocultural Heritage in Valle Nacional, Oaxaca: Women's Contributions and Community Resilience

Gema Lugo-Espinosa [1], Marco Aurelio Acevedo-Ortiz [1,*], Yolanda Donají Ortiz-Hernández [2,*], Fernando Elí Ortiz-Hernández [3] and María Elena Tavera-Cortés [4]

1 Secretaría de Ciencias, Humanidades, Tecnología e Innovación—Instituto Politécnico Nacional, CIIDIR Unidad Oaxaca, Santa Cruz Xoxocotlán 71230, Mexico; glugoe@ipn.mx
2 Instituto Politécnico Nacional, Centro Interdisciplinario de Investigación para el Desarrollo Integral Regional, Unidad Oaxaca, Santa Cruz Xoxocotlán 71230, Mexico
3 Instituto Politécnico Nacional, Escuela Superior de Ingeniería Mecánica y Eléctrica Unidad Culhuacán, Ciudad de Mexico 04440, Mexico; fortizh@ipn.mx
4 Instituto Politécnico Nacional, Unidad Profesional Interdisciplinaria de Ingeniería y Ciencias Sociales y Administrativas, Ciudad de Mexico 08400, Mexico; mtavera@ipn.mx
* Correspondence: macevedoo@ipn.mx (M.A.A.-O.); yortiz@ipn.mx (Y.D.O.-H.); Tel.: +52-9511726274 (M.A.A.-O.); +52-9515170610 (ext. 82723) (Y.D.O.-H.)

Abstract: Territorial transformations in Indigenous regions are shaped by intersecting ecological, political, and cultural dynamics. In San Juan Bautista Valle Nacional, Oaxaca, the construction of the Cerro de Oro dam disrupted river flows, displaced livelihoods, and triggered the decline of irrigated agriculture. This study examines the long-term impacts of these changes on land use, demographics, and cultural practices, emphasizing women's contributions to community resilience. Using a mixed-methods approach, the study integrates geospatial analysis (1992–2021), census data (2000–2020), documentary review, and ethnographic fieldwork, including participatory mapping. Results show a shift toward seasonal rainfed agriculture, fluctuating forest cover, and a rise in female-headed households. Women have emerged as central actors in adapting to change through practices such as seed saving, agroforestry, and backstrap-loom weaving. These spatially grounded practices, enacted across varied socio-ecological zones, sustain food systems, preserve biodiversity, and reinforce biocultural memory. Although often overlooked in formal governance, women's territorial agency plays a vital role in shaping land use and community adaptation. This research highlights the need to recognize Indigenous women's roles in managing change and sustaining territorial heritage. Acknowledging these contributions is essential for building inclusive, culturally grounded, and sustainable development pathways in regions facing structural and environmental pressures.

Keywords: community-based adaptation; gender and territory; indigenous governance

1. Introduction

Contemporary territorial movements are complex processes that reflect both structural transformations and symbolic disputes over the use, control, and meaning of space [1,2]. These changes are shaped by external drivers such as market expansion, infrastructure megaprojects, and state-led development policies [2,3], which, individually or collectively, generate structural transformations that deeply impact community life, economic dynamics, and cultural heritage [4,5]. In the Papaloapan regions of Oaxaca, Mexico, territory represents more than a productive base; it is a social construct shaped by customary practices, traditional knowledge, and communal governance expressed through collective labor, community assemblies, and the defense of biocultural heritage [6,7].

Territorial governance and development are deeply connected to the people who inhabit the land [8]. Land use is influenced by cultural values, economic activities, and social structures, including gender roles [9]. Territorial transformation often depends on the degree of social organization, the sense of belonging, and the historical memory maintained by the community. In many Indigenous localities, spatial knowledge is transmitted orally and preserved through symbolic markers such as boundary stones, rivers, or ancestral footpaths [9,10]. However, climate variation, economic pressures, and cultural shifts are gradually altering the landscape and erasing these longstanding territorial references [11].

In this context, the knowledge of elders plays a central role in maintaining the memory of how territories have changed over time [5,12–14]. Their guidance is key in assemblies, where collective decisions are made regarding land use, cultural preservation, and local development. However, demographic changes caused by migration, social conflicts, or the emergence of dominant groups have transformed the composition of rural populations [15]. In many cases, women and older adults remain and assume increasing responsibilities in land and resource management [16].

In Oaxaca, territorial transformation has also been shaped by state-led modernization processes that often fail to consider the ecological and cultural particularities of Indigenous territories [17,18]. Infrastructure expansion, extractive policies, and centralized spatial planning have fragmented landscapes, reduced access to agricultural land, and weakened traditional systems of land control [18,19]. These changes have intensified precarious working conditions and led to the depopulation of rural areas, significantly altering the dynamics of community participation and land use [19].

The continued presence of women in these territories has given rise to new forms of agency [20]. Women engage in productive and care work while also participating in community organization [21]. They are key actors in sustaining local economies, often through agroecological initiatives, food systems, and artisanal production such as textiles [22–26]. These contributions, though often invisible, support household livelihoods and are central to the cultural continuity and resilience of the territory [5,20,22,27].

The Papaloapan region of Oaxaca, commonly known as Chinantla, provides a relevant case for analyzing these interactions. Development planning in this region operates across three levels: local, municipal, and regional. At the local level, decision-making occurs through assemblies and is based on available resources and community involvement. At the municipal level, public infrastructure projects are coordinated across wider areas. At the regional level, economic development policies are often implemented without sufficient alignment with local priorities and social structures. The interaction or disconnection between these levels has a significant impact on the success of conservation and sustainable development efforts.

The main research gap addressed by this study lies in the limited empirical evidence on how community-based territorial management, supported by women's participation and intergenerational knowledge, influences land use transitions while preserving biocultural heritage. Unlike conventional top-down planning approaches, this study demonstrates how collective decision-making can promote equitable access to natural resources, strengthen local governance, and support the regeneration of cultural identity through sustainable territorial practices.

This study aims to contribute to the understanding of cultural landscapes as living territorial constructions shaped by both ecological conditions and social processes. It offers a critical perspective of land use change in Indigenous rural contexts, with particular attention to the role of women as agents of cultural continuity, community adaptation, and sustainable development. These dynamics are explored through the case of San Juan Bautista Valle Nacional, located in the Chinantla region of northern Oaxaca, where Indigenous women play a central role in shaping biocultural resilience amidst ongoing territorial transformations.

2. Theoretical Background

This research is grounded in an interdisciplinary theoretical framework that integrates feminist political ecology, territorial governance, and Indigenous epistemologies rooted in biocultural memory. These perspectives provide critical tools to understand the interplay between gender, land use change, cultural continuity, and socio-ecological resilience in Indigenous communities.

2.1. Feminist Political Ecology and Territorial Governance

Feminist political ecology (FPE) offers a lens through which to analyze how socio-environmental transformations differentially impact men and women, particularly regarding access to resources, decision-making authority, and ecological knowledge systems [21–23,25]. It underscores the central—yet often undervalued—role of women in sustaining agri-food systems, conserving biodiversity, and transmitting intergenerational knowledge, especially under conditions shaped by male outmigration and institutional neglect [16,20,22].

Territorial governance complements FPE by emphasizing local agency in shaping land use trajectories and resisting top-down impositions. Moving beyond technocratic approaches, it focuses on the symbolic, affective, and cultural dimensions of territory. In regions like Valle Nacional, state-led interventions such as the construction of the Cerro de Oro dam are not passively accepted [17–19]; rather, they are reinterpreted, resisted, or adapted through grassroots organizing, customary land management practices, and collective mobilization [6–8,28].

From a broader perspective, territorial transformation is also shaped by dynamic and contested processes of access, exclusion, claim-making, and territorialization [29]. These processes reflect historical struggles and complex agrarian structures, often reproducing inequalities through institutional arrangements that govern land ownership, use, and control. Thus, governance is not neutral; it reflects and reproduces power dynamics.

In Valle Nacional, governed by customary law, community-based institutions play a central role. Through assemblies and collective labor, residents demonstrate strong organizational capacities for natural resource management and conflict resolution. Meanwhile, municipal and state authorities influence territorial dynamics through social and productive development programs. The intersection of these governance layers creates complex negotiation arenas that shape land access and decision-making processes.

2.2. Indigenous Epistemologies and Biocultural Memory

Indigenous epistemologies deepen the understanding of territory by framing it as a living and relational entity imbued with cultural meaning, historical continuity, and collective responsibility [13,14]. Knowledge is place-based and transmitted orally through direct engagement with the landscape, linking spiritual beliefs, daily practices, and environmental stewardship.

Practices such as seed saving, home gardening, and textile weaving are deeply embedded in this worldview. Beyond their practical utility, they serve as symbolic expressions that sustain identity and preserve ecological balance [16,27,30–38]. In particular, weaving encapsulates narratives of ecological and cultural memory, with motifs that reference mountains, rivers, and native species—bridging the tangible landscape with its symbolic representation. These actions embody biocultural memory—the cumulative reservoir of ecological and cultural knowledge, values, and practices that co-evolved with specific environments.

Biocultural memory encompasses localized agricultural calendars, traditional ceremonies, and artisanal production such as the huipil chinanteco, all of which reflect embedded knowledge systems. These expressions serve as strategies for continuity and resilience, particularly in the face of marginalization or environmental degradation. They also represent epistemological resistance to development paradigms that externalize or commodify natural and cultural resources.

Acknowledging these epistemologies not only legitimizes Indigenous ways of knowing but also reveals the crucial role of cultural continuity in sustaining socio-ecological systems. Their relational approach to land and stewardship offers critical insights for biodiversity conservation, climate adaptation, and territorial governance in contexts of rapid change. For this reason, it is essential to promote the autonomy of Indigenous peoples, ensuring that their self-determination is respected as a foundational principle in the design and implementation of sustainable land use practices.

2.3. Biocultural Resilience and Adaptive Socio-Ecological Systems

Biocultural resilience refers to the adaptive capacity of communities and ecosystems to respond to environmental, social, and economic disruptions while maintaining core functions, cultural identity, and social cohesion. Rooted in reciprocal human–environment relationships, this resilience depends on flexibility, intergenerational learning, and innovation. Nonetheless, it requires balance: overexploitation of natural resources can lead to negative outcomes when ecological thresholds are exceeded [39]. This concept underscores the interdependence of social and ecological systems and the importance of preserving both biodiversity and cultural integrity. It is particularly relevant in contexts of overlapping vulnerabilities, where hybrid knowledge systems—blending ancestral practices and adaptive innovations—enable continuity and well-being.

Biocultural diversity [40], understood as the co-evolution of biological and cultural systems, reinforces resilience by maintaining heterogeneity and context-specific practices. Its erosion through homogenizing pressures can lead to cultural dislocation and ecological degradation. In contrast, reinforcing biocultural resilience strengthens a community's capacity to navigate change.

2.4. Biocultural Heritage as a Tool for Territorial Continuity and Cultural Resistance

Biocultural heritage refers to dynamic, place-based knowledge systems, practices, values, and memories that connect communities to their landscapes and sustain local livelihoods [41]. It includes ecological knowledge, as well as spiritual, linguistic, and symbolic dimensions embedded in daily life. As such, it embodies both continuity and adaptability in response to environmental transformations.

Home gardening, seed conservation, and textile production are not merely productive strategies; they are acts of cultural affirmation. These practices reinforce community identity while resisting cultural homogenization and ecological degradation. Weaving, in particular, operates as a cultural cartography—encoding symbolic representations of territory, kinship, and ancestral knowledge—thus serving as a textile form of territoriality. Analyzing landscape multifunctionality through the lens of biocultural heritage enables the identification of vulnerable areas, conservation priorities, and sustainable development opportunities [42].

In this study, biocultural resilience is defined as the collective capacity of Indigenous communities to respond to socio-ecological uncertainty through cultural cohesion, adaptive stewardship, and the revitalization of ancestral practices. It manifests in actions that promote food sovereignty, defend territorial rights, and maintain spiritual and material relationships with the land.

Together, these frameworks provide an integrated lens to examine how gender, cultural practices, Indigenous knowledge, and collective action shape resilience and sustainable land governance in Indigenous territories undergoing transformation.

3. Materials and Methods

3.1. Study Site Location

The municipality of San Juan Bautista Valle Nacional (commonly referred to as Valle Nacional) is situated in the northern region of Oaxaca, Mexico, between 17°46′ N latitude and 96°08′ to 96°31′ W longitude, at an average elevation of 60 m above sea level [43]. According to the 2020 Population and Housing Census, it has a total population of 23,067 inhabitants, of whom 47.6% are men and 52.4% are women [44]. The presence of Chinantec Indigenous communities is highly significant, with 51.8% of the population speaking an Indigenous language and 41.48% self identifying as Indigenous [45]. Valle Nacional's local economy is predominantly based on primary sector activities, particularly agriculture, forestry, and fishing, which together engage 46.97% of the economically active population [45].

Land tenure in Valle Nacional is governed predominantly by social property regimes. The municipality comprises 33 ejidos and two communities with communal land tenure systems, each managed by local agrarian authorities [46]. In total, the municipality encompasses 18 police precincts, 10 municipal precincts, and 27 rural settlements.

Valle Nacional is part of the broader Chinantla region, renowned for its exceptional biodiversity, humid tropical forests, and rich sociolinguistic diversity. Sites such as Cerro Baúl, Cerro de Uva, Montebello, Cuchara, and San Mateo Yetla are located within the municipality and form part of the "Ruta Chinantla," a cultural and ecological circuit promoted by the Oaxaca State Secretariat of Tourism [47].

In addition to its natural wealth, the municipality is a repository of intangible cultural heritage, particularly the traditional weaving of huipiles on backstrap looms. This artisanal practice, passed down through generations, remains a vital expression of Chinantec women's identity and territorial memory [34,37]. Recent documentation efforts have drawn attention to the decline of this tradition due to economic pressures and generational shifts [31], making Valle Nacional a critical site for examining the intersections between land use, gender, and cultural resilience.

3.2. Methodological Approach and Data Collection

This study adopts a qualitative, diachronic approach aimed at understanding territorial dynamics from a socio-spatial and cultural perspective. A multiscale and longitudinal design was employed to examine changes in land use and demographic structure over a thirty-year period.

Spatial analysis was conducted using ArcGIS Pro software 3.5.2 [48], based on official land use and vegetation maps from the National Land Use and Vegetation Inventory for the years 1992, 2001, 2005, 2009, 2013, 2017, and 2021 [49–55]. Eight land use categories were digitized and analyzed: irrigated agriculture, seasonal rainfed agriculture, pine-oak forest, montane cloud forest, permanent cultivated grassland, tall evergreen rainforest, secondary shrubby vegetation of tall evergreen rainforest, and secondary arboreal vegetation of tall evergreen rainforest. This allowed the identification of spatial trends and territorial transitions in San Juan Bautista Valle Nacional.

In parallel, demographic data and gender-disaggregated indicators were gathered from the National Institute of Statistics and Geography (INEGI) Population and Housing

Censuses for 2000, 2005, 2010, and 2020 [56]. Variables such as marginalization levels, educational lag, lack of access to basic services, and population structure by sex and age were used to contextualize territorial transformations through a socio-economic and gender lens.

The spatial and demographic analyses were complemented by archival research and review of press articles documenting the impacts of the Cerro de Oro dam and other state-led infrastructure projects [57]. To understand women's spatial practices across different socio-ecological zones, the research incorporated participatory mapping and field-based observations conducted between 2020 and 2024. These interactions enabled the identification of gendered land use practices related to rainfed agriculture, home gardens, milpa–agroforestry systems, and permanent grasslands. The resulting typology was developed by cross-referencing observed practices with geo-referenced landscape features using ArcGIS, supported by iterative field visits. Although not based on a quantitative survey, this approach provides a grounded framework for analyzing spatial agency from a feminist political ecology and biocultural heritage perspective, shedding light on knowledge systems and territorial engagement often excluded from formal land classifications.

4. Results

4.1. Land Use Transitions and Vegetation Cover Change (1992–2021)

Geospatial and temporal analysis of San Juan Bautista Valle Nacional reveals a profound transformation in land use and vegetation cover from 1992 to 2021. One of the most significant changes was the expansion of seasonal rainfed agriculture, which increased from just 78 hectares in 1992 to 10,290 hectares in 2021. This growth coincided with the complete disappearance of irrigated agriculture after 2001, a land use that previously occupied 3934 hectares. The transition toward less water-dependent agriculture reflects the cumulative effect of altered hydrological regimes and local adaptation strategies (Figure 1).

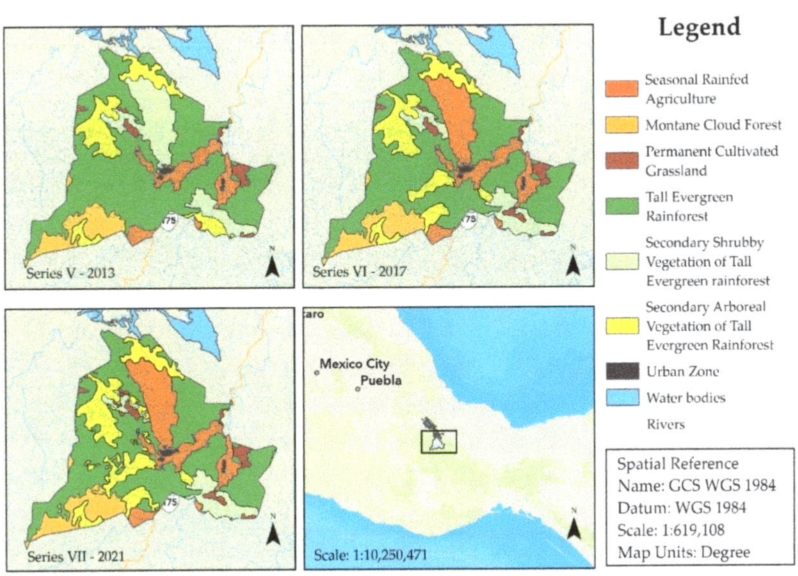

Figure 1. Land use transition and vegetation cover change (2013–2021) (source: own elaboration based on geospatial information from INEGI [53–55]).

Earlier dynamics are illustrated in Figure 2, showing how forested areas declined and were replaced by secondary vegetation types between 1992 and 2009. This period marked a turning point toward a more fragmented, yet flexible, territorial configuration.

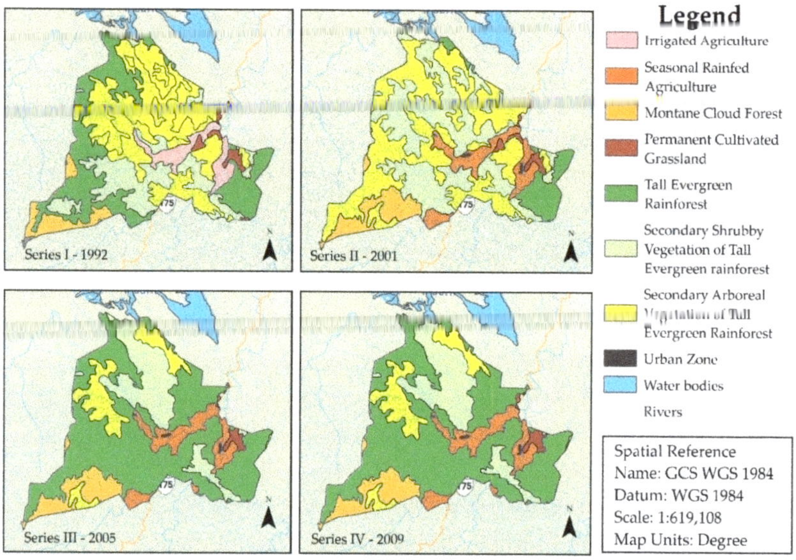

Figure 2. Land use change transition and vegetation cover change (1992–2009) (Source: own elaboration based on geospatial information from INEGI [49–52]).

A comprehensive overview is provided in Table 1, which summarizes changes across all land use classes. These data show not only linear expansions or reductions but also ecological cycles—degradation followed by partial regeneration—across multiple ecosystems and productive zones.

Table 1. Land use changes in San Juan Bautista Valle Nacional, Oaxaca (1992–2017) in hectares.

Land Use	1992	2001	2005	2009	2013	2017	2021
Irrigated agriculture	3934	0	0	0	0	0	0
Seasonal rainfed agriculture	78	4811	5191	5082	4781	10,283	10,290
Pine-oak Forest	131	819	0	0	0	0	0
Montane cloud forest	2961	4100	4034	4034	4047	4047	4047
Permanent cultivated grassland	1043	819	622	706	1221	1515	1631
Tall evergreen rainforest	22,398	8406	40,526	40,622	41,924	39,339	37,627
Secondary shrubby vegetation of tall evergreen rainforest	23,606	22,088	10,669	10,598	9067	4198	3713
Secondary arboreal vegetation of tall evergreen rainforest	14,289	28,005	7188	7188	6344	8656	10,616
Urban zone	0	147	147	147	356	356	456
Water bodies	0	66	66	66	66	49	63

Source: own elaboration based on geospatial information from INEGI [49–55].

Among forested ecosystems, the tall evergreen rainforest experienced the most dramatic changes. A sharp decline of 62% occurred between 1992 and 2001, driven by intensified land use. However, partial recovery was observed starting in 2005, with the area reaching a peak of 41,924 hectares in 2013, followed by a gradual reduction to 37,627 hectares by 2021 (Figure 3).

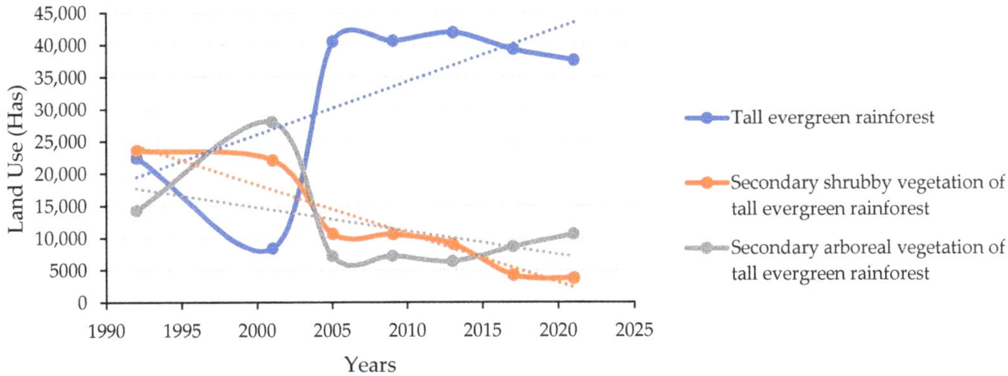

Figure 3. Trends in tall evergreen rainforest and secondary vegetation (1992–2021). Solid lines show actual land use values (in hectares), while dotted lines indicate linear trends for each land use type (source: own elaboration based on geospatial information from INEGI [49–55]).

Secondary shrubby vegetation steadily declined by 84% during the study period, indicating both degradation and forest recovery processes. Secondary arboreal vegetation, on the other hand, followed a non-linear trajectory—initially expanding, then declining, and finally stabilizing above 10,000 hectares.

The montane cloud forest remained relatively stable over time, with minimal surface loss. Permanent cultivated grassland showed early reduction but experienced recovery after 2013. These combined trends, including modest urban expansion, are illustrated in Figure 4.

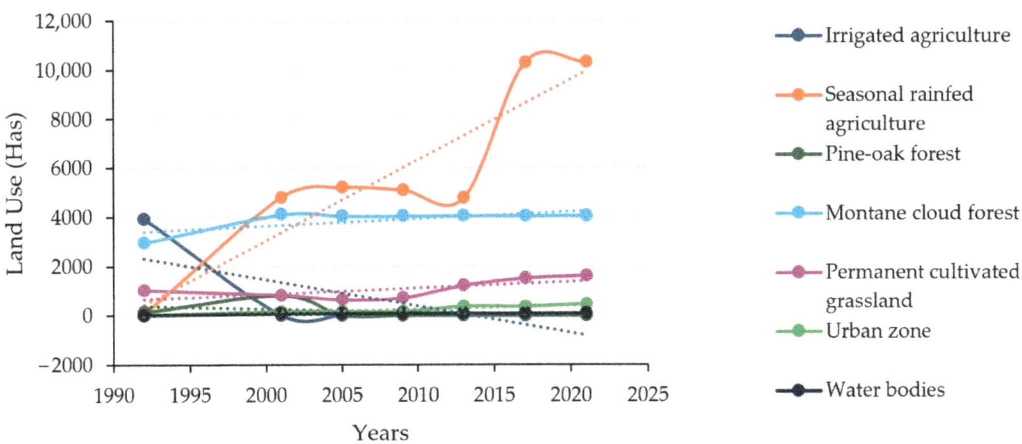

Figure 4. Trends in agriculture, forest, and urban land use (1992–2021). Solid lines show actual land use values (in hectares), while dotted lines indicate linear trends for each land use type (source: own elaboration based on geospatial information INEGI [49–55]).

4.2. Socioeconomic Dynamics and Gender Roles

Demographic and household-level data suggest an ongoing transformation in the gender composition of population and land management. Between 2000 and 2020, the total population of San Juan Bautista Valle Nacional rose slightly from 22,886 to 23,067. Male population steadily increased from 11,140 to 12,081, while the female population peaked in 2010 (11,758) and declined to 10,986 by 2020 (Figure 5).

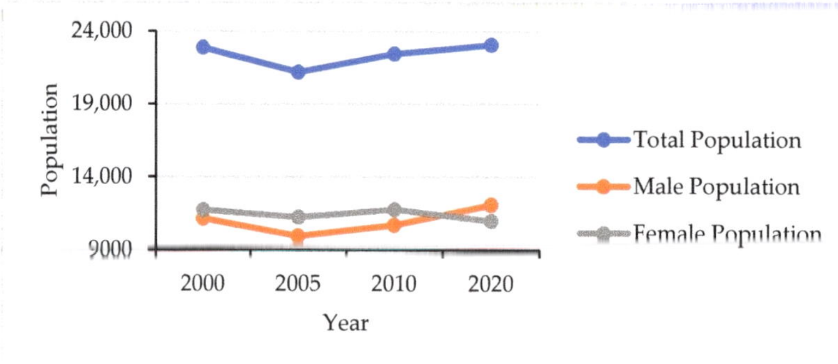

Figure 5. Total, male, and female population in San Juan Bautista Valle Nacional (2000–2020) (source: own elaboration based on demographic and social information from INEGI [56]).

Household dynamics changed more noticeably. Total households increased from 4586 to 6027 between 2000 and 2020. Female-headed households more than doubled, from 819 to 1724, while male-headed households remained relatively constant (Figure 6).

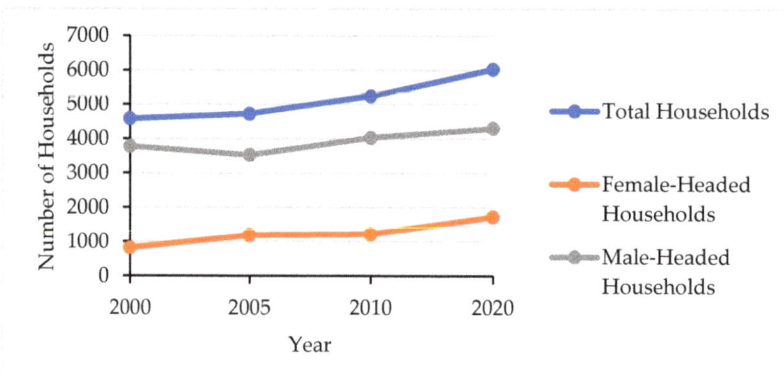

Figure 6. Household composition by gender of head (2000–2020) (source: own elaboration based on demographic and social information from INEGI [56]).

Further analysis (Figure 7) shows that the number of individuals living in female-headed households grew from 3219 to 5680, underscoring women's growing responsibility for domestic and economic resilience.

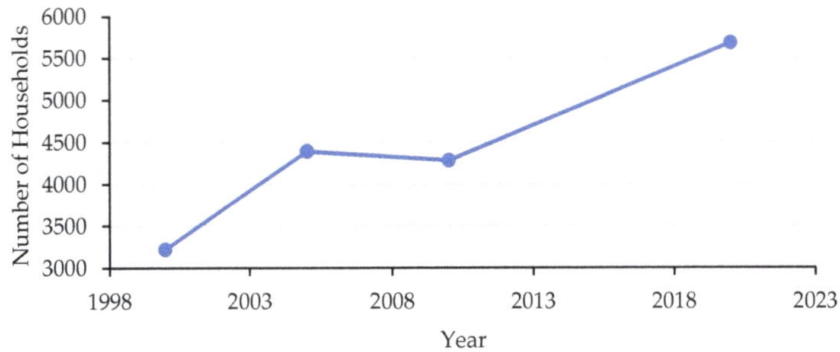

Figure 7. Family members in female-headed households (2000–2020) (source: own elaboration based on demographic and social information from INEGI [56]).

These findings indicate a redistribution of household authority and rural labor roles, suggesting a gendered transition in the spatial and productive organization of the territory [44,46,58].

4.3. Gendered Territorial Practices Across Socio-Ecological Zones

Women's engagement in land management in San Juan Bautista Valle Nacional reveals a profound spatial awareness, expressed through daily practices across diverse ecological zones. These zones—rainfed plots, home gardens, milpa–agroforestry systems, and grasslands—are not neutral or interchangeable spaces. Rather, they constitute socio-ecological niches where women navigate topographic constraints, resource access, and household needs, transforming these environments through situated knowledge and care.

Each ecosystem type reflects distinct gendered practices. On sloped or marginal terrain, women cultivate rainfed plots through seasonal intercropping and soil conservation strategies. Adjacent to their homes, gardens are used for seed saving, cultivation of vegetables and medicinal plants, and raising poultry. In the milpa–agroforestry systems [59], women foster biodiversity by integrating fruit trees, coffee, native timber, and shade crops—often within landscapes that border secondary forests. Coyuchi cotton (*Gossypium hirsutum*)—from the Nahuatl word coyōichcatl, meaning "coyote-colored cotton" (light brown or beige)—is a native variety of Oaxaca, Mexico. It is intercropped in these plots in small quantities and used for weaving huipiles, napkins, and other handcrafted items [60]. Although its production has declined, it held substantial economic value in the past, representing 34% of the municipality's total agricultural production in 1980 [61]. In lower riverine areas, pasturelands are managed through fodder planting and rotational grazing, sometimes collectively organized.

This diversity of spatial practices underscores women's territorial agency in adapting to ecological constraints and shaping multifunctional landscapes. Their decisions about where to grow, graze, or regenerate ecosystems reflect both environmental knowledge and cultural continuity, contributing to the creation of resilient and productive territories (Table 2).

Table 2. Gendered management practices and spatial characteristics of ecosystems.

Ecosystem Type	Management Practices by Women	Spatial Characteristics
Rainfed agricultural plots	Seasonal cropping, maize–bean intercropping, soil conservation	Hillsides, former irrigation zones, marginal terrain
Home gardens	Seed saving, vegetable cultivation, medicinal plants, poultry raising	Adjacent to homes, fenced patios or backyards
Milpa–agroforestry systems	Integration of fruit trees, coffee, native timber, shade crops, and small-scale cotton cultivation for artisanal use	Edges of secondary forests, partially shaded areas, interspersed crops
Permanent cultivated grasslands	Fodder cultivation, rotational grazing, water access planning	Lowland pastures, individual or communal plots, near rivers

Source: Own elaboration. This typology reflects women's differentiated ecological roles and their contribution to food systems, textile production, and the resilience of biocultural landscapes.

4.4. Adaptive Land Management in Response to Territorial Change

Women's land practices have responded dynamically to environmental shifts, particularly following the hydrological changes caused by the dam's construction. As irrigated plots became less productive or entirely unviable, women intensified the use of rainfed areas through techniques such as soil recovery, organic composting, and integrating short-cycle crops such as maize (*Zea mays*), beans (*Phaseolus vulgaris*), squash (*Cucurbita pepo*), radishes (*Raphanus sativus*), tomato (*Solanum lycopersicum*), and edible greens such as *Opuntia* spp. and *Piper auritum*, adapted to rainfall variability, demonstrating how these strategies, often applied on sloped and marginal terrains, reflect a deliberate shift toward agroecological restoration.

In the milpa–agroforestry system [59], women's adaptive responses include adjusting planting calendars, selecting drought-resistant seed varieties, and diversifying with shade-tolerant crops such as coffee and fruit trees. The continued cultivation of coyuchi cotton within these plots illustrates not only ecological adaptation but also the maintenance of ancestral production systems, even at a reduced scale. This spatial reorganization and knowledge application has led to visible transformations: abandoned lands were re-integrated into productive mosaics, increasing both ecological functionality and household autonomy.

4.5. Spatial Agency and Biocultural Resilience

The land use decisions made by women in Valle Nacional are not only practical but also symbolic, rooted in the transmission of cultural identity and collective memory. Home gardens, for example, serve as living archives of edible and medicinal biodiversity. Their arrangement often follows ancestral knowledge, including planting by lunar cycles or slope orientation.

These practices embody what might be termed invisible geographies—forms of spatial agency that are underrepresented in cartographic data yet crucial for ecological and cultural resilience. By managing the edges of forests and watersheds, interplanting tree species, and maintaining soil fertility through organic methods, women ensure continuity in food security and biodiversity retention. Their spatial agency bridges the tangible (soil, water, species) with the intangible (rituals, calendar, identity), making them pivotal actors in sustaining territorial integrity.

4.6. Weaving Practices and Cultural Continuity

The weaving of huipiles and other traditional textiles represents a territorial expression grounded in women's ecological practices. Though often excluded from land use analysis due to its domestic and artisanal character, weaving is intrinsically linked to territorial management. The raw material—coyuchi cotton—is cultivated within milpa systems; the natural dyes derive from local flora; and the motifs themselves encode ecological and cosmological knowledge.

Motifs such as mazorcas of maize, birds, or rivers are not merely decorative—they serve as symbolic references that evoke territorial knowledge, narrating the cyclical rhythms and sacred features of the landscape. In the context of increasing outmigration and generational shifts, textile production becomes a medium of resilience, maintaining a tangible connection between domestic life and the surrounding environment. Women thus act as cultural cartographers, weaving territorial narratives that sustain biocultural heritage across generations.

4.7. Historical Drivers of Territorial Transformation

Territorial change must also be understood within its historical context. In the 1960s, flooding prompted the construction of the Cerro de Oro dam, which dramatically reduced water availability and led to the decline of irrigated agriculture between the 1970s and 1990s [18,62,63]. This large-scale infrastructure significantly reduced river flows, resulting in the loss of irrigated agricultural areas and a gradual shift toward seasonal rainfed systems starting in 1996 [63,64]. Women's territorial engagement—through the management of marginal lands, diversification of crops, and adaptation to rainfed conditions—must be understood as an integral part of this broader adaptive process. These transformations in land use were further reinforced by national conservation programs, such as the Payment for Ecosystem Services (PES), launched in 2003 [65].

Concurrently, women upheld the tradition of textile weaving, which gained renewed institutional visibility after 2000 through cultural programs and legal protections [30–33,64,66]. Although textile production had long served as a form of cultural expression, this period marked a turning point in its external recognition. In 2019, the government advanced the promotion of collective intellectual property rights for Indigenous communities [67], further reinforcing its political visibility. However, this institutional validation did not emerge spontaneously—it was the result of sustained efforts by Indigenous women to defend and assert the originality of their textile designs, a process that exemplifies biocultural resilience [39,42,68]. However, the COVID-19 pandemic in 2020 disrupted the intergenerational transmission of this knowledge, contributing to demographic shifts that also impacted territorial practices (Figure 8).

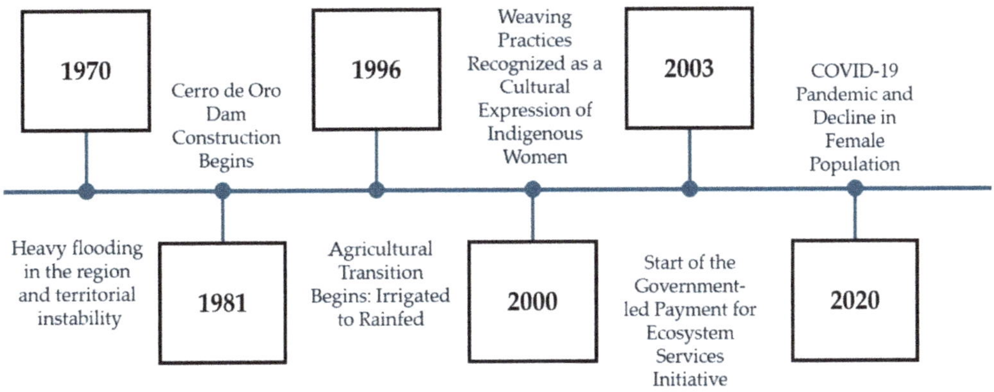

Figure 8. Timeline of key socio-environmental and cultural events influencing territorial transformation in San Juan Bautista Valle Nacional, Oaxaca.

5. Discussion

5.1. Territorial Change as a Historical and Socioecological Process

The historical trajectory outlined in the previous section underscores how territorial change in Valle Nacional has been shaped by overlapping processes of environmental intervention, state-driven infrastructure, and community adaptation. Rather than isolated events, the construction of the dam, the decline of irrigated agriculture, and subsequent conservation efforts reflect a continuum of imposed and negotiated transformations. These pivotal moments not only reshaped the physical landscape but also redefined social roles and community structures over time [1,2,19].

Such transformations are not merely technical or environmental in nature; they are also deeply symbolic, influencing cultural practices, gender roles, and territorial meanings. In the Chinantla region, women engage actively in the use and transformation of territory—through agriculture, textile production, and environmental stewardship—yet they often remain excluded from formal land tenure and decision-making structures. This mirrors similar cases in Africa, where women farmers continue to seek recognition of their land rights and struggle for equitable participation in governance [21].

5.2. Migration, Depopulation, and the Reconfiguration of Rural Life

Migration—particularly of men—intensified in the early 2000s, largely in response to declining agricultural productivity, land scarcity, and insufficient institutional support. One of the key drivers was the region's loss of commercial importance following the construction of the Cerro de Oro dam, prompting residents to seek alternative employment in other areas [64]. Between 2015 and 2020, 34,644 people from Oaxaca migrated abroad, with 2% of this figure originating from the Chinantla region [69]. This trend reflects broader patterns of structural displacement in Indigenous territories affected by state-led development interventions [2,5,17].

According to INEGI, the primary reasons for migration from Valle Nacional were family reunification (67.3%), followed by employment (18.2%), education (6%), insecurity (2.7%), and other causes (5.8%) [43–45]. This demographic restructuring has resulted in the visible feminization of rural life. Women have assumed not only caregiving and subsistence roles but also increasing responsibility for agricultural plots and household economies.

These roles are grounded in specific territorial practices. Women in female-headed households cultivate rainfed plots in marginal lands and former pastures, manage forest patches through agroforestry systems, and oversee communal grazing areas for small livestock. In areas where irrigation is no longer viable due to the dam's impact, women have reorganized production strategies around seasonal cycles and microclimates, relying on local knowledge to optimize land suitability and resource access. This spatial adaptation constitutes a form of territorial management that redefines how land is occupied and utilized.

The sharp rise in female-headed households—from 819 in 2000 to 1724 in 2020—indicates a broader gendered realignment in land and resource management. This demographic shift has had a direct impact on land use. Between 2000 and 2005, the expansion of tall evergreen rainforest—from 8406 ha to 40,526 ha—suggests a temporary reduction in agricultural pressure, likely linked to rural depopulation and the abandonment of cultivation zones [49–51]. This pattern supports the hypothesis that while depopulation is socially disruptive, it may enable ecological regeneration when not followed by extractive activities [7,9].

5.3. Repopulation, Institutional Support, and the Return to Territory

By 2010, repopulation trends began to emerge, coinciding with the stabilization of rainfed agriculture and the rise of community-led conservation initiatives. The increase in the total number of households during this period can be attributed to several factors.

First, although many residents had emigrated for economic reasons, favorable local conditions encouraged their return. These conditions included renewed opportunities for sustainable livelihoods supported by public policy interventions from the Oaxaca state government and federal initiatives such as the Payment for Environmental Services program promoted by the National Forestry Commission (CONAFOR), alongside support from civil society organizations [29,65]. Return migration was also driven by a strong sense of belonging and cultural attachment, particularly among Indigenous residents who view land as both a material and symbolic space.

Second, demographic changes contributed to household growth. Many local offspring who had reached adulthood formed new nuclear households, either within the community or nearby, often maintaining strong ties to their extended families [46,58].

Third, repopulation was not limited to returning residents. In-migration from neighboring communities also played a role, drawn by local productive initiatives, collective land management practices, and territorial governance structures that enabled access to land and social integration [44,46,69].

Importantly, this repopulation process was shaped by the intersection of state support and strong local governance. Customary institutions played a pivotal role in reinforcing communal land rights, enabling collective decision-making, and facilitating community participation. These governance structures enhanced the appeal of returning, as they offered real opportunities for dignified livelihoods anchored in cultural continuity and ecological sustainability.

In Indigenous communities, attachment to territory is not only utilitarian but deeply symbolic. Migration is often a necessity rather than a choice, and the possibility of returning is conditioned by the potential to reestablish one's role within a culturally meaningful and ecologically viable setting. In this context, repopulation reflects more than demographic recovery—it signals the resilience of territorial identity and the capacity for adaptive self-determination within socio-ecological systems.

The COVID-19 pandemic further encouraged return migration, particularly among men who lost employment in urban centers. These dynamics reflect the complex interplay between external shocks, social responses, and territorial memory—mediated by cultural values and collective agency [4,6,12].

5.4. Women's Agency, Community Governance, and Biocultural Resilience

Throughout this period, community-based governance has played a pivotal role in maintaining territorial resilience. Local committees have responded to pressing threats such as wildfires, illegal resource extraction, and land degradation. The introduction of the national Payment for Environmental Services (PES) program in 2003 helped institutionalize conservation efforts by providing economic incentives for the protection of ecosystem services [6,28,65]. These decentralized and participatory mechanisms contrast sharply with earlier technocratic approaches, which frequently excluded Indigenous perspectives from land management and spatial planning decisions [8,19].

By 2020, 28.6% of households in Valle Nacional were headed by women. While this figure captures a demographic shift shaped by factors such as male outmigration, widowhood, and aging, it also provides an entry point for examining how women navigate rural life under changing conditions. In some cases, women become the primary economic providers and take on decision-making roles within their households. However, female-headed households do not inherently signal greater autonomy or collective power; rather, agency emerges through broader socio-cultural dynamics and support networks.

In the Chinantla region, long-term participatory research has documented how Indigenous women, regardless of household structure, contribute decisively to community life, subsistence farming, and ecological stewardship. Some women face structural barriers to public engagement due to restrictive gender norms and unequal domestic burdens. Others, however, are actively involved in local governance and collective initiatives—especially when supported by family networks and community-based institutions.

Women across these contexts are central to the preservation of agroecological knowledge. They cultivate a diversity of crops [16,70,71] that sustain food sovereignty and agrobiodiversity [10,19,22,24]. These contributions are not confined to female-headed households but are part of a broader gendered division of labor that shapes biocultural resilience.

These gendered roles are not only symbolic or cultural; they have material consequences for land use patterns and territorial dynamics. Women's participation in agroforestry, rainfed agriculture, and cattle management—particularly in response to male outmigration—has directly influenced land occupation strategies. As such, the transformations documented in Table 1 are not solely driven by environmental or demographic pressures, but also by the adaptive agency of women who sustain food systems and shape community resilience.

In addition, women safeguard ancestral knowledge through agroecological practices such as seed saving, intercropping, and home garden management. These techniques—often overlooked or undervalued by mainstream rural development programs—are essential for maintaining ecological integrity and enhancing community resilience [12–14,16,38]. Beyond their practical function, they facilitate intergenerational knowledge transfer and reinforce cultural identity and a deeply rooted sense of place. Figure 9 illustrates this multifaceted role, highlighting how women act as custodians of both ecological systems and cultural continuity.

(**a**) (**b**)

Figure 9. Home Gardens as Spaces of Agroecological Practice and Cultural Continuity: (**a**) Intercropped plot with squash (*Cucurbita pepo*) and radish (*Raphanus sativus*); (**b**) Intercropped plot with cilantro (*Coriandrum sativum*), radish (*Raphanus sativus*), and tomato (*Solanum lycopersicum*), cultivated by women in the Chinantla region. These home gardens represent one facet of women's broader involvement in land stewardship, which also includes rainfed agriculture, agroforestry systems, and livestock care, especially in contexts of male outmigration.

5.5. Cultural Heritage, Symbolism, and the Paradox of Recognition

This stewardship also extends to the realm of textile production—particularly the weaving of huipiles on backstrap looms—which has long been recognized within Chinantec communities as a vital cultural expression. Since the mid-20th century, these garments have obtained national visibility through their presence in rituals and dances such as the Flor de Piña at the Guelaguetza festival. Weaving traditions encode cosmological knowledge and territorial memory through specific motifs—birds, rivers, trees, and maize—that symbolically represent ecological elements and social values [27,32,33,35–38,47].

However, it was not until the early 2000s that institutional policies began to formally support and promote the revitalization of these practices. For example, Mexico's Alternative Tourism Program in Indigenous Zones (PTAZI), launched in 2007 by the former National Commission for the Development of Indigenous Peoples (CDI), fostered the preservation of Indigenous cultural heritage by promoting artisanal production, community tourism, and environmental stewardship [72]. In this context, weaving has not "suddenly" gained recognition but has instead become more publicly valorized through initiatives such as storytelling, educational workshops, and local exhibitions.

Today, huipiles continue to serve as living archives of identity, worn in communal festivities and proudly exhibited as embodiments of collective memory and cultural pride [11,27,31,38,73,74] (Figure 10). Yet, the practice faces increasing threats due to generational disinterest, economic hardship, and migration to urban areas [20,25,75]. Outside the community, these garments are often abandoned as young women face pressures to assimilate or avoid ethnic discrimination [75].

(a) (b)

Figure 10. Public Exhibition of Chinantec Huipiles from San Juan Bautista Valle Nacional as Part of Local Cultural Revitalization Efforts: (**a**) Huipil displayed at the 2023 Guelaguetza Artisan Fair. (**b**) Huipil displayed at the 2025 Guelaguetza Artisan Fair. Both garments feature traditional backstrap-loom weaving and intricate iconography, including motifs of birds, trees, maize, and cosmological patterns. These elements embody territorial memory, Chinantec identity, and the symbolic relationship between women, land, and cultural resilience.

Despite their vital cultural and ecological contributions, women remain marginalized within formal governance structures. Community decision-making continues to be shaped by customary law, which often limits women's participation to auxiliary roles in church groups, school committees, or supportive tasks [20–22,46]. Leadership positions are predominantly held by men, reflecting persistent gender hierarchies embedded in both traditional norms and institutional frameworks [23,26,74]. This underscores a structural paradox in which women sustain critical territorial practices yet remain politically invisible. In Valle Nacional, this invisibility is most striking in the cultural domain: while the Chinantec huipil has gained national recognition, the social systems that support its production—land access, gender equity, and intergenerational learning—remain fragile.

6. Conclusions

This study shows that land use change in San Juan Bautista Valle Nacional is shaped not only by ecological and infrastructural shifts—such as the construction of the Cerro de Oro dam—but also by community-based responses rooted in cultural memory, gender roles, and local governance. The decline of irrigation agriculture prompted transitions to rainfed systems and contributed to demographic restructuring, including a notable rise in female-headed households due to male outmigration.

Amid these transformations, Indigenous women have played a central role in sustaining community life and biocultural resilience. Their contributions—ranging from agroecological food production and seed saving to textile weaving—reflect adaptive strategies grounded in ancestral knowledge and daily territorial decision-making. These practices enhance food security, strengthen cultural identity, and help maintain biodiversity across complex ecosystems.

The analysis of women's spatial practices—including in rainfed plots, home gardens, agroforestry systems, and pastures—reveals a gendered geography of land management. Through decisions related to planting, water use, soil restoration, and ecological regeneration, women have reconfigured land use patterns to respond to environmental and socio-economic pressures. This spatial agency remains largely invisible within formal land use planning frameworks, yet it plays a crucial role in ecological recovery and the continuity of cultural landscapes.

Recognizing these practices as valid forms of territorial governance is essential for developing inclusive and resilient strategies. Women's agency—though often operating outside formal institutions—shapes reterritorialization processes that are simultaneously social, ecological, and symbolic. Territory, as shown in this case, is not governed solely through infrastructure or state policy but through the embodied, everyday decisions of women who preserve and transmit biocultural knowledge.

Nevertheless, a paradox persists: women are vital to maintaining land, culture, and biodiversity, yet they remain politically marginal in both customary and state governance structures. This institutional gap limits the recognition of gendered contributions to territorial transformation. In Valle Nacional, land use change, cultural heritage, and gender dynamics must be understood as interconnected—not isolated—dimensions of rural transformation. Women's agency is not simply reactive to environmental crises; it represents a force of continuity that preserves memory, resists marginalization, and shapes future trajectories.

While this study integrates participatory mapping and field observation to document women's land practices, it is not exhaustive. Limitations include the lack of high-resolution temporal data to track short-term change and the absence of systematically collected oral histories, which would enrich the symbolic and historical reconstruction of the territory. Future research should strengthen participatory approaches, engage younger generations, and pursue comparative analysis to better understand intergenerational resilience and spatial memory.

This research underscores the importance of designing public policies that explicitly include women in the conservation of cultural heritage and territorial management. Such policies should not only increase economic opportunities for women but also promote a more equitable distribution of household responsibilities—enabling greater participation in collective and community-based initiatives.

Author Contributions: Conceptualization, G.L.-E. and F.E.O.-H.; data curation, M.A.A.-O.; formal analysis, M.A.A.-O. and Y.D.O.-H.; investigation, F.E.O.-H. and M.E.T.-C.; methodology, G.L.-E.; resources, Y.D.O.-H. and M.E.T.-C.; supervision, Y.D.O.-H. and M.E.T.-C.; validation, M.A.A.-O., Y.D.O.-H., F.E.O.-H. and G.L.-E.; writing—original draft, G.L.-E.; writing—review and editing, G.L.-E., M.A.A.-O. and Y.D.O.-H. All authors have read and agreed to the published version of the manuscript.

Funding: This research was funded by the Instituto Politécnico Nacional through the project SIP-IPN 20241375.

Institutional Review Board Statement: Not applicable.

Data Availability Statement: The data on which this study is based are openly available in the Population and Housing Censuses and Surveys provided by the Demographic and Social Information Subsystem of INEGI at https://www.inegi.org.mx/programas/ccpv/2020/ (accessed on 20 July 2025) and public data from San Juan Bautista Valle Nacional on the Oaxaca Planning System Platform id 559 at https://sisplade.oaxaca.gob.mx/bm_sim_services/PlanesMunicipales/2011_2013/ (accessed on 20 July 2025).

Acknowledgments: We acknowledge the backing of the EcoLogic Development Fund, Fondo Ambiental Regional de la Chinantla, Oaxaca, A.C., the artisan women of San Juan Bautista Valle Nacional for preserving and sharing their cultural heritage, and the Instituto Politécnico Nacional for their support; as well as the Secretaría de Ciencia, Humanidades, Tecnología e Innovación (SECIHTI) and Sistema Nacional de Investigadoras e Investigadores (SNII-SECIHTI).

Conflicts of Interest: The authors declare no conflicts of interest. The funders had no role in the design of the study; in the collection, analyses, or interpretation of data; in the writing of the manuscript; or in the decision to publish the results.

Abbreviations

The following abbreviations are used in this manuscript:

CDI	National Commission for the Development of Indigenous Peoples
CONAFOR	National Forestry Commission
INEGI	National Institute of Statistics and Geography
PTAZI	Alternative Tourism Program in Indigenous Zones

References

1. Marino, D.; Barone, A.; Marucci, A.; Pili, S.; Palmieri, M. The Integrated Analysis of Territorial Transformations in Inland Areas of Italy: The Link between Natural, Social, and Economic Capitals Using the Ecosystem Service Approach. *Land* **2024**, *13*, 1455. [CrossRef]
2. Abramson, S.F.; Carter, D.B.; Ying, L. Historical Border Changes, State Building, and Contemporary Trust in Europe. *Am. Political Sci. Rev.* **2022**, *116*, 875–895. [CrossRef]
3. Damayanti, R.W.; Hartono, B.; Wijaya, A.R. Clarifying megaproject complexity in developing countries: A literature review and conceptual study. *Int. J. Eng. Bus. Manag.* **2021**, *13*, 18479790211027414. [CrossRef]
4. De Nito, E.; Caccialanza, A.; Canonico, P.; Favari, E. Analyzing the role of social value in megaprojects: Toward a new performance framework. *Meas. Bus. Excell.* **2024**, *28*, 243–254. [CrossRef]
5. Pérez Velasco Pavón, J.C. Economic behavior of indigenous peoples: The Mexican case. *Lat. Am. Econ. Rev.* **2014**, *23*, 12. [CrossRef]
6. García-Padilla, E.; DeSantis, D.L.; Rocha, A.; Fucsko, L.A.; Johnson, J.D.; Lazcano-Villarreal, D. Biological and Cultural Diversity in the state of Oaxaca, Mexico: Strategies for Conservation among Indigenous Communities. *Biol. Soc.* **2022**, *5*, 48–72. [CrossRef]
7. Lugo-Espinosa, G.; Acevedo-Ortiz, M.A.; Aquino-Bolaños, T.; Ortiz-Hernández, Y.D.; Ortiz-Hernández, F.E.; Pérez-Pacheco, R.; López-Cruz, J.Y. Cultural Heritage, Migration, and Land Use Transformation in San José Chiltepec, Oaxaca. *Land* **2024**, *13*, 1658. [CrossRef]

8. Torre, A. Contribution to the theory of territorial development: A territorial innovations approach. *Reg. Stud.* **2025**, *59*, 2193218. [CrossRef]
9. Živojinović, I.; Elomina, J.; Pülzl, H.; Calanasan, K.; Dabić, I.; Ólafsdóttir, R.; Siikavuopio, S.; Iversen, A.; Robertsen, R.; Bjerke, J.; et al. Exploring land use conflicts arising from economic activities and their impacts on local communities in the European Arctic. *J. Land Use Sci.* **2024**, *19*, 186–210. [CrossRef]
10. Sepie, A.J. More than Stories, More than Myths: Animal/Human/Nature(s) in Traditional Ecological Worldviews. *Humanities* **2017**, *6*, 78. [CrossRef]
11. Aktürk, G.; Dastgerdi, A.S. Cultural Landscapes under the Threat of Climate Change: A Systematic Study of Barriers to Resilience. *Sustainability* **2021**, *13*, 9974. [CrossRef]
12. Dei, G.J.S.; Karanja, W.; Erger, G. The Role of Elders and Their Cultural Knowledges in Schools. In *Elders' Cultural Knowledges and the Question of Black/ African Indigeneity in Education*; Springer International Publishing: Cham, Switzerland, 2022; pp. 127–151.
13. García, H.D.; García, W.A.; Curcio, C.L. Aging in Indigenous Communities: Perspective from Two Ancestral Communities in the Colombian Andean-Amazon Region. *J. Cross-Cult. Gerontol.* **2024**, *39*, 189–205. [CrossRef] [PubMed]
14. Ford, J.D.; King, N.; Galappaththi, E.K.; Pearce, T.; McDowell, G.; Harper, S.L. The Resilience of Indigenous Peoples to Environmental Change. *One Earth* **2020**, *2*, 532–543. [CrossRef]
15. Adger, W.N.; Fransen, S.; Safra de Campos, R.; Clark, W.C. Migration and sustainable development. *Proc. Natl. Acad. Sci. USA* **2024**, *121*, e2206193121. [CrossRef] [PubMed]
16. Acevedo-Ortiz, M.A.; Lugo-Espinosa, G.; Ortiz-Hernández, Y.D.; Pérez-Pacheco, R.; Ortiz-Hernández, F.E.; Granados-Echegoyen, C.A. Women's Leadership in Sustainable Agriculture: Preserving Traditional Knowledge Through Home Gardens in Santa Maria Jacatepec, Oaxaca. *Sustainability* **2024**, *16*, 9513. [CrossRef]
17. Martínez Casas, R. Todo sea para el Desarollo: El reacomodo de comunidades indígenas por la construcción de represas en el Alto Papaloapan. *Público Priv.* **2022**, *20*, 35–59. [CrossRef]
18. Dominguez, J. The Construction of Dams in Mexico: Evolution, Current Situation and New Approaches for the Viability to Water Infrastructure. *Gest. Politica Publica* **2019**, *28*, 3–37. [CrossRef]
19. Estrada, A.; Garber, P.A.; Gouveia, S.; Fernández-Llamazares, Á.; Ascensão, F.; Fuentes, A.; Garnett, S.T.; Shaffer, C.; Bicca-Marques, J.; Fa, J.E.; et al. Global importance of Indigenous Peoples, their lands, and knowledge systems for saving the world's primates from extinction. *Sci. Adv.* **2022**, *8*, eabn2927. [CrossRef]
20. Muñoz, J.d.L.Q. Territory, Memory and Gender: The meanings of women's political participation in Atenco, México. *Ambiente Soc.* **2019**, *22*, e01161. [CrossRef]
21. Maybom, C.; Funder, M.K. Women's agency in environmental interventions: Navigating "resilient livelihoods" in Kenya. *Geoforum* **2025**, *164*, 104343. [CrossRef]
22. Durán-Díaz, P.; Armenta-Ramírez, A.; Kurjenoja, A.K.; Schumacher, M. Community Development through the Empowerment of Indigenous Women in Cuetzalan Del Progreso, Mexico. *Land* **2020**, *9*, 163. [CrossRef]
23. Gustavsson, M. Women's changing productive practices, gender relations and identities in fishing through a critical feminisation perspective. *J. Rural. Stud.* **2020**, *78*, 36–46. [CrossRef]
24. Visser, J.; Wangu, J. Women's dual centrality in food security solutions: The need for a stronger gender lens in food systems' transformation. *Curr. Res. Environ. Sustain.* **2021**, *3*, 100094. [CrossRef]
25. Vujko, A.; Karabašević, D.; Cvijanović, D.; Vukotić, S.; Mirčetić, V.; Brzaković, P. Women's Empowerment in Rural Tourism as Key to Sustainable Communities' Transformation. *Sustainability* **2024**, *16*, 10412. [CrossRef]
26. Brogan, G.; Dooley, K. Weaving together social capital to empower women artisan entrepreneurs. *Int. J. Gend. Entrep.* **2023**, *16*, 69–88. [CrossRef]
27. Klein, K. *El Hilo Continuo: La Conservación de las Tradiciones Textiles de Oaxaca*; Getty Conservation Institute: Los Angeles, CA, USA, 1997.
28. Acevedo-Ortiz, M.A.; Lugo-Espinosa, G.; Ortiz-Hernández, Y.D.; Ortiz-Hernández, F.E. Comunidades en la Chinantla Oaxaca conservando bosques y selvas: Impulso de iniciativas locales. *Cienc. Agronómicas Apl. Biotecnol.* **2023**, *3*, 110–115.
29. Lee-Peluso, N.; Lund, C. New frontiers of land control: Introduction. *Peasant. Stud.* **2011**, *38*, 667–681. [CrossRef]
30. Abad, C. El Huipil de Valle Nacional, en Peligro de Extinción; Solo Ocho Personas Luchan por su Rescate. Available online: https://www.youtube.com/watch?v=4V3S2WclA_4&t=1s (accessed on 15 July 2025).
31. BBVA México. BBVA en las Actividades de la Guelaguetza 2024. Available online: https://www.youtube.com/watch?v=AZ7wD7k0yhI (accessed on 2 July 2025).
32. FAHHO. Huipil Chinanteco de Valle Nacional. Available online: https://repositorio.fahho.mx/handle/123456789/36336 (accessed on 14 July 2025).
33. Jerónimo Santiago, M. *Huipiles de Valle Nacional: Arte y Mitología*; Secretaría de Cultura del Gobierno del Estado de Oaxaca: Oaxaca, Mexico, 2006.
34. Rodríguez, L. Valle Nacional a Través de ss Huipiles. Available online: https://www.facebook.com/watch/?v=10155499828726954 (accessed on 15 July 2025).

35. Secretaría de las Culturas y Artes de Oaxaca (SECULTA). Bordado y Huipil Chinanteco. Available online: https://www.youtube.com/watch?v=FM2-sWBcsJg (accessed on 17 July 2025).
36. Secretaría de las Culturas y Artes de Oaxaca (SECULTA). Explicación del Significado de las Partes del Huipil Capsula 14. Available online: https://www.youtube.com/watch?v=j_qmkHdtAbw (accessed on 14 July 2025).
37. SiCmexico. Textilería de San Juan Bautista Valle Nacional. Available online: https://sic.gob.mx/ficha.php?table=frpintangible&table_id=368 (accessed on 2 July 2025).
38. Nance, D.; Lopez, O. The Emotional Meaning of Dress for Older Oaxacan Women: Identity, Pride, Racism and Nostalgia in Twenty-First Century Mexico. *J. Dress Hist.* **2022**, *6*, 24–48.
39. Ungar, M.; McRuer, J.; Liu, X.; Theron, L.C.; Blais, D.; Schnurr, M.A. Social-ecological resilience through a biocultural lens: A participatory methodology to support global targets and local priorities. *Ecol. Soc.* **2020**, *25*, 8. [CrossRef]
40. Martínez-Reyes, J.E. *Biocultural Diversity Conservation: A Global Sourcebook*; Francis, T., Ed.; Earthscan: New York, NY, USA, 2012; Volume 3.
41. Swiderska, K.; Argumedo, A.; Wekesa, C.; Ndalilo, L.; Song, Y.; Rastogi, A.; Ryan, P. Indigenous Peoples' Food Systems and Biocultural Heritage: Addressing Indigenous Priorities Using Decolonial and Interdisciplinary Research Approaches. *Sustainability* **2022**, *14*, 11311. [CrossRef]
42. Costea, A.; Gligor, V.; Nicula, E.-A.; Domnița, M.; Alexe, M. The Biocultural Resilience Index (BRI): A New Tool for Assessing the Sustainability of Multifunctional Landscapes. *Preprint* **2024**. [CrossRef]
43. Instituto Nacional de Estadística y Geografía (INEGI). Panorama Sociodemográfico de México. Available online: https://www.inegi.org.mx/programas/ccpv/2020/ (accessed on 2 July 2025).
44. Gobierno de México, D.M. San Juan Bautista Valle Nacional. Available online: https://www.economia.gob.mx/datamexico/es/profile/geo/san-juan-bautista-valle-nacional (accessed on 10 June 2025).
45. Gobierno de México. Informe Anual Sobre la Situación de la Pobreza y Rezago Social 2024. San Juan Bautista Valle Nacional. Available online: https://www.gob.mx/cms/uploads/attachment/file/889133/20559SanJuanBautistaValleNacional2024.pdf (accessed on 3 February 2025).
46. Mora-Díaz, I.N. Plan Municipal de Desarrollo San Juan Bautista Valle Nacional 2011–2013. Available online: https://sisplade.oaxaca.gob.mx/bm_sim_services/PlanesMunicipales/2011_2013/ (accessed on 2 July 2025).
47. Secretaría de Turismo de Oaxaca. Ruta de la Chinantla. Available online: https://tuxtepecturismo.com/que-ver-y-hacer/ruta-chinantla/ (accessed on 2 July 2025).
48. ESRI. ArcGIS. Available online: www.esri.com (accessed on 20 August 2025).
49. Instituto Nacional de Estadística y Geografía (INEGI). Conjunto de Datos Vectoriales de la Carta de Uso del Suelo y Vegetación. Escala 1:250 000. Serie I. Continuo Nacional. Available online: https://www.inegi.org.mx/app/biblioteca/ficha.html?upc=702825007020 (accessed on 24 April 2025).
50. Instituto Nacional de Estadística y Geografía (INEGI). Conjunto de Datos Vectoriales de la Carta de Uso del Suelo y Vegetación. Escala 1:250 000. Serie II. Continuo Nacional. Available online: https://www.inegi.org.mx/app/biblioteca/ficha.html?upc=702825007021 (accessed on 24 April 2025).
51. Instituto Nacional de Estadística y Geografía (INEGI). Conjunto de Datos Vectoriales de la Carta de Uso del Suelo y Vegetación. Escala 1:250 000. Serie III. Continuo Nacional. Available online: https://www.inegi.org.mx/app/biblioteca/ficha.html?upc=702825007022 (accessed on 24 April 2025).
52. Instituto Nacional de Estadística y Geografía (INEGI). Conjunto de Datos Vectoriales de la Carta de Uso del Suelo y Vegetación. Escala 1:250 000. Serie IV. Conjunto Nacional. Available online: https://www.inegi.org.mx/app/biblioteca/ficha.html?upc=702825007023 (accessed on 24 April 2025).
53. Instituto Nacional de Estadística y Geografía (INEGI). Conjunto de Datos Vectoriales de la Carta de Uso del Suelo y Vegetación. Escala 1:250 000. Serie V. Conjunto Nacional. Available online: https://www.inegi.org.mx/app/biblioteca/ficha.html?upc=702825007024 (accessed on 24 April 2025).
54. Instituto Nacional de Estadística y Geografía (INEGI). Conjunto de datos Vectoriales de la Carta de Uso del Suelo y Vegetación. Escala 1:250 000. Serie VI. Conjunto Nacional. Available online: https://www.inegi.org.mx/app/biblioteca/ficha.html?upc=889463598459 (accessed on 24 April 2025).
55. Instituto Nacional de Estadística y Geografía (INEGI). Conjunto de Datos Vectoriales de la Carta de Uso del Suelo y Vegetación. Escala 1:250 000. Serie VII. Conjunto Nacional. Available online: https://www.inegi.org.mx/app/biblioteca/ficha.html?upc=889463842781 (accessed on 24 April 2025).
56. Instituto Nacional de Estadística y Geografía (INEGI). Datos Abiertos. Available online: https://www.inegi.org.mx/datosabiertos/ (accessed on 10 June 2025).
57. Lima, P.T.; Valdiviezo, A.C. La Chinantla oaxaqueña: La historia de un patrimonio perdido. *Diseño Soc.* **2016**, *1*, 54–61.
58. Vargas-Romero, A. Plan Municipal de Desarrollo San Juan Bautista Valle Nacional 2008–2010. Available online: https://sisplade.oaxaca.gob.mx/bm_sim_services/PlanesMunicipales/2008_2010/ (accessed on 2 July 2025).

59. Benrey, B.; Bustos-Segura, C.; Grof-Tisza, P. The mesoamerican milpa system: Traditional practices, sustainability, biodiversity, and pest control. *Biol. Control* **2024**, *198*, 105637. [CrossRef]
60. Enríquez-Nolasco, S.G. *Impacto Territorial del Cultivo del Algodón en el Estado de Oaxaca, México*; Universidad Nacional Autónoma de México: Ciudad de México, Mexico, 2020.
61. de Teresa, A.P. Del porfiriato a la Revolución: La propiedad de la tierra en la parroquia de Valle Nacional, Oaxaca, 1884-1915. *Cuad. Sur* **2020**, *1*, 61–83.
62. McCormick, G. The Forgotten Jaramillo: Building a Social Base of Support for Authoritarianism in Rural Mexico. In *Dictablanda: Politics, Work, and Culture in Mexico, 1938–1968*; Gillingham, P., Smith, B., Eds.; Duke University Press: Durham, NC, USA, 2014.
63. Núñez Núñez, M. El agua en la región de la Chinantla, México. Estudio comparativo de una cosmovisión chinanteca a partir de su tradición oral. *Boletín Lit. Oral* **2019**, *9*, 121–142. [CrossRef]
64. De Teresa, A.P. Población y recursos en la región chinanteca de Oaxaca. *Desacatos* **1999**, *1*, 1–24.
65. Comisión Nacional Forestal. Conservación—Programa de Pago por Servicios Ambientales. Available online: https://snif.cnf.gob.mx/conservacion/ (accessed on 15 July 2025).
66. Stephen, L. Women's Weaving Cooperatives in Oaxaca. *Crit. Anthropol.* **2005**, *25*, 253–278. [CrossRef]
67. Andrés-Rosales, R.; de Jesús-Almonte, L.; Carbajal-Suárez, Y. Empleo, producción y salario manufacturero en México ante la pandemia por la COVID-19. Un análisis de VAR espacial. *Cienc. Administrativas. Teoría Prax.* **2022**, *17*, 32–49. [CrossRef]
68. Caillon, S.; Cullman, G.; Verschuuren, B.; Sterling, E.J. Moving beyond the human–nature dichotomy through biocultural approaches: Including ecological well-being in resilience indicators. *Ecol. Soc.* **2017**, *22*, 27. [CrossRef]
69. Gobernación, S.d. Movilidad Humana Oaxaca. Available online: https://portales.segob.gob.mx/work/models/PoliticaMigratoria/CPM/foros_regionales/estados/centro/info_diag_F_centro/info_Oaxaca.pdf (accessed on 15 June 2025).
70. Lugo Espinosa, G.; Acevedo Ortiz, M.A.; Ortiz Hernández, Y.D. Conservación de semilla criolla y control biológico, resguardo del patrimonio biocultural de la Chinantla Oaxaca. In *Nuevas Territorialidades-Gestión de los Territorios y Recursos Naturales con Sustentabilidad Ambiental*; Sarmiento-Franco, J.F., Ed.; UNAM-AMECIDER: Ciudad de México, Mexico, 2023; Volume 1, pp. 221–236.
71. Lugo-Espinosa, G.; Acevedo-Ortiz, M.A.; Ortiz-Hernández, Y.D.; Aquino-Bolaños, T. Huertos familiares y participación comunitaria para la generación de medios de vida sostenibles. In *Los Objetivos del Desarrollo Sostenible versus la Pandemia de la COVID-19*, 1st ed.; Tavera-Cortés, M.E., Ed.; ASMIIA: Texcoco, Mexico, 2023; pp. 125–138.
72. Comisión Nacional para el Desarrollo de los Pueblos Indígenas. Programa Turismo Alternativo en Zonas Indígenas (PTAZI). Available online: https://www.coneval.org.mx/Informes/Evaluacion/Consistencia/2007_2008/CDI/Programa%20Turismo%20Alternativo%20en%20Zonas%20Ind%C3%ADgenas%20(PTAZI).pdf (accessed on 17 August 2025).
73. Githapradana, W.; Raharja, G.; Ratna, T.; Pebryani, N. Gender and Culture Inclusivity in the Story of Hope Fashion Collection. *ARCHive-SR* **2024**, *8*, 43–51. [CrossRef]
74. Anaya, S.; Zepeda, L. Vestir la Resistencia: Mujeres Indígenas Frente a la Violencia Patriarcal. Available online: https://www.zonadocs.mx/2022/03/22/vestir-la-resistencia-mujeres-indigenas-frente-a-la-violencia-patriarcal/ (accessed on 15 July 2025).
75. Gracia, M.A.; Horbath, J.E. Condiciones de vida y discriminación a indígenas en Mérida, Yucatán, México. *Estud. Sociológicos* **2019**, *37*, 277–307. [CrossRef]

land

Article

Who Is Most Responsible for the Mitigation of Climate Change? An Intercultural Study in Central Europe, Central Asia, and the Middle East

Anna Kwiatkowska [1,*], Magdalena Mosanya [2], Patrycja Uram [1], Dilbar Urazbayeva [3], Baxtigul Nurullayeva [3], Vita Mikuličiūtė [4], Rūta Sargautytė [4] and Konstantin Karpinskij [3,5]

1 Institute of Psychology, Polish Academy of Sciences, 1 Jaracza, 00-378 Warsaw, Poland; patrycja.uram@sd.psych.pan.pl
2 Faculty of Psychology, Murdoch University, Level 1&2, Block 18, Dubai Knowledge Park, Dubai P.O. Box 500700, United Arab Emirates; magdalena.mosanya@murdoch.edu.au
3 Department of Psychology, Manun University, Building 3, 110 Islam Karimov Str., Urgench 220100, Uzbekistan; dilbarurazbayeva8@gmail.com (D.U.); nurullayevabaxtigul73@gmail.com (B.N.); karpkostia@gmail.com (K.K.)
4 Faculty of Philosophy, Institute of Psychology, Vilnius University, Universiteto g., 01131 Vilnius, Lithuania; vita.mikuliciute@fsf.vu.lt (V.M.); ruta.sargautyte@fsf.vu.lt (R.S.)
5 Department of Exprimental and Applied Psychology, Faculty of Psychology, Yanka Kupala State University of Grodno, 22 Ozheshko Str., 230023 Hrodna, Belarus
* Correspondence: akwiatkowska@psych.pan.pl

Abstract: Researchers observed that even if one's environmental concern was high, people would delegate others to take responsibility for climate change mitigation and undertake sustainable actions. In this study, we explored how citizens of different countries in Central Europe, Central Asia, and the Middle East perceived responsibility for mitigating and reducing climate change consequences of various collective and individual agents. Also, we asked about the role of cultural values, environmental worldviews, and beliefs in the intractability of climate change in the prediction of the responsibility distribution. The total sample consisted of n = 1267 participants from Belarus, Lithuania, Poland, the United Arab Emirates, and Uzbekistan. We created the list of 11 collective and individual entities as accountable for mitigation and sustainable activities. We used the Collindex scale to measure collectivistic and individualistic values, the NEP scale to measure anthropocentric and ecocentric worldviews, and two questions concerning the intractability beliefs. Results showed that participants attributed more responsibility to collective agents than to individuals across countries. The predictors' patterns indicated that ecocentric worldviews consistently increased perceived responsibility of both collective and individual actors, whereas anthropocentrism reduced the attribution of collectives' responsibility. Collectivistic values appeared to foster greater expectations of individuals. Also, differences between national samples were observed.

Keywords: climate change; mitigation; cultural values; new environmental paradigm; intractability; cross-cultural study

1. Who Is Most Responsible for the Mitigation of Climate Change?

An Intercultural Study in Central Europe, Central Asia, and the Middle East

Climate change mitigation refers to any action taken by governments, businesses, and people to reduce the flow of heat-trapping greenhouse gases into the atmosphere to make the impact of climate change less severe. Also, mitigation includes enhancing carbon sinks that remove greenhouse gases from the atmosphere by protecting and restoring forests and critical ecosystems. Reducing and avoiding emissions requires changing almost every sphere of people's lives, and at every level of functioning of societies and individuals, from how economies work, to what everyday choices are made by individuals [1]. Mitigation efforts include transitioning to renewable energy sources, reducing energy consumption, changing agricultural practices, and changing individuals' habits concerning transportation or diet [2]. The climate change issue is fundamentally collective because everyone acts in ways that benefit themselves but produce greenhouse gas emissions. Therefore, efforts to address the impacts of climate change must also be collective [3], as the problem is too vast and complex to be solved by a single individual or entity [4]. There is considerable psychological literature on personal beliefs, emotions, and attitudes toward climate change [2,5–8]. There is a significant gap in research regarding how people perceive the responsibility for climate change action among different stakeholders. Understanding these perceptions is crucial for fostering effective collaboration and accountability in the fight against climate change.

There are many reasons why we should inquire about how people view responsibility for acting on climate change. To start with, although individuals are aware of climate change, they are often hesitant to alter their behaviors significantly. These activities, such as waste sorting, reducing shower duration, and unplugging electronics when not in use, are somewhat beneficial; however, they are unlikely to induce substantial mitigation effects [1,9,10]. Nevertheless, many people find it difficult to include these habits in their daily routines. Furthermore, convincing people to change their lifestyle—such as eating less meat, driving cars less, or avoiding extra flights, which have the greatest potential to cut greenhouse emissions—is much more difficult, if not impossible. Naturally, adopting the most effective behaviors requires greater effort and can be psychologically and financially demanding. Therefore, we should not be surprised that people tend to avoid deeper mitigation efforts, reject personal responsibility, and prefer to pass the task onto others. In this study, we aimed to examine how citizens of countries in Central Europe, Central Asia, and the Middle East attribute responsibility for climate change mitigation to various entities. Also, we are interested in the role of cultural values, ecological worldviews, and beliefs in the intractability of climate change in shaping opinions about responsibility distribution and sustainable practices.

Responsible agents. First of all, it should be underscored that before ascribing responsibility to someone, one must admit that a climate change problem exists, and it should be resolved by taking proper actions. For people who deny or minimize the consequences of climate change, responsibility issues are of less importance or even irrelevant to their lives. Nevertheless, research shows that responsibility for mitigation is the major concern for most individuals and for collective bodies, such as corporations or governments [3,9,11].

It is reasonable to expect that responsibility for climate change mitigation should be shared across different actors, such as governments, corporations, international organizations, and individuals, within their means. For example, governments have the means and power to implement policies and regulations appropriate to achieve mitigation, and sustainability aims, and to cooperate on climate issues with their counterparts at the international level [10,12]. Stoddart and his colleagues explored environmental group members' opinions on responsibility and discovered that among the top four responsibilities, the government was perceived as the most responsible, followed by individuals, "everyone," and corporations [13]. Becker and Sparks's qualitative study, in which many participants agreed that the government should play a crucial role in addressing climate change, though along with members of the public and corporations [14], confirmed these findings.

Nowadays, corporations have even more power, which enables them to invest in the research and development of cleaner technologies, transition to sustainable business models, and disseminate best practices across the globe [3,11,15]. The list of collective agents can be quite extensive, comprising numerous organizations, such as universities, cities, sub-national governments, and businesses [3].

Also, individuals can contribute to mitigation efforts by adopting sustainable practices in their daily lives, such as reducing energy consumption, using public transportation, consuming less meat, and minimizing waste [9]. Some people may help to achieve mitigation goals indirectly by increasing public and individual awareness of environmental threats and by teaching and supporting sustainable behaviors in children, community members, and social media users. These may be parents [16], teachers [10,17], religious leaders [18,19], social media influencers [20,21], as well as community leaders, respected by others.

In addition, individuals can also play a role in advocating for climate action by supporting policies that promote sustainability and checking on how governments and businesses operate within climate issues [3,22].

In the study on a German representative sample, the following actors were assessed as responsible for climate protection measures: politicians, economists, individuals, the society, industrialized countries, emerging economies, and developing countries. It occurred that every actor was considered highly responsible. However, developing countries were held the least responsible compared to other actors, while the highest responsibility was attributed to industrialized countries. The individuals were assumed to be less responsible than politicians, economists, or society in general [23]. The study carried out in Australia also showed that people placed more responsibility on groups and organizations, and less on individuals, for both causing and responding to climate change, regardless of their opinion on the causes of climate change [24].

If people believe that such a large number of entities are responsible for climate change action, it may generate diffusion of responsibility, which in turn will interfere with mitigation efforts [9]. Individuals may think that their actions are meaningless and pointless to politicians or businesses; politicians may place responsibility back onto individuals, etc. [12]. Indeed, the study with focus groups carried out in Switzerland revealed that individuals tend to justify their lack of engagement in mitigation efforts by blaming others for inaction, including governments [23].

The distribution of responsibility among numerous actors may depend on how individuals perceive the actors: their intention to act in the name of the right cause, their ability to act efficiently, their power to make a substantial impact, and their means to do so. Perception of responsible others may be entangled in social, political, and cultural contexts, as well as in psychological factors [10,12,14,25,26]. For example, in democratic states, the government is trusted more than in less democratic countries, so citizens may grant governmental institutions more rights to act in the environmental domain [12]. On the other hand, in countries ruled by authoritarian leaders, people may accept strict government controls as given and comply with governmental regulations [27]. It is worth noticing that interpersonal trust is higher in democratic societies, so people tend to believe that others are involved in the battle against climate change as much as everybody is [22].

Cultural context. Studies examining the influence of cultural value orientations on environmental activities frequently concentrate on postmaterialist values, usually relating them to increased environmental concern [4]. In this study we defined cultural context in terms of collectivistic and individualistic value orientations. There are mixed research results concerning the relationship between collectivism and individualism and attitudes toward the environment. Cultural values may support or diminish efforts undertaken to alleviate climate change and work for sustainability. Preference for collectivistic values may result in higher concern for climate change, seen as a common good—belonging to the community—degradation; therefore, people may attribute more responsibility to various agents fighting for the restoration of the environment for the benefit of society. Positive links between collectivistic orientation and sustainable behaviors are evidenced in many studies [28–30]. However, there are also ambiguous findings [31]. Furthermore, some studies deliver contradictory results [32–34], showing negative relationships between collectivism and attitudes toward climate change concern.

Individualistic values promote egocentric interests, a lifestyle framed by seeking pleasure, excitement, and self-fulfillment, regardless of others' needs. So, people of individualistic orientation may be less interested in the community's goals and be indifferent to responsibility issues [26], except for being responsible for taking care of themselves [35]. Xue and his colleagues found in Chinese participants that higher scores on individualism were associated with reduced support for policies to mitigate climate change [27]. However, more pro-environmental activities are observed in individualistic societies than in collectivistic cultures [1,31–36]. The widespread presence of environmental groups is more likely in individualistic societies rather than in collectivistic societies due to the individualists' ability to form temporary groups aimed at achieving a specific goal rather than basing their activity on stable, long-lived groups [35]. Also, there are studies showing the lack of a relationship between the individualism–collectivism dimension and environmental performance [11].

Such ambivalent and even contradictory findings make it difficult to predict who people of collectivistic or individualistic orientation delegate to bear responsibility for climate change mitigation. Obviously, the concept of culture is more complex, and collectivistic–individualistic dimensions could not be treated as sufficient proxy factors for the whole cultural context.

Ecological worldviews. The concept of the New Ecological Paradigm introduced by Dunlap and his colleagues [37–39] captured the dilemma of human–nature relationships well. The NEP approach is based on the belief that human beings depend on a natural world that has its own intrinsic value, regardless of its usefulness to humans. The NEP scale has been widely used to measure pro-ecological worldviews understood as a multifaceted concept, comprising the five hypothesized facets: acknowledgement of the reality of limits to growth, anti-anthropocentrism, the fragility of nature's balance, rejection of human exemptionalism, and the possibility of an ecocrisis [38].

However, studies show that the NEP scale can be applied as a measurement of a one-dimensional construct, representing a general pro-environmental concern, or multidimensional constructs [38]. The latter may reflect at least bipolar beliefs as pro-environmental and anti-environmental views, which may be called ecocentric and anthropocentric views, respectively [40,41]. Anthropocentric beliefs claim human domination over nature and rights to use natural resources satisfying human needs, without regard to harm made to nature. Although distinction between these two ways of understanding humans' relations with nature is the most popular approach in the literature, some authors offer more nuanced views on this matter. Unai Pascual and his team [42] suggest that anthropocentric worldviews can be connected to the ways of living described as "living from nature" or "living in nature", which underscore instrumental values of nature; while ecocentric and pluricentric views relate to the concepts framed as "living with nature" and "living as nature". The two latter ideas emphasize living beings, including other-than-human beings, or nature's processes as a whole (i.e., living with) *and prioritizes embodying and perceiving nature as a physical, mental, and spiritual part of oneself, emphasizing broad values of oneness, kinship, and interdependence* (i.e., living as) ([40] p. 813).

Following this way of reasoning, we assume that anthropocentrism puts into question the human role in the climate disaster. The tension between anthropocentrism and ecocentrism lies in the lack of balance between how much humans take from nature and how much nature can offer to humans. Such a balance requires taking into account concerns both for the environment and for humans.

Looking at the ecological worldviews as related to responsibility issues, we may expect that ecocentrism can generate a high tendency to attribute responsibility for climate change mitigation to collective bodies as much as to individuals, because ecocentrism predicts pro-environmental engagement and willingness to protect nature [43]. In contrast, anthropocentrism, in ignoring human responsibility in causing and resolving problems, can make people indifferent to mitigation, even to refuse to participate in mitigation and sustainable activities.

Intractability. Intractability of climate change refers to the challenges posed by a multitude of various factors, which make mitigation actions ineffective. The array of factors includes the scientific complexity of the climate issues, economic short-term interests, lack of global consensus, political polarization, cultural norms and habits, lack of public awareness and engagement, and other [44].

The perception of the intractability of climate change may hamper people's willingness to undertake actions to reduce greenhouse gas emissions. Perceived intractability refers to one's belief that climate change cannot be addressed by individual action [26]; therefore, a single person cannot be obliged to act. People feel they lack the ability or sense of empowerment to undertake actions that will "make the difference" on climate change. It could be seen as one "dragon"—among other "dragons"—of inaction [25]. According to Gifford, environmental or climate-related inaction has many reasons, which may be categorized as lack of proper knowledge on environmental and climate issues; as a variety of psychological processes interfering with effective action; and action can be inadequate because it makes too little of a difference. As Jennifer Kent noted: *In the US and UK research 75% (consumers) stated that they were concerned about global warming "but challenged to see how their action could make a difference" and only 9% indicated both concern and willingness to take action. In the Australian research, an equal number expressed concern but not willingness to act (75%)* ([9] p. 142). Belinda Xie and her colleagues used the term "mitigation response inefficacy" for the perceived inability of the response to effectively reduce or control a threat. They found high correlation scores between mitigation response inefficacy and unwillingness to act for climate benefits [45].

Intractability of climate change, perceived by individuals, may have an impact on the distribution of responsibility in such a way that people would consider collective bodies more responsible for tackling climate issues than individuals. Although they may assume, quite rightly, that the global problem of climate change requires the combined actions of many individuals working together, they more likely call on governments and other public institutions to act [46,47]. Likewise, putting too much of the responsibility on governments may diminish individuals' motivations to be involved personally; such a tendency was observed in the sample of young Australians [48].

The present study.

The current study aimed to explore the perception of responsibility for climate change mitigation ascribed to a number of collective and individual actors. Research on responsibility distribution usually focuses on responsibility attributed to collective entities such as governments, corporations, organizations, and to individual entities presented simply as unspecific individuals. We expanded the list of responsible agents by including more collective actors, e.g., countries-polluters, and rich countries, and by including individuals with a special status or playing special roles in society: scientists, rich people, social activists, religious leaders, and teachers. The larger set of actors may provide a better insight into the process of responsibility distribution.

We intended to explore the cultural and psychological contexts in which the distribution of climate change responsibility took place. Shifting responsibility to others may be caused by morally dubious intentions of avoiding personal responsibility, but it also may be a perfectly rational act, as the best way of tackling mitigation problems in given circumstances. Thus, we wanted to learn how cultural context sets the stage for deciding to whom responsibilities should be transferred: collectives or individuals. We focus on collectivistic and individualistic values as the most relevant to the dilemma of choosing between collective and individual actors.

As for psychological factors shaping opinions about responsibility for mitigation, we proposed cognitive variables: beliefs about human nature relationships and beliefs in climate change intractability. Beliefs about human—nature relationships, i.e., ecological worldviews, reflect one's concern for the environment in general, while intractability of climate change—concern for climate. Both concepts may have an impact on a tendency to attribute responsibility, regardless of whether agents represent collectives or individuals.

Lastly, we addressed these objectives by studying different samples coming from countries understudied in environmental psychology. We chose five countries, in three geographical regions: Belarus, Lithuania, and Poland in Central Europe; the United Arab Emirates in the Middle East; and Uzbekistan in Central Asia. Table 1 provides basic regional and cultural characteristics.

Table 1. Basic regional and cultural characteristics of Belarus, Lithuania, Poland, the UAE, and Uzbekistan.

Country	Geographical Region	GDP per Capita/Rank *	Democracy Index **	EPI 2024 /Rank ***	Individualism ****	Cognitive Style	Dominant Religion
Belarus	Central Europe	$22,591/70	1.99	58.1 /32	48	Analytical	Christianity/ Orthodox
Lithuania	Central Europe	$48,397/35	7.59	63.9 /22	55	Analytical	Christianity/ Catholic
Poland	Central Europe	$43,269/39	7.4	64.4/19	47	Analytical	Christianity/ Catholic
UAE	Middle East	$87,729/6	3.07	52.0/53	36	Holistic/ Analytical	Islam
Uzbekistan	Central Asia	$9533/116	2.1	42.9/104	*Kazakhstan—20*	Holistic/ Analytical	Islam

* World Bank 2022; number of countries = 177 [1]. ** Democracy Index 2024; range 0–10 (most democratic); made of sub-indices of electoral pluralism, civil lberties, political participation, democratic culture, and functioning of government [2]. *** The 2024 EPI combines 58 indicators across 11 issue categories, ranging from climate change mitigation and air pollution to waste management, sustainability of fisheries and agriculture, deforestation, and biodiversity protection [3]. **** After Hofstede scores [4].

The study was approved by the Ethics Committee of the Psychology Institute, Polish Academy of Sciences.

Three European countries shared a common history in the XVI-XVII century, creating the Polish Lithuanian Commonwealth, which included contemporary Belarus, and recent history as post-Soviet states. The dominant religion is Christianity: Roman Catholic in Poland and Lithuania, and Russian Orthodox in Belarus. Also, according to Hofstede scores, they share a similar position on the individualistic dimension, with Lithuania as the most individualistic country out of the three. Belonging to the Western cultural zone, the prevailing cognition style is analytical. They differ from each other as regards the ranking of the GDP per capita, since Belarus occupies the lowest position in the group; and the ranking of the EPI, with—again—Belarus having the lowest rank.

The United Arab Emirates share Islamic religion and low scores on individualism with Uzbekistan. However, we can infer the position of Uzbekistan on an individualistic dimension from the scores of its closest neighbor, Kazakhstan, because there is no data on Uzbekistan in the Hofstede research. Also, both countries may share the same cognition style, i.e., holistic, though Uzbekistan people may be familiar with analytic thinking as well. Regarding rankings in GDP per capita, the UAE and Uzbekistan occupy opposite positions: the UAE as the richest country, and Uzbekistan as the poorest one. The difference in the EPI is also quite large; however, both countries are located quite low in this ranking.

As regards the Democracy Index, produced by the Economist Intelligence Unit, we have two democratic states, which are Lithuania and Poland, and three states—the UAE, Uzbekistan, and Belarus—with a low Democracy Index.

According to a recent analysis by Kitayama and Salvador [49], the holistic cognitive style, typical for Arab and Far East cultures, is complemented by Western analytical, argumentative thinking. It is especially true for countries that *have historically served as a hub for trade and cultural exchanges between Eastern and Western regions* ([49] p. 516). Living where they do, residents of the UAE and Uzbekistan likely possess access to either cognitive style. It is worth noting that similarity of certain characteristics of studied countries aligns with the thesis of values convergence along geographical and religious lines [50].

We posed the following research questions:

RQ1. How do citizens of different countries distribute responsibility for climate change mitigation across various actors?

RQ2. Which cultural values: collectivistic or individualistic, best predict an attribution of responsibility for mitigation to collective and individual actors?

RQ3. Do cognitive beliefs about human–nature relationships and beliefs in the intractability of climate change predict attribution of climate change mitigation to collective and individual actors?

Since the study has an exploratory character, we have planned no specific hypotheses.

2. Method

Participants

The total sample consisted of N = 1268 participants, recruited through convenience sampling at the local universities; 30% were males, 72% were students, with a mean age of approximately 23 years. In the UAE, foreigners made up 88.1% of the population as of 2020, reflecting the unique characteristics of the country. The largest group of non-UAE nationals is South Asians, who account for 59.4% of the foreign population. The demographic structure of our sample for the UAE closely mirrors these population characteristics.

Table 2 provides characteristics of national samples.

Table 2. National samples' characteristics.

Country	N	Age M (SD)	N & % of Females	N & % of Students	Dominant Religion	Dominant Ethnicity
Belarus	207	23,16 (5,18)	140 (67.63)	87 (42.03)	Christianity Orthodox	Belarusian (90%)
Lithuania	225	23.97 (5.63	184 (81.78)	112 (49.78)	Christianity Catholic	Lithuanian (90%)
Poland	303	22.60 (5.79)	206 (67.76)	261 (85.86)	Christianity Catholic	Polish (95%)
UAE	251	23.30 (8.30)	163 (64.94)	189 (75.30)	Islam	South Asia and Arab (50%)
Uzbekistan	281	22.04 (5.86)	239 (85.05)	267 (95.02)	Islam	Turkish (100%)
Total	1267	22.97 (6.30)	932 (73.50)	916 (72.24)		

3. Measures

The study used a demographic questionnaire that collected participants' information on their age, gender, occupation, confession, and ethnicity, listed in the participants section, followed by the subsequent questionnaires.

Responsibility for Climate Change Mitigation. We made a list of 12 entities: everyone, social activists, teachers, religious leaders, scientists, political leaders, businesses, governments, the richest people, rich countries, countries-polluters, and media, as those who should to take responsibility for climate change mitigation. We gave examples of responsible activities for each agent (Table 3). We asked participants to *choose the option that best corresponds with how you feel about the statement on a scale from 1—the least responsibility, to 5—the greatest responsibility*.

Table 3. The list of the responsible agents.

1	Every person (e.g., sorting waste, recycling, saving energy, less consumerist lifestyle)
2	Social activists (e.g., organizing protests, sending appeals to politicians)
3	Teachers (e.g., teaching about sustainability, climate change, etc.)
4	Religious leaders (e.g., promoting respect towards nature through their teachings)
5	Scientists (e.g., technology development)
6	Political leaders (e.g., lobbying and environmentally friendly decision-making)
7	Businesses (e.g., introducing environmentally friendly technological solutions)
8	Governments (e.g., introduction of legal regulations)
9	The richest people, billionaires (e.g., donating part of their income to the environment, giving up a luxurious lifestyle)
10	Rich countries (e.g., as environmental protection is expensive, poor countries cannot afford it, hence, the responsibility lies on countries with high GDP)
11	Countries with the highest carbon footprint (e.g., it is imperative for these countries to transition their economies to be more environmentally friendly)
12	Media (e.g., news reporting, investigation, publication)

To investigate the factor structure of the measure for the total sample, we performed an exploratory factor analysis (EFA) using the principal components method with Varimax rotation on the measure. Bartlett's test of sphericity was significant, $\chi^2(66) = 5629$, $p < 0.001$, and the Kaiser–Meyer–Olkin measure of sampling adequacy was 0.90. The analysis revealed two-factor solution that explained 54.6% of the total variance. Factor loadings showed that seven items were loaded on the first factor, which explained 31.80% of variance, and five items on the second factor, which explained 22.80% of variance. One item (media) had similarly high loadings on two factors (0.443 and 0.572), so we excluded it from subsequent analyses.

Based on the first factor items, we created the variable "Collective responsible agents". *(Resp_coll)*. This category included businesses, governments, politicians, countries-polluters, rich people, rich countries, and scientists. Reliability was satisfactory: Cronbach's α coefficients: 0.86 (total); 0.89 (Belarus); 0.84 (Lithuania); 0.86 (Poland); 0.81 (the UAE); 0.80 (Uzbekistan). Based on the second factor items, we created the variable "Individual responsible agents" *(Resp_Individ)*. This category included everyone, teachers, activists, and religious leaders. Reliability was satisfactory. Cronbach's α coefficients: 0.75 (total); 0.75 (Belarus); 0.78 (Lithuania); 0.68 (Poland); 0.69 (the UAE); 0.70 (Uzbekistan).

Cultural Values. We used the COLINDEX scale [51] to measure collectivistic and individualistic values. The scale comprises 13 items reflecting collectivistic (6 items) and individualistic (7 items) values. Participants were asked to evaluate how important the values were to them as principles guiding their lives on the five-point scale, where 1—completely unimportant, 5—very important.

We performed a confirmatory factor analysis (CFA) on the total sample using a two-factor structure. The two-structure solution was not supported as having an adequate fit in this sample, as indicated by model fit indices: $\chi^2(64) = 613.84$, $p < 0.000$; Comparative Fit Index (CFI) = 0.849; root mean square error of approximation (RMSEA) = 0.082, 90% confidence interval (CI) = [0.077, 0.088], standardized root mean squared residual (SRMR) = 0.059. An exploratory factor analysis (principal components method with Varimax rotation; Bartlett's test $\chi^2(78) = 3700$, $p < 0.001$, KMO = 0.835) revealed a three-factor solution, which explained 51.43% of total variance. Factor loadings showed that items were loaded on the first factor, which explained 20.67% of variance; four items on the second factor, which explained 17.37% of variance; and three items on the third factor, which explained 13.39% of variance.

We created three variables based on the three-factor solution. The first variable represents collectivistic values (Values_collect): obedience, respect for elders, politeness, security, self-discipline, order; Cronbach's α coefficients: 0.77 (total); 0.83 (Belarus); 0.72 (Lithuania); 0.71 (Poland); 0.77 (the UAE); 0.74 (Uzbekistan). The second variable represents individualistic values manifesting a tendency to have an indulgent life (Values_Ind_indulg): exciting life, pleasure, varied life, being daring; Cronbach's α coefficients: 0.72 (total); 0.65 (Belarus); 0.73 (Lithuania); 0.70 (Poland); 0.66 (the UAE); 0.62 (Uzbekistan). The third variable also represents individualistic values, but with a focus on self-realization (Values_Ind_real): creativity, freedom, independence; Cronbach's α coefficients: 0.53 (total); 0.58 (Belarus); 0.61 (Lithuania); 0.53 (Poland); 0.48 (the UAE); 0.53 (Uzbekistan).

Ecological Worldviews. We employed the New Ecological Paradigm (NEP) Scale [39] to measure ecological worldviews. We added one item to the original 15-item scale, referring to the last pandemic situation across the globe: *The fundamental reason for the emergence of pandemics is that we failed to live in harmony with nature.* Respondents were asked to indicate how much they agree with the statements on the five-point scale, where 1—strongly disagree, 5—strongly agree.

Following EFA solution (principal components method with Varimax rotation; Bartlett's test $\chi^2(120) = 4087$, $p < 0.001$, KMO = 0.828), we identified two factors that explained 38,11% of total variance: the first factor explained 19,60% of variance, and the second factor explained 18,51% of variance.

The first factor (9 items) represents ecocentric worldview (Cronbach's α coefficients: 0.74 (total); 0.73 (Belarus); 0.65 (Lithuania); 0.81 (Poland); 0.69 (the UAE); 0.74 (Uzbekistan). The second factor (7 items) represents an anthropocentric worldview; Cronbach's α coefficients: 0.75 (total); 0.66 (Belarus); 0.66 (Lithuania); 0.71 (Poland); 0.70 (the UAE); 0.58 (Uzbekistan). Ecocentric worldview examples: *Plants and animals have as much right as humans to exist. The balance of nature is very delicate and easily upset.* Anthropocentric worldview examples: *Humans have the right to modify the natural environment to suit their needs. Humans were meant to rule over the rest of nature.*

Intractability of climate change. To measure the belief in intractability of climate change, we asked participants to indicate how much they agreed with the two following statements: *My individual action would likely do little to aid the fight against climate change* and *Climate change is an unstoppable process; we cannot do anything about it.* They used a five-point scale, where 1—strongly disagree, 5—strongly agree.

Each scale was back-translated from English into Russian for Belarusian participants, and into Lithuanian, Polish, and Uzbek for relevant national samples. The questionnaire for

UAE participants was provided in English, which is informally the most common spoken language of the country.

Analytical Strategy

We ran an invariance measurement analysis by comparing the factor structure, factor loading, and intercepts of three measures across five national samples [52,53] with AMOS v.29 [54] (Table S1).

Invariance of three measurement instruments is supported by the data only at the metric level. Changes in fit indices between the configural and the metric invariance models are smaller than cutoff criteria for RMSEA (Δ < 0.015) and SRMR (Δ < 0.03) indices, except Δ CFI, which is close to criterion (0.014; 0.031; 0.018). This implies that data across national samples are comparable only in terms of covariances and regression coefficients, so the only meaningful analyses that can be performed in the total sample and across national samples are correlational and regression analyses. Because of the lack of scalar invariance, we cannot compare mean scores between national samples.

Therefore, we will analyze the distribution of mean scores of the responsibility for climate change mitigation within national samples. Then, we will perform several hierarchical regression analyses in the total sample and in national samples to find predictors of the tendency to attribute responsibility for mitigation to various actors. We used the SPSS v. 29.

4. Results

Descriptive Statistics

Means, standard deviations, and bivariate correlations between the examined variables in the total sample and national sub-samples are presented in Table 4.

Table 4. Descriptive statistics in the total sample and national sub-samples.

Variable		Country	M (SD)	2	3	4	5	6	7	8
1	Collectivistic Values	BL	4.29 (0.66)	0.39 **	0.48 **	0.24 **	0.30 **	−0.04	0.06	0.23 **
		LT	4.04 (0.58)	0.30 **	0.25 **	0.05	0.22 **	−0.11	0.02	0.24 **
		PL	3.82 (0.66)	0.14 *	0.19 **	0.32 **	−0.01	−0.01	−0.12 *	0.16 **
		UAE	4.23 (0.62)	0.13 *	0.15 *	0.35 **	0.04	0.01	−0.01	0.19 **
		UZ	4.40 (0.58)	0.50 **	0.60 **	0.27 **	0.30 **	0.02	0.471 **	0.38 **
		Total	**4.14 (0.66)**	**0.36 ****	**0.33 ****	**0.36 ****	**0.13 ****	**0.04**	**0.02**	**0.28 ****
2	Individualistic Self-Indulgent Values I	BL	4.10 (0.65)		0.48 **	0.29 **	0.14 *	−0.03	0.01	0.13
		LT	3.77 (0.72)		0.38 **	0.13	0.21 **	−0.03	0.02	0.15 *
		PL	3.51 (0.71)		0.30 **	0.02	0.07	−0.04	0.04	0.04
		UAE	3.91 (0.71)		0.31 **	0.16 *	0.05	0.17 **	0.04	0.05
		UZ	4.20 (0.72)		0.53 **	0.30 **	0.30 **	0.02	0.29 **	0.15 *
		Total	**3.89 (0.75)**		**0.39 ****	**0.29 ****	**0.12 ****	**0.05**	**0.04**	**0.16 ****
3	Individualistic Self-Realization Values I	BL	4.41 (0.58)			0.14 *	0.26 **	−0.08	0.13	0.15 *
		LT	4.34 (0.58)			−0.02	0.28 **	−0.15 *	0.19 **	0.11
		PL	4.30 (0.59)			0.05	0.22 **	−0.20 **	0.16 **	0.28 **
		UAE	4.44 (0.52)			0.05	0.18 **	−0.03	0.09	0.15 *
		UZ	4.34 (0.67)			0.24 **	0.35 **	0.07	0.30 **	0.24 **
		Total	**4.36 (0.60)**			**0.07 ***	**0.27 ****	**−0.07 ***	**0.18 ****	**0.21 ****

Table 4. *Cont.*

Variable		Country	M (SD)	2	3	4	5	6	7	8
4	Anthropocentrism	BL	3.19 (0.59)				*0.02*	0.17 *	*−0.08*	*0.05*
		LT	2.56 (0.57)				−0.32 **	0.17 *	−0.25 **	−0.25 **
		PL	2.78 (0.70)				−0.56 **	0.26 **	−0.35 **	−0.27 **
		UAE	3.06 (0.72)				*−0.10*	0.30 **	−0.22 **	*0.01*
		UZ	3.81 (0.63)				0.51 **	0.16 **	0.34 **	0.29 **
		Total	**3.09 (0.78)**				**−0.17 ****	**0.26 ****	**−0.23 ****	***−0.01***
5	Ecocentrism	BL	3.98 (0.52)					−0.15 *	0.37 **	0.45 **
		LT	4.09 (0.47)					−0.22 **	0.32 **	0.26 **
		PL	3.79 (0.68)					−0.28 **	0.32 **	0.47 **
		UAE	4.11 (0.55)					*−0.11*	0.27 **	0.37 **
		UZ	3.64 (0.74)					*0.06*	0.43 **	0.35 **
		Total	**3.90 (0.64)**					**−0.16 ****	**0.40 ****	**0.40 ****
6	Intractability	BL	3.07 (0.87)						*0.05*	−0.24 **
		LT	2.53 (0.79)						−0.19 **	−0.22 **
		PL	2.97 (0.95)						−0.21 **	−0.41 **
		UAE	2.83 (0.99)						*−0.05*	*−0.06*
		UZ	3.06 (0.88)						*0.03*	*−0.03*
		Total	2.90 (0.92)						**−0.16 ****	**−0.22 ****
7	Responsibility of Collective Entities	BL	4.04 (0.73)							0.54 **
		LT	4.49 (0.61)							0.43 **
		PL	4.07 (0.80)							0.40 **
		UAE	4.35 (0.61)							0.53 **
		UZ	3.79 (0.77)							0.51 **
		Total	**4.13 (0.76)**							**0.48 ****
8	Responsibility of Individual Entities	BL	3.50 (0.79)							
		LT	3.90 (0.84)							
		PL	3.14 (0.87)							
		UAE	4.09 (0.67)							
		UZ	3.68 (0.85)							
		Total	**3.64 (0.88)**							

Note: * $p < 0.01$; ** $p < 0.001$; italics are for $p > 0.01$. Bold are for statistics in the total sample. Abbreviations: BL—Belarus; LT—Lithuania, PL—Poland, UAE—the United Arab Emirates, UZ—Uzbekistan.

When considering cultural values mean scores, self-realization individualistic values are rated highest in each national sample, with collectivistic values second, and self-indulgent individualistic values after that (Table 4). The differences between the pair of values are statistically significant ($p < 0.001$), but the effect sizes of the differences are small to medium. Only the Polish sample showed a large effect size for the difference between individualistic values (Cohen's d = 1.02) (Table S2). Similarly, each country's sample reveals significant mean score differences between anthropocentric and ecocentric worldviews, showing higher scores of ecocentric views than scores of anthropocentric views. Yet, here we have effect sizes that are large or very large, except for Uzbekistan, where the effect size is very small (Cohen's d = 0.12). Concerning responsibility for mitigation variables, their difference is statistically significant across all national samples, with a medium-to-large effect size, except in Uzbekistan, where Cohen's d is near 0. It is worth mentioning that the mean scores for the intractability variable in each national sample are near the midpoint of the five-point scale (2.83; 2,97; 3,06; 3,07) or below the midpoint (2.53).

RQ1. Distribution of responsibility for climate change mitigation across various actors

Findings revealed that, concerning both the total and national samples, participants felt all entities mentioned should bear responsibility for climate change mitigation (almost all scores were above the midpoint of the five-point scale) (Figure 1a–e). But, as noted earlier, participants assigned higher responsibility to collective bodies compared to individuals (Tables 5 and 6). In the Belarusian sample, the relevant means are: 4.04 vs. 3.50; Cohen's d = 0.73; in the Lithuanian sample: 4.49 vs. 3.90; Cohen's d = 0.74; in the Polish sample: 4.07 vs. 3.14; Cohen's d = 1.00; in the UAE sample: 4.35 vs. 4.09; Cohen's d = 0.43; in the Uzbekistan sample: 3.79 vs. 3.68; Cohen's d = 0.02.

Regarding mean scores of attributions of the responsibility assigned to collective agents (Figure 1a–e; Tables S3 and S4) scientists (M = 4.20) were ranked highest in Belarus, businesses (4.70) in Lithuania and Poland (M = 4.26), governments (M = 4.60) in the UAE, and politicians (M = 3.92) in Uzbekistan. The lowest rankings were assigned to rich people by Belarus (M = 3.92) and the Emirates (M = 4.17), and to rich countries by Lithuania (M = 4.14), Poland (M = 3.82), and Uzbekistan (M = 3.60).

Teachers in Lithuania (M = 4.20), Poland (M = 3.61), and the Emirates (M = 4.14) ranked highest among individual agents responsible for mitigation, while Belarusians (M = 3.85) and Uzbeks (M = 3.88) ranked highest the agent named "everyone". Social activists had the lowest ranking in the Lithuanian (M = 3.62), Polish (M = 2.49), UAE (M = 3.71), and Uzbek (3.71) samples, while Belarusians ranked religious leaders lowest (M = 3.00).

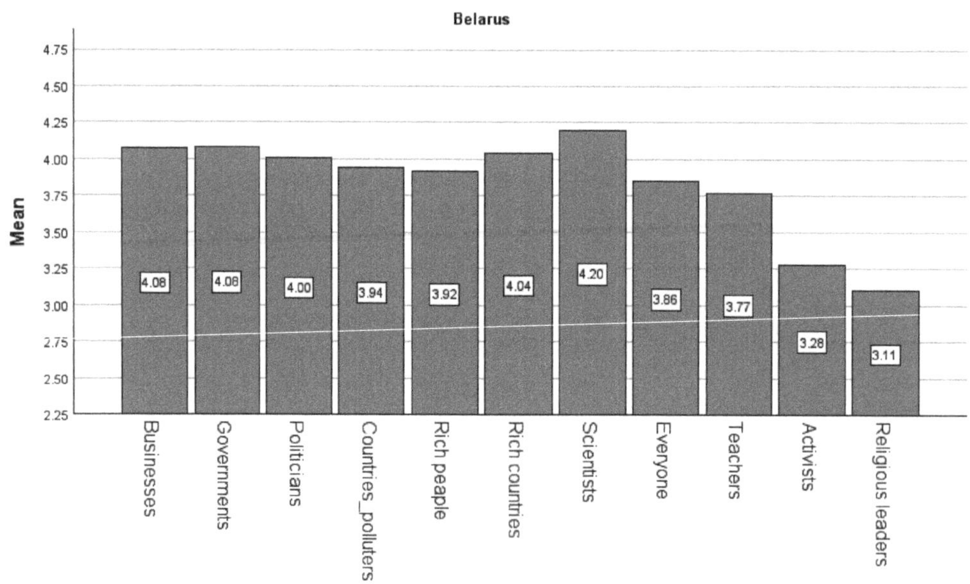

(a) Belarus

Figure 1. *Cont.*

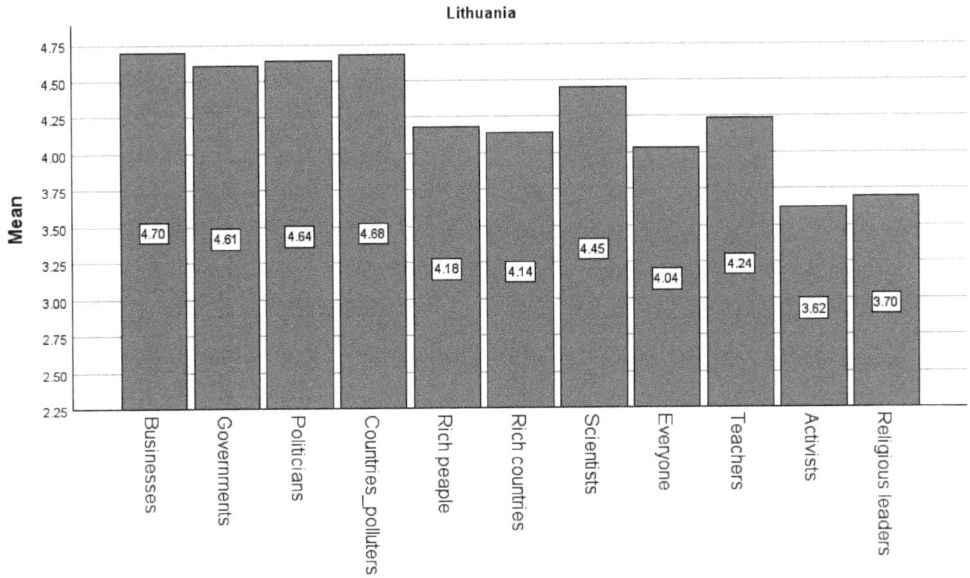

(**b**) Lithuania

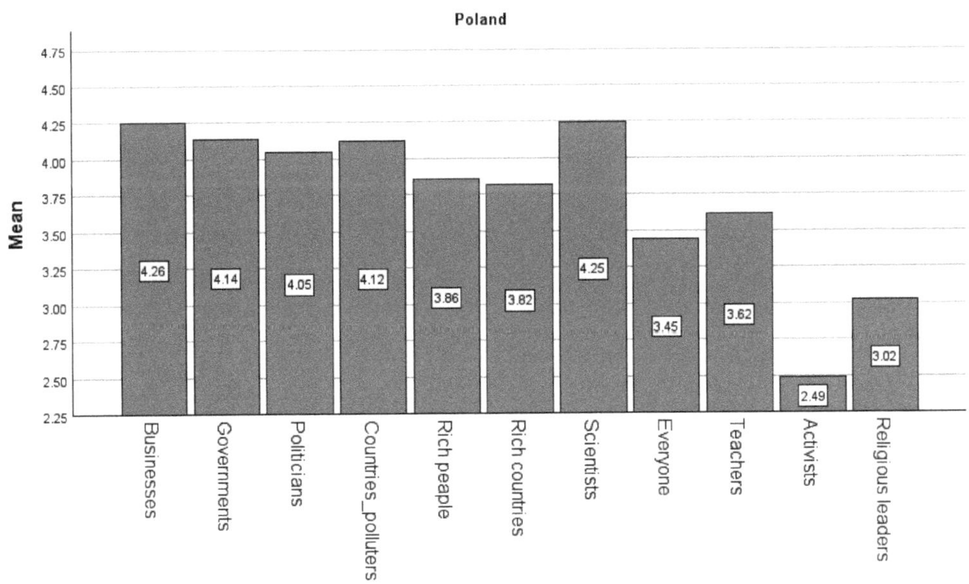

(**c**) Poland

Figure 1. *Cont.*

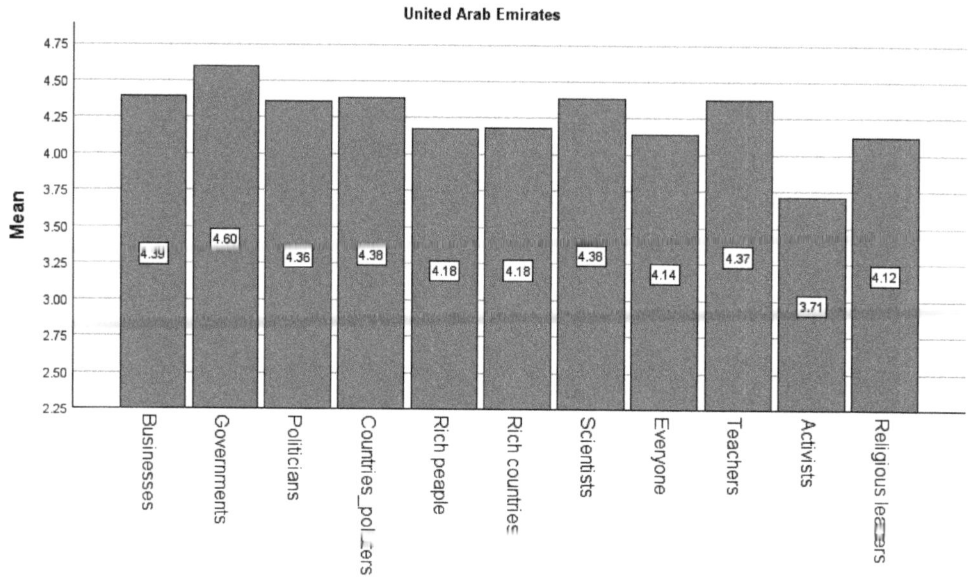

(**d**) The United Arab Emirates

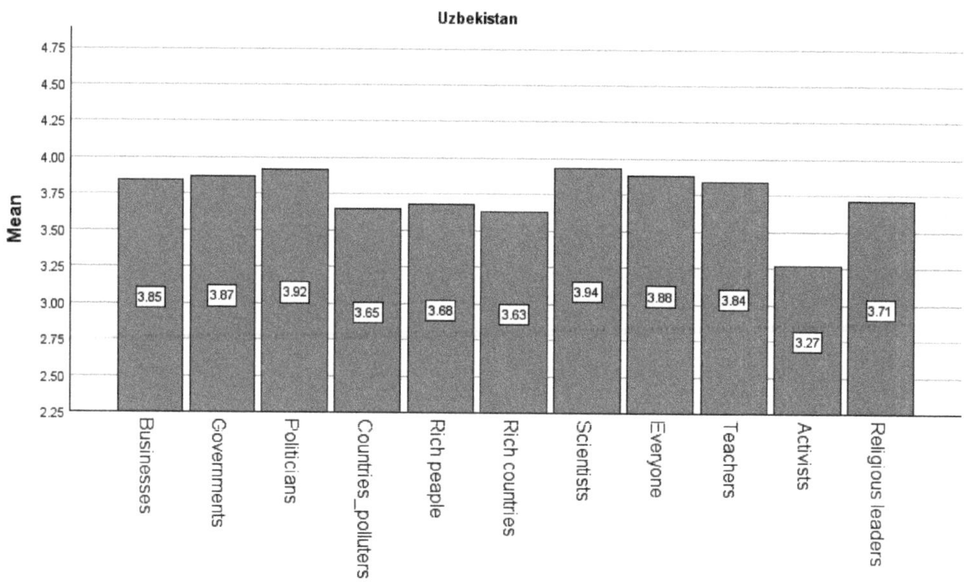

(**e**) Uzbekistan

Figure 1. Mean scores of responsibility distribution in the national samples.

RQ2 and RQ3. Predictors of attribution of responsibility for climate change mitigation.

We performed hierarchical multiple regression analyses in the total sample and in national samples to examine the extent to which cultural values, ecological worldviews, and belief in the intractability of climate change predict attribution of mitigation responsibility to collective and individual actors. Each of these two variable sets and a single variable (i.e., intractability) were entered into the regression equation on separate steps. In Table 5, we report regression analysis results for attribution of responsibility to collective agents in the total sample.

Table 5. Hierarchical regression analysis results for responsibility of collective agents in the total sample.

Predictors	Model 1		Model 2		Model 3	
	B [CI]	**β**	**B [CI]**	**β**	**B [CI]**	**β**
Val_Collect	−0.04 [−0.11; 0.03]	−0.03	0.01 [−0.05; 0.08]	0.01	0.01 [−0.05; 0.08]	0.01
Val_Ind_Indulg	−0.03 [−0.10;03]	−0.03	0.01 [−0.05; 0.06]	0.01	0.01 [−0.05; 0.06]	0.01
Val_Ind_Realiz	0.26 [0.19; 34]	0.21 ***	0.13 [0.06; 0.20]	0.10 ***	0.12 [0.05; 0.19]	0.10 ***
Anthropocentrism			−0.18 [−0.23;0.12]	0.18 ***	−0.16 [−0.22; −0.11]	−0.17 ***
Ecocentrism			0.41 [34; 0.47]	0.34 ***	0.40 [0.34; 0.46]	0.34 ***
Intractability					−0.05 [−0.09; −0.01]	−0.05 *
F (df)	15.97 (3;1263) ***		62.42 (5;1261) ***		52.90 (6;1260) ***	
R^2 (Adj R^2)	0.037 (0.034)		0.198 (0.195)		0.201 (0.197)	
ΔR^2	0.037 ***		0.162 ***		0.003 *	

Note: *** $p < 0.001$; * $p < 0.05$ [CI]—confidence interval. Val_Collect—collectivistic values; Val_Ind_Indulg—individualistic self-indulgent values; Val_Ind_Realiz—individualistic self-realization values.

Table 6. Hierarchical regression analysis results for collective agents in the national samples: Beta coefficients in the final model.

	Belarus	**Lithuania**	**Poland**	**Emirates**	**Uzbekistan**
Val_Collect	−0.07	−0.06	−0.07	0.04	0.28 ***
Val_Ind_Indulg	−0.04	−0.05	0.01	0.06	0.04
Val_Ind_Realiz	0.09	0.14 *	0.10	0.08	−0.01
Anthropocentrism	−0.06	−0.15 *	−0.21 **	−0.23 ***	0.11
Ecocentrism	0.36 ***	0.23 ***	0.15 *	0.25 ***	0.28 ***
Intractability	−0.07	−0.10	−0.10	0.03	0.01
F (df)	6.02 (6;200) ***	6.59 (6:218) ***	9.80 (6;296) ***	5.54 (6;244) ***	17.91 (6;274) ***
R^2	0.15	0.15	0.17	0.12	0.28

Note: *** $p<.001$; ** $p < 0.01$; * $p < 0.05$. Val_Collect—collectivistic values; Val_Ind_Indulg—individualistic self-indulgent values; Val_Ind_Realiz—individualistic self-realization values.

Model 1 included three cultural factors. Individualistic self-realization values were the only significant predictor, which accounted for 3.7% of variance. Model 2 added ecological worldview factors to examine whether worldviews explain additional variance in attribution of responsibility beyond that explained by cultural values. Both variables, i.e., anthropocentrism and ecocentrism, are significant predictors (anthropocentrism is a negative predictor) and additionally explain 16.2% of variance. Model 3 added one more variable: belief in climate change intractability, which is significant only on $p < 0.05$ and added a very small amount of variance—0.3%. All predictors collectively explain 20.1% of total variance in attribution of responsibility to collective agents.

So, the higher the preference for individualistic self-realization values (creativity, independence, freedom), the lower the anthropocentric orientation, and the higher the ecocentric orientation, the higher the tendency to ascribe mitigation responsibility to collective agents. In other words, a person who adheres to self-realization values and prefers ecocentric to anthropocentric worldviews tends to put mitigation responsibility on collective agents.

Table 6 reports the results of regression analyses on predicting collective agents' responsibility in national samples. The table contains beta coefficients calculated in Model 3. In each sample, the models are significant and explain from 12% (UAE) to 28% (Uzbekistan) of variance. However, we can only observe the pattern of the three significant predictors found in the total sample in the Lithuanian sample. In Poland and the United Arab Emirates, responsibility of collective agents can be predicted solely by anthropocentrism and ecocentrism, while in Belarus only by ecocentrism. In the Uzbekistan sample, in addition to ecocentric orientation, a new predictor occurred, which is collectivistic values.

We repeated the above hierarchical regression analyses for responsibility ascribed to individual agents (Tables 7 and 8). As previously, Model 1 included cultural values. Collectivistic and individualistic self-realization values are significant predictors and accounted for 9.2% of variance in attribution of responsibility to individuals. Model 2 added anthropocentrism and ecocentrism to cultural values and additionally explained 21.7% of total variance, despite the fact that individualistic values lost their significance. It turned out that collectivistic values remained a significant predictor in this model, along with an ecocentric orientation. The belief in climate change intractability, included in Model 3, became a significant predictor and added 2.8% of variance. In sum, the amount of total variance explained in the final model was 24.6%.

Table 7. Hierarchical regression analysis results for individual agents' responsibility in the total sample.

Predictors	Model 1		Model 2		Model 3	
	B [CI]	β	B [CI]	β	B [CI]	β
Val_Collect	0.30 [0.22;0.38]	0.23 ***	0.30 [0.22;0.38]	0.23 ***	0.29 [0.22; 0.37]	0.22 ***
Val_Ind_Indulg	0.03 [−0.03; 0.10]	0.03	0.04 [−0.02; 0.11]	0.03	0.04 [−0.02; 0.11]	0.04
Val_Ind_Realiz	0.18 [0.09; 26]	0.12 ***	0.04 [−0.04; 0.12]	0.03	0.02 [−0.06; 0.10]	0.01
Anthropocentrism			−0.05 [−0.11; 0.01]	−0.04	0.00 [−0.06; 0.06]	0.00
Ecocentrism			0.49 [0.42; 0.56]	0.36 ***	0.46 [0.39; 0.53]	0.34 ***
Intractability					−0.17 [−0.22; −0.12]	−0.18 ***
F (df)	42.70 (3; 1263) ***		69.99 (5; 1261) ***		68.41 (6; 1260) ***	
R^2 (Adj R^2)	0.092 (0.090)		0.217 (0.214)		0.246 (0.242)	
ΔR^2	0.092 ***		0.125 ***		0.028 ***	

Note: *** $p < 0.001$. [CI]—confidence interval. Val_Collect—collectivistic values; Val_Ind_Indulg—individualistic self-indulgent values; Val_Ind_Realiz—individualistic self-realization values.

Table 8. Hierarchical regression analysis results for individual agents in the national samples: Beta coefficients in the final model.

	Belarus	Lithuania	Poland	Emirates	Uzbekistan
Val_Collect	0.09	0.19 **	0.16 **	0.18 **	0.33 ***
Val_Ind_Indulg	0.04	0.11	−0.05	0.08	−0.13
Val_Ind_Realiz	−0.04	−0.03	0.13 **	0.04	0.01
Anthropocentrism	0.05	−0.21 **	−0.06	−0.02	0.14 *
Ecocentrism	0.40 ***	0.11	0.33 ***	0.34 ***	0.22 ***
Intractability	−0.19**	−0.14 *	−0.28 ***	−0.06	−0.07
F (df)	10.92 (6;200) ***	7.44 (6;218) ***	25.98 (6;296) ***	8.80 (6;244) ***	13.58 (6;274) ***
R^2	0.25	0.17	0.34	0.18	0.23

Note: *** $p < 0.001$; ** $p < 0.01$; * $p < 0.05$. Val_Collect—collectivistic values; Val_Ind_Indulg—individualistic self-indulgent values; Val_Ind_Realiz—individualistic self-realization values.

Therefore, valuing collectivism more, having a more ecocentric view, and perceiving climate change intractability as less strong are associated with a higher tendency to place mitigation responsibility on individual actors. Those with collectivistic values, an ecocentric nature perspective, and a belief that climate change is not intractable may be inclined to put mitigation efforts on individuals.

Collectivistic values were significant predictors of responsibility attribution across national samples, as evidenced by the hierarchical regression analyses presented in Table 8. Notably, the Belarusian sample exhibits a weaker, yet still significant, association for this predictor. Furthermore, the ecocentric worldview emerges as a primary predictor across most national samples, with the exception of Lithuania. In the Lithuanian sample, an anthropocentric perspective significantly and negatively predicts responsibility attribution, diverging from the pattern observed in other countries. Additionally, the belief that climate change is intractable predicts a decreased sense of responsibility in Poland, Belarus, and Lithuania; however, this relationship is not observed in the United Arab Emirates and Uzbekistan. Interestingly, in Uzbekistan, both anthropocentric and ecocentric orientations serve as significant positive predictors of responsibility attribution.

5. Discussion

Studies often examine how avoiding personal responsibility is connected to assigning responsibility to others, mainly seeking psychological explanations [9,25,26,48,55]. Schmitt et al. [15] argue that focusing only on psychological causes of climate inaction, while ignoring context, overlooks the *complex interplay between psychological tendencies, social relations, and social structures* (p.123), particularly power dynamics. Authors claim that individuals in societies often lack the power to act meaningfully, even within pro-environmental groups. On the other hand, certain people or organizations, especially those in power, counteract climate efforts, striving to sustain their profitable status quo, even with environmental and human consequences [15]. Individuals may realize their inability to influence global change through their personal agency. Feelings of powerlessness may turn people to those who have more power to act. These include governments, international corporations, businesses, and other collective institutions which have the capacity to make decisions and have the means to implement them [3,48]. In particular, governments play a crucial role in addressing climate change. For example, governments can set regulations concerning the reduction in emissions, invest in environmentally friendly public infrastructure, promote public awareness on climate issues, and ensure international cooperation in the environmental domain. So, it seems reasonable to hold governments and other collective bodies responsible for climate change mitigation.

Similarly, our study results showed that respondents attributed greater responsibility to collective actors than to individual actors, confirming findings from other researchers [12,23,24]. However, the government was only ranked highest in the UAE. Politicians and businesses received high rankings in Lithuania, Poland, and Uzbekistan. The quite interesting finding is that scientists (perceived as a collective body—a scientific community) were those who received the highest ranking in Belarus, Poland, and Uzbekistan. Evidently, respondents delegate taking action on climate change to those who are the most able to do something meaningful: politicians and governments, as having the power to set regulations and elaborate strategies; scientists, as capable of developing technologies; and businesses responsible for implementing climate-friendly solutions.

Respondents had to assess the responsibility of four individual entities: teachers, religious leaders, social activists, and "everyone". Responsibility of "everyone" was described in the questionnaire as performing typical individual activities such as sorting waste or saving water in the household, showing personal responsibility. Teachers, religious leaders, and activists are those who can affect people indirectly by spreading knowledge, convincing them to live more sustainably, increasing awareness, or triggering politicians. Attributing responsibility mostly to teachers suggests that respondents recognized the importance of ecological education. Personal responsibility for climate change mitigation (admitting that everyone is responsible) was estimated as high as teachers' responsibility, but much lower than collective agents' responsibility.

What is most striking is how low social activists were placed in the ranking. In the Polish sample, their responsibility scores were put below the midpoint of the scale, which suggests that respondents had low regard for their actions. Religious leaders achieved the lowest rankings in the Christian countries, i.e., Belarus, Lithuania, and Poland, while in the Islamic countries, Uzbekistan and the Emirates, their responsibility was estimated as relatively high. These results mirror the current role of religion and religious institutions in more secular European cultures and, on the other hand, in more traditional Islamic cultures. Notably, environmental literature is showing a trend of relating religion with sustainability issues, a trend seen both in Islamic authors and Western authors [56–60]. In summary, our data showed how people distribute responsibility for climate change mitigation across collective and individual actors. People believe collective entities bear more responsibility than individuals, and the most accountable for this are scientists, businesses, and governments. They also recognize teachers as key agents advocating for change, and they do not hesitate to assign accountability to individual persons, though not as heavily as to collective agents.

We considered cultural values, ecological worldviews, and belief in the intractability of climate change, seeking predictors of the tendency to attribute mitigation responsibility to collective and individual agents. As cultural values (collectivistic and individualistic) provide more lifestyle insight, in this study, we prefer them over environmental values (e.g., biospheric) [43]. It was found that collectivistic values did not relate to the allocation of responsibility to collectives, but instead significantly predicted the responsibility of individual agents, including personal responsibility. Inclination to see responsible actors in individuals rather than in collectives might be linked to collectivistic views on how a community operates. An efficient community is based on strong community leaders and common efforts of community members [61], so leaders and every member are accountable for vital decisions, particularly in addressing climate change. On the other hand, collective agents as defined in this study (governments, businesses, politicians) may be perceived as too distant from community life to be trusted sufficiently and depended on.

When it comes to individualistic values, the split into self-indulgence and self-realization values was important and had some consequences for outcomes. Results showed that the first type of values had no relevance to the distribution of responsibility, either across collectives or across individuals, while self-realization values proved to be a significant predictor of responsibility assigned to collective agents. The way that two types of individualistic values work differently with environmental issues clarifies the mixed results seen when studying culture's connection to environmental attitudes [27,33,34,62]. If one's life is guided by values aimed at pleasure, stimulation, and instant gratification, worrying about the grim future is unlikely, so there is not much interest in the mitigation of climate change responsibility issues. On the contrary, when meaning of life is in the center of values, individuals are more inclined to care for the planet and look to collective agents who can alter the future. Similar ideas are expressed in the notion of post-materialistic values, which correlate strongly with environmental concern [4].

The ecocentric worldview is a strong positive predictor of assigning responsibility for mitigation both to collective and individual actors. An ecocentric viewpoint signifies a person's environmental concern and worry about the harms of climate change on the environment [63]. So, ecocentricity means that all actors should participate in mitigation activities. Conversely, anthropocentrism is a strong negative predictor of mitigation accountability of collective bodies. It seems that there is a similar tendency regarding individual agents; however, the positive link between anthropocentrism and individual responsibility, found in the Uzbekistan sample, works against this tendency and complicates the results (beta coefficient in the total sample is zero). Still, individuals with an anthropocentric view often minimize climate change damage [37], making them less likely to believe in the need to involve significant actors like governments or companies.

The mean scores of beliefs in the intractability of climate change in national samples are around/below the midpoint of the scale, i.e., from $M = 2.53$ to $M = 3.07$. It implies that respondents are not completely sure that individual actions are pointless or climate change is unstoppable. Nevertheless, those who think in that way tend to diminish accountability for mitigation placed on collective agents and, to some extent, on individual agents. The negative relationship between intractability beliefs and responsibility for mitigation occurred only in the Belarusian, Lithuanian, and Polish samples. In the Emirates and Uzbekistan samples, intractability beliefs are irrelevant to opinions on who should bear responsibility. The difference between countries needs to be explained in future studies.

Limitations

Several limitations of the present study should be acknowledged.

Firstly, the sampling method relied on convenience sampling at local universities, which resulted in a large proportion of young, highly educated participants (over 70% students, mean age 23, mostly women). This restricts the generalizability of the results to the broader populations of the studied countries, particularly older age groups, people outside higher education, and rural residents. However, the study functions as an important pilot study, guiding future research directions.

Further, the measurement invariance analysis showed only metric invariance across the five national samples, meaning that mean-level comparisons between countries are not meaningful. The lack of scalar invariance restricts the extent of cross-national conclusions and indicates that some of the variation in responses may stem from cultural or linguistic interpretation differences rather than genuine attitudinal disparities. Also, it is likely that we did not detect and answer properly to response-bias problems in some samples, such as an acquiescent responding (i.e., uncritical acceptance of an item) [64], which may suppress or inflate associations between variables.

Moreover, although the study used a relatively broad list of responsible entities, the selection of actors and their descriptions may have influenced the responses. Providing examples of typical activities for each entity could have anchored participants' perceptions, possibly inflating or lowering responsibility attributions depending on the salience or perceived feasibility of those activities.

Furthermore, the Uzbek sample warrants special consideration. The cognitive style in Uzbekistan can be described as holistic/analytical [49]. Holistic thinking means one can hold conflicting beliefs on the same topic and not feel cognitive dissonance [65]. So, maybe this is why there was high positive correlation between anthropocentric and ecocentric beliefs in the Uzbekistan sample. Also, the country has relatively low exposure to public climate change discourse compared to Western samples. Limited personal experience with climate-related activism, coupled with strong community orientation, might have shaped responses in ways not fully captured by the current predictors. In particular, the positive association of anthropocentrism with responsibility attribution in Uzbekistan diverges from patterns in other countries, which may reflect unique socio-cultural interpretations of responsibility and human–nature relations, as well as some aspects of cognitive style.

Similarly, the multi-ethnic population in the UAE presents a challenge when interpreting culture-related aspects of this study's findings. The country is home to many international youths considered residents, and these individuals often embody a hybrid cross-cultural identity known as "Third Culture Kids" [66,67]. Multicultural individuals may perceive themselves as "global citizens", which could lead to a stronger connection to and sense of responsibility for global issues like climate change, as suggested by research [68]. Our sample, however, reflected the demographic characteristics of the UAE, with most participants being of South Asian and Arab descent, thus making it somewhat representative of the youth living in the region. Nonetheless, the generalizability is limited due to the factors mentioned earlier. Importantly, both non-Western samples from Uzbekistan and the UAE are valuable because prior research on these groups is limited. Future research should delve deeper into the cultural contexts of these groups, as these factors may influence their beliefs regarding responsibility for climate change mitigation.

Finally, the study did not address potential social desirability bias, especially given that climate change is a socially sensitive and morally charged topic. Respondents might have over-reported support for pro-environmental responsibility distribution to align with perceived social norms.

6. Conclusions

Despite these limitations, the study provides meaningful insights into how responsibility for climate change mitigation is attributed in under-researched cultural contexts, including Central Asia and the Middle East. Across all countries, collective entities—particularly scientists, businesses, and governments—were perceived as more responsible than individuals. However, teachers and the notion of "everyone" also received high responsibility scores, suggesting that individual agency is not dismissed entirely. The predictors' patterns indicate that ecocentric worldviews consistently increase perceived responsibility across both collective and individual actors, whereas anthropocentrism reduces the attribution of collectives' responsibility. Collectivistic values appear to foster greater expectations from individuals, possibly reflecting a cultural emphasis on community participation in problem-solving.

These findings underline the complex interplay between cultural values, cognitive orientations, and perceptions of climate change intractability. They suggest that public communication and policy strategies should be tailored to cultural contexts: in societies where personal efficacy is low, strengthening trust in both collective and individual action may be crucial; in others, emphasizing the moral and practical capacities of both community members and institutions could foster broader engagement. In sum, while the research cannot provide conclusive cross-national comparisons, it highlights important attitudinal patterns and cultural nuances that can inform climate change mitigation campaigns and further cross-cultural studies.

Supplementary Materials: The following supporting information can be downloaded at: https://www.mdpi.com/article/10.3390/land14091914/s1, Table S1: Results of invariance measurement analyses for the Responsibility for Climate Change Mitigation scale, the Cultural Values scale, and the New Ecological Paradigm scale; Table S2: Means comparisons in the national samples. Results of t-paired tests; Table S3: Means and standard deviations of the responsibility of collective agents in national samples; Table S4: Means and standard deviations of the responsibility on individual agents in national samples.

Author Contributions: Conceptualization: A.K., M.M., P.U., R.S., D.U., B.N. and K.K.; Methodology: A.K., M.M., P.U., R.S. and D.U.; Data Collection: M.M., P.U., B.N., V.M., R.S. and K.K.; Formal Analysis: A.K., M.M., P.U. and V.M.; Data Curation: A.K.; Writing—original draft preparation: A.K.;

Writing—review and editing: M.M. and P.U. All authors have read and agreed to the published version of the manuscript.

Funding: The authors received no external funding.

Informed Consent Statement: Informed consent was obtained from all individual participants included in the study.

Data Availability Statement: Data is available on reasonable request by sending an email to the corresponding author. Code availability: Not applicable.

Conflicts of Interest: The authors declare no conflicts of interest.

Notes

1 https://www.worldometers.info/gdp/gdp-by-country (accessed on 25 July 2025)
2 https://ourworldindata.org/grapher/democracy-index-eiu (accessed on 25 July 2025)
3 https://epi.yale.edu/measure/2024/EPI (accessed on 25 July 2025)
4 https://www.hofstede-insights.com/country-comparison-tool (accessed on 25 July 2025)

References

1. Stoddard, I.; Anderson, K.; Capstick, S.; Carton, W.; Depledge, J.; Facer, K.; Gough, C.; Hache, F.; Hoolohan, C.; Hultman, M.; et al. Three Decades of Climate Mitigation: Why Haven't We Bent the Global Emissions Curve? *Annu. Rev. Environ. Resour.* **2021**, *46*, 653–689. [CrossRef]
2. Swim, J.; Clayton, S.; Doherty, T.; Gifford, R.; Howard, G.; Reser, J.; Stern, P.; Weber, E. *Psychology & Global Climate Change. Addressing a Multifaceted Phenomenon and Set of Challenges. A Report of the APA Task Force on the Interface Between Psychology and Global Climate Change*; American Psychological Association: Washington, DC, USA, 2009.
3. Hormio, S. Collective Responsibility for Climate Change. *WIREs Clim. Change* **2023**, *14*, e830. [CrossRef]
4. Pokharel, P.R.; Alqahtani, M.S.; Nandy, M.; Lodh, S. Exploring Global Environmental Engagement: The Role of Willingness and Membership in Environmental Action. *Sustainability* **2025**, *17*, 3611. [CrossRef]
5. Clayton, S.; Manning, C. (Eds.) *Pychology and Climate Change. Human Perceptions, Impacts, and Responses*; Elsevier Academic Press: London, UK, 2019; ISBN 9781119130536.
6. Tam, K.P.; Milfont, T.L. Towards Cross-Cultural Environmental Psychology: A State-of-the-Art Review and Recommendations. *J. Environ. Psychol.* **2020**, *71*, 101474. [CrossRef]
7. Milfont, T.L.; Schultz, P.W. The Role of Attitudes in Environmental Issues. In *The handbook of attitudes*; Taylor and Francis Ltd.: Abingdon, UK, 2018; pp. 337–364.
8. Steg, L.; Van Den Berg, A.E.; de Groot, J.I.M. (Eds.) *Environmental Psychology. An Introduction*; BPS & Wiley and Sons, Ltd.: Chichester, UK, 2013; ISBN 978-0-470-97638-8.
9. Kent, J. Individualized Responsibility: "If Climate Protection Becomes Everyone's Responsibility, Does It End Up Being No-One's?". *Cosmop. Civ. Soc. Interdiscip. J.* **2009**, *1*, 132–149. [CrossRef]
10. Wynes, S.; Nicholas, K.A. The Climate Mitigation Gap: Education and Government Recommendations. *Enviro. Res. Lett.* **2017**, *12*, 074024. [CrossRef]
11. Ringov, D.; Zollo, M. Corporate Responsibility from a Socio-Institutional Perspective: The Impact of National Culture on Corporate Social Performance. *Corp. Gov.* **2007**, *7*, 476–485. [CrossRef]
12. Falck, R. How Politicians and the Population Attribute Responsibility for Climate Change Mitigation: No Indication of a 'Governance Trap' in Norway. *Environ. Polit.* **2024**, *33*, 699–726. [CrossRef]
13. Stoddart, M.C.J.; Tindall, D.B.; Greenfield, K.L. "Governments Have the Power"? Interpretations of Climate Change Responsibility and Solutions Among Canadian Environmentalists. *Organ. Environ.* **2012**, *25*, 39–58. [CrossRef]
14. Becker, S.; Sparks, P. Talking about Climate Change Mitigation: People's Views on Different Levels of Action. *Sustainability* **2018**, *10*, 1357. [CrossRef]
15. Schmitt, M.T.; Neufeld, S.D.; Mackay, C.M.L.; Dys-Steenbergen, O. The Perils of Explaining Climate Inaction in Terms of Psychological Barriers. *J. Soc. Issues* **2020**, *76*, 123–135. [CrossRef]
16. Nche, G.C.; Achunike, H.C.; Okoli, A.B. From Climate Change Victims to Climate Change Actors: The Role of Eco-Parenting in Building Mitigation and Adaptation Capacities in Children. *J. Environ. Educ.* **2019**, *50*, 131–144. [CrossRef]
17. Hermans, M.; Korhonen, J. Ninth Graders and Climate Change: Attitudes towards Consequences, Views on Mitigation, and Predictors of Willingness to Act. *Int. Res. Geogr. Environ. Educ.* **2017**, *26*, 223–239. [CrossRef]

18. Chryssavagis, J. Ecumenical Patriarch Bartholomew: Insights into an Orthodox Christian Worldview. *Int. J. Environ. Stud.* **2007**, *64*, 9–18. [CrossRef]
19. Johnston, L. The Religious Dimensions of Sustainability: Institutional Religions, Civil Society, and International Politics since the Turn of the Twentieth Century. *Relig. Compass* **2010**, *4*, 176–189. [CrossRef]
20. Maiella, R.; La Malva, P.; Marchetti, D.; Pomarico, E.; Di Crosta, A.; Palumbo, R.; Cetara, L.; Di Domenico, A.; Verrocchio, M.C. The Psychological Distance and Climate Change: A Systematic Review on the Mitigation and Adaptation Behaviors. *Front. Psychol.* **2020**, *11*, 568899. [CrossRef]
21. Ortega-Egea, J.M.; García-de-Frutos, N.; Antolín-López, R. Why Do Some People Do "More" to Mitigate Climate Change than Others? Exploring Heterogeneity in Psycho-Social Associations. *PLoS ONE* **2014**, *9*, e106645. [CrossRef]
22. Cologna, V.; Siegrist, M. The Role of Trust for Climate Change Mitigation and Adaptation Behaviour: A Meta-Analysis. *J. Environ. Psychol.* **2020**, *69*, 101428. [CrossRef]
23. Stoll-Kleemann, S.; O'Riordan, T.; Jaeger, C.C. The Psychology of Denial Concerning Climate Mitigation Measures: Evidence from Swiss Focus Groups. *Glob. Environ. Change* **2001**, *11*, 107–117. [CrossRef]
24. Leviston, Z.; Walker, I. The Influence of Moral Disengagement on Responses to Climate Change. *Asian J. Soc. Psychol.* **2021**, *24*, 144–155. [CrossRef]
25. Gifford, R. The Dragons of Inaction: Psychological Barriers That Limit Climate Change Mitigation and Adaptation. *Am. Psychol.* **2011**, *66*, 290–302. [CrossRef]
26. Xiang, P.; Zhang, H.; Geng, L.; Zhou, K.; Wu, Y. Individualist-Collectivist Differences in Climate Change Inaction: The Role of Perceived Intractability. *Front. Psychol.* **2019**, *10*, 187. [CrossRef]
27. Xue, W.; Hine, D.W.; Marks, A.D.G.; Phillips, W.J.; Zhao, S. Cultural Worldviews and Climate Change: A View from China. *Asian J. Soc. Psychol.* **2016**, *19*, 134–144. [CrossRef]
28. Arısal, İ.; Atalar, T. The Exploring Relationships between Environmental Concern, Collectivism and Ecological Purchase Intention. *Procedia—Soc. Behav. Sci.* **2016**, *235*, 514–521. [CrossRef]
29. McCarty, J.A.; Shrum, L.J. The Influence of Individualism, Collectivism, and Locus of Control on Environmental Beliefs and Behavior. *J. Public Policy Mark.* **2001**, *20*, 93–104. [CrossRef]
30. Segev, S. Modelling Household Conservation Behaviour among Ethnic Consumers: The Path from Values to Behaviours. *Int. J. Consum. Stud.* **2015**, *39*, 193–202. [CrossRef]
31. Kim, Y.; Choi, S.M. Antecedents of Green Purchase Behavior: An Examination of Collectivism, Environmental Concern, and PCE. *Adv. Consum. Res.* **2005**, *32*, 592–599.
32. Lahuerta-Otero, E.; González-Bravo, M.I. Can National Culture Affect the Implementation of Common Sustainable Policies? A European Response. *Cross-Cultural Res.* **2018**, *52*, 468–495. [CrossRef]
33. Onel, N.; Mukherjee, A. The Effects of National Culture and Human Development on Environmental Health. *Environ. Dev. Sustain.* **2014**, *16*, 79–101. [CrossRef]
34. Vachon, S. International Operations and Sustainable Development: Should National Culture Matter? *Sustain. Dev.* **2010**, *18*, 350–361. [CrossRef]
35. Dangelico, R.M.; Fraccascia, L.; Nastasi, A. National Culture's Influence on Environmental Performance of Countries: A Study of Direct and Indirect Effects. *Sustain. Dev.* **2020**, *28*, 1773–1786. [CrossRef]
36. Leong, L.Y.C.; Fischer, R.; McClure, J. Are Nature Lovers More Innovative? The Relationship between Connectedness with Nature and Cognitive Styles. *J. Environ. Psychol.* **2014**, *40*, 57–63. [CrossRef]
37. Dunlap, R.E.; Jones, R.E. Environmental Concern: Conceptual and Measurement Issues. In *Handbook of Environmental Sociology*; Greenwood Press: Westport, CT, USA, 2002.
38. Dunlap, R.E.; Van Liere, K.D.; Mertig, A.G.; Jones, R.E. Measuring Endorsement of the New Ecological Paradigm: A Revised NEP Scale. *J. Soc. Issues* **2000**, *56*, 425–442. [CrossRef]
39. Dunlap, R.E.; Van Liere, K.D. The "New Environmental Paradigm": A Proposed Measuring Instrument and Preliminary Results. *J. Environ. Educ.* **1978**, *9*, 10–19. [CrossRef]
40. Gagnon Thompson, S.C.; Barton, M.A. Ecocentric and Anthropocentric Attitudes toward the Environment. *J. Environ. Psychol.* **1994**, *14*, 149–157. [CrossRef]
41. Sockhill, N.J.; Dean, A.J.; Oh, R.R.Y.; Fuller, R.A. Beyond the Ecocentric: Diverse Values and Attitudes Influence Engagement in pro-Environmental Behaviours. *People Nat.* **2022**, *4*, 1500–1512. [CrossRef]
42. Pascual, U.; Balvanera, P.; Anderson, C.B.; Chaplin-Kramer, R.; Christie, M.; González-Jiménez, D.; Martin, A.; Raymond, C.M.; Termansen, M.; Vatn, A.; et al. Diverse Values of Nature for Sustainability. *Nature* **2023**, *620*, 813–823. [CrossRef]
43. de Groot, J.I.M.; Steg, L. Value Orientations to Explain Beliefs Related to Environmental Significant Behavior: How to Measure Egoistic, Altruistic, and Biospheric Value Orientations. *Environ. Behav.* **2008**, *40*, 330–354. [CrossRef]
44. Thomas, V. Intractability of Climate Change. In *Risk and Resilience in the Era of Climate Change*; Palgrave Macmillan: Singapore, 2024; pp. 113–142.

45. Xie, B.; Brewer, M.B.; Hayes, B.K.; McDonald, R.I.; Newell, B.R. Predicting Climate Change Risk Perception and Willingness to Act. *J. Environ. Psychol.* **2019**, *65*, 101331. [CrossRef]
46. Lorenzoni, I.; Pidgeon, N.F. Public Views on Climate Change: European and USA Perspectives. *Clim. Change* **2006**, *77*, 73–95. [CrossRef]
47. Bickerstaff, K.J.; Simmons, P.; Pidgeon, N. Constructing Responsibilities for Risk: Negotiating Citizen—State Relationships. *Environ. Plan. A* **2008**, *40*, 1312–1330. [CrossRef]
48. Fielding, K.S.; Head, B.W. Determinants of Young Australians' Environmental Actions:The Role of Responsibility Attributions, Locus of Control, Knowledge and Attitudes. *Environ. Educ. Res.* **2012**, *18*, 171–186. [CrossRef]
49. Kitayama, S.; Salvador, C.E. Cultural Psychology: Beyond East and West. *Annu. Rev. Psychol.* **2024**, *75*, 495–526. [CrossRef] [PubMed]
50. Jackson, J.C.; Medvedev, D. Worldwide Divergence of Values. *Nat. Commun.* **2024**, *15*, 2650. [CrossRef]
51. Kwan-Shing Chan, D.; COLINDEX. A Refinement of Three Collectivism Measure. In *Individualism and Collectivism. Theory, Method, and Applications*; Kim, U., Triandis, H.C., Kagitcibasi, C., Choi, S.-C., Yoon, G., Eds.; SAGE Publications: Thousand Oakes, CA, USA, 1994; pp. 200–210.
52. Byrne, B.M.; Shavelson, R.J.; Muthén, B. Testing for the Equivalence of Factor Covariance and Mean Structures: The Issue of Partial Measurement Invariance. *Psychol. Bull.* **1989**, *105*, 456–466. [CrossRef]
53. Cieciuch, J.; Davidov, E.; Schmidt, P.; Algesheimer, R. How to Obtain Comparable Measures for Cross-National Comparisons. *Kolner Z. Soz. Sozpsychol.* **2019**, *71*, 157–186. [CrossRef]
54. Byrne, B.M. *Structural Equation Modeling with AMOS. Basic Concepts, Applications, and Programming*, 2nd ed.; Routledge: New York, NY, USA, 2010.
55. Gifford, R.; Nilsson, A. Personal and Social Factors That Influence Pro-Environmental Concern and Behaviour: A Review. *Int. J. Psychol.* **2014**, *49*, 141–157. [CrossRef]
56. El Jurdi, H.A.; Batat, W.; Jafari, A. Harnessing the Power of Religion: Broadening Sustainability Research and Practice in the Advancement of Ecology. *J. Macromarketing* **2017**, *37*, 7–24. [CrossRef]
57. Konisky, D.M. The Greening of Christianity? A Study of Environmental Attitudes over Time. *Environ. Polit.* **2018**, *27*, 267–291. [CrossRef]
58. Mo, Y.; Zhao, J.; Tang, T.L.P. Religious Beliefs Inspire Sustainable HOPE (Help Ourselves Protect the Environment): Culture, Religion, Dogma, and Liturgy—The Matthew Effect in Religious Social Responsibility. *J. Bus. Ethics* **2023**, *184*, 665–685. [CrossRef]
59. Preston, J.L.; Baimel, A. Towards a Psychology of Religion and the Environment. *Curr. Opin. Psychol.* **2021**, *40*, 145–149. [CrossRef] [PubMed]
60. York, M. Religion and the Environmental Crisis. In *Ecotheology—Sustainability and Religions of the World*; Hufnagel, L., Ed.; Intechopen: London, UK, 2022.
61. Kemmelmeier, M.; Kemmelmeier, M.; Burnstein, E.; Burnstein, E.; Krumov, K.; Krumov, K.; Genkova, P.; Genkova, P.; Kanagawa, C.; Kanagawa, C.; et al. Individualism, Collectivism and Authoritarianism in Seven Societies. *J. Cross. Cult. Psychol.* **2003**, *34*, 304–322. [CrossRef]
62. Shields, T.; Zeng, K. The Reverse Environmental Gender Gap in China: Evidence from "The China Survey". *Soc. Sci. Q.* **2012**, *93*, 1–20. [CrossRef] [PubMed]
63. Dunlap, R.E. The New Environmental Paradigm Scale: From Marginality to Worldwide Use. *J. Environ. Educ.* **2008**, *40*, 3–18. [CrossRef]
64. Johnson, T.P.; Shavitt, S.; Holbrook, A.L. Survey Response Syles across Cultures. In *Cross-Cultural Research and Methods in Psychology*; Matsumoto, D., Van De Vijver, F.J.R., Eds.; Cambridge University Press: Cambridge, UK, 2011; pp. 130–175.
65. Nisbett, R.E.; Choi, I.; Peng, K.; Norenzayan, A. Culture and Systems of Thought: Holistic versus Analytic Cognition. *Psychol. Rev.* **2001**, *108*, 291–310. [CrossRef]
66. Dillon, A.; Ali, T. Global Nomads, Cultural Chameleons, Strange Ones or Immigrants? An Exploration of Third Culture Kid Terminology with Reference to the United Arab Emirates. *J. Res. Int. Educ.* **2019**, *18*, 77–89. [CrossRef]
67. Habeeb, H.; Hamid, A.A.R.M. Exploring the Relationship between Identity Orientation and Symptoms of Depression among Third Culture Kids College Students. *Int. J. Instr.* **2021**, *14*, 999–1010. [CrossRef]
68. Mosanya, M.; Kwiatkowska, A. New Ecological Paradigm and Third Culture Kids: Multicultural Identity Configurations, Global Mindset and Values as Predictors of Environmental Worldviews. *Int. J. Psychol.* **2023**, *58*, 103–115. [CrossRef]

land

MDPI

Article

Generative AI for Biophilic Design in Historic Urban Alleys: Balancing Place Identity and Biophilic Strategies in Urban Regeneration

Eun-Ji Lee and Sung-Jun Park *

Department of Architectural Engineering, Keimyung University, Daegu 42601, Republic of Korea; eunjilee@kmu.kr
* Correspondence: sjpark@kmu.ac.kr

Abstract: Historic urban alleys encapsulate cultural identity and collective memory but are increasingly threatened by commercialization and context-insensitive redevelopment. Preserving their authenticity while enhancing environmental resilience requires design strategies that integrate both heritage and ecological values. This study explores the potential of generative artificial intelligence (AI) to support biophilic design in historic alleys, focusing on Daegu, South Korea. Four alley typologies—path, stairs, edge, and node—were identified through fieldwork and analyzed across cognitive, emotional, and physical dimensions of place identity. A Flux-based diffusion model was fine-tuned using low-rank adaptation (LoRA) with site-specific images, while a structured biophilic design prompt (BDP) framework was developed to embed ecological attributes into generative simulations. The outputs were evaluated through perceptual and statistical similarity indices and expert reviews (n = 8). Results showed that LoRA training significantly improved alignment with ground-truth images compared to prompt-only generation, capturing both material realism and symbolic cues. Expert evaluations confirmed the contextual authenticity and biophilic effectiveness of AI-generated designs, revealing typology-specific strengths: the path enhanced spatial legibility and continuity; the stairs supported immersive sequential experiences; the edge transformed rigid boundaries into ecological transitions; and the node reinforced communal symbolism. Emotional identity was more difficult to reproduce, highlighting the need for multimodal and interactive approaches. This study demonstrates that generative AI can serve not only as a visualization tool but also as a methodological platform for participatory design and heritage-sensitive urban regeneration. Future research will expand the dataset and adopt multimodal and dynamic simulation approaches to further generalize and validate the framework across diverse urban contexts.

Keywords: historic urban alleys; urban regeneration; place identity; biophilic design; generative AI; LoRA fine-tuning; AI-assisted visualization

1. Introduction

1.1. Motivation and Purpose

Urban streets and alleys embody the history, culture, and communal life of a region, encapsulating distinctive identities and collective memories. In particular, historic–cultural landscapes shaped during the modernization process represent complex layers that extend beyond physical structures, incorporating accumulated social events and architectural styles over time. These layers provide a unique sense of placeness, organizing urban life, mediating social relations, and serving as cultural landscapes that reproduce regional identity [1,2].

The cultural value of historic alleys has been increasingly reappraised in recent years, establishing them as strategic assets for tourism, regeneration, and urban branding. In particular, Daegu, South Korea, has developed a distinctive street culture within its historic core, shaped by governmental offices, religious institutions, and heritage sites associated with nationalist movements during the Japanese colonial period [3]. In response, the city has introduced cultural tourism initiatives, such as the Historical and Cultural Streets Tours, which strategically leverage visual traces and historical placeness as urban marketing resources [4].

However, the original character of Daegu's alleys has been increasingly threatened by tourism-oriented commercialization and uncontextual visual interventions, often leading to the erosion of authenticity and rupture of historical continuity. As critics note, preservation alone is insufficient to ensure the sustainability and resilience of these spaces [5]. Place identity is not merely a function of visual elements or building materials, but is formed through layered sensory and experiential dimensions encompassing memory, emotion, and behavior [6]. Consequently, urban environments face a dual tension: the simultaneous demands of preserving identity and enhancing environmental quality.

Against this backdrop, biophilic design has emerged as a promising approach to reintegrate human–nature relationships into design and promote restoration [7]. Biophilic design refers to the intentional incorporation of natural elements, patterns, and processes into the built environment to enhance human well-being and foster emotional connections with nature [8,9]. Yet a critical question remains: how does nature operate within historic and cultural urban landscapes? Traditionally framed in planning as a source of comfort or decorative value [7], nature in historic and cultural contexts fulfills multiple, interdependent functions:

- Memory and emotional restoration: Elements such as greenery, water, shade, and wind evoke collective memories and facilitate emotional recovery at both individual and communal levels [10,11]. In historic contexts, natural features reinforce temporal continuity by evoking past experiences [12].
- Place attachment and sensory immersion: Seasonal vegetation, water features, and variations in natural light foster sensory immersion and reinforce emotional attachment to place [13,14].
- Harmonization of landscape tensions: Vegetation-based interventions, including green façades and permeable pavements, can mediate tensions between conservation and contemporary development, thereby maintaining visual coherence and contextual integration [15,16].

Thus, biophilic design in historic and cultural landscapes should extend beyond functional comfort, operating as a strategic design intervention that preserves place identity while enhancing emotional resilience. This necessitates a contextual approach grounded in memory, form, and symbolism.

To address this complexity, generative artificial intelligence (AI) offers new opportunities for predictive experimentation and visual simulation in design. Generative image models translate textual prompts into visual outputs after being trained on large-scale datasets [17]. With additional fine-tuning through low-rank adaptation (LoRA) [18], these models can be specialized for specific themes or contexts. Integrating generative AI with LoRA fine-tuning enables efficient learning of typological visual data and facilitates the simulation of design interventions in urban environments. This process serves both as a predictive design tool and as a visual platform for public engagement and feedback.

However, recent studies caution that mainstream AI models such as ChatGPT and Midjourney often fail to adequately capture cultural diversity and place-specific identities, particularly in non-Western contexts [19,20]. Campo-Ruiz [19], for instance, argues that these models tend to generate imagery biased toward iconic landmarks or commercialized urban culture, thereby diminishing cultural authenticity and reinforcing urban inequalities. Similarly, Jang et al. [20] emphasized that although generative AI holds potential for visualizing place identity, the quality of representation is contingent upon the cultural scope of training data, which often excludes or oversimplifies the particularities of non-Western cities.

Therefore, using AI to preserve place identity in historic and cultural landscapes should not remain a purely technical exercise. It requires a design approach that integrates urban and architectural contexts with cultural identity and restorative spatial experiences. The historic alleys of Daegu, shaped by modernization and enriched with both physical features and sensory memories, provide a clear example of where such an approach is both necessary and timely.

The present study analyzes the elements of place identity embedded in Daegu's "Historic Modern Culture Alley" and explores their integration with biophilic design attributes through generative AI simulations. By doing so, it aims to reconcile identity preservation with emotional resilience, providing new insights into the design and regeneration of culturally significant urban environments.

1.2. Research Aim and Questions

This study investigates design strategies that balance place preservation and restorative potential in Daegu's historic alleys by analyzing their cognitive, emotional, and physical elements and integrating these with biophilic design attributes through generative simulations. Specifically, a Flux-based LoRA model was employed to generate alley landscape images, followed by the application of biophilic design prompts to create intervention scenarios. The resulting outputs were evaluated through quantitative metrics and expert assessments, offering empirical insights into reconciling identity preservation with emotional restoration in historic urban landscapes.

The study addresses the following research questions (RQs):

(1) How can the cognitive, emotional, and physical components of place identity in Daegu's historic alleys be represented through visual imagery?

(2) What biophilic design strategies are most appropriate for historic cultural landscapes in Daegu, and how do these strategies contribute to both the preservation of identity and the realization of restorative experiences?

(3) To what extent can generative AI serve as an effective tool for designing biophilic alley landscapes grounded in place identity, and what are its limitations?

2. Theoretical Background

2.1. Place Identity of Cultural Landscape and Urban Alley

The concept of place identity was first articulated by environmental psychologist Proshansky et al. [21], who defined it as "cognitions about the physical world in which the individual lives." This perspective highlights that place identity is not merely a reflection of subjective awareness, but emerges from the interplay of physical, symbolic, institutional, and social dimensions [22]. Relph [12] and Paasi [23] broadened the concept, viewing place identity as a composite structure that integrates nature, culture, and everyday life to differentiate one locality from another. Similarly, Groote and Haartsen [24] conceptualize place identity as the outcome of the interaction between physical and artificial structures, systems of meaning, and social signifiers. Collectively, these perspectives emphasize that place identity is not a fixed entity but a fluid and multi-layered cultural construct that is continually reconstituted through time and context.

Synthesizing prior scholarship, place identity can be understood across three interrelated dimensions:

- Cognitive elements: Spatial perception, sense of orientation, and interpretation of symbolic meanings.
- Emotional elements: Attachment, familiarity, belonging, and psychological mechanisms tied to identity formation.
- Physical elements: Landscape structures, architectural styles, materials, spatial organization, and visual markers.

These three dimensions do not operate independently but interact dynamically. In historic urban places, past experiences and collective memories intersect with present emotional and physical environments, generating a multi-layered and evolving reproduction of place identity.

Place identity is also closely tied to the spatial composition of urban landscapes. Lynch [25] identified five key elements contributing to the imageability of cities—paths, edges, nodes, districts, and landmarks—illustrating how citizens perceive and structure urban form. Among these, paths, edges, and nodes constitute critical components in alley landscapes, mediating place experience through cognitive mapping, perception of boundaries, and recognition of transitional points. In this study, alleys are interpreted as district units with a unique identity, further distinguished by surrounding landmarks that enhance visibility and character. Within these districts, the configuration of paths (e.g., non-linear flows, directional guidance), edge elements (e.g., walls, vertical differences), and nodes (e.g., corners, small squares, spaces of interaction) serve as key visual cues shaping the place identity of alleys.

2.2. Biophilic Design Theory and Urban Landscape Application

Biophilic design is a design paradigm grounded in the premise that emotional and psychological connections with nature can enhance human health and well-being [7]. It offers human-centered principles for urban environments, moving beyond conventional approaches that focus on the physical incorporation of natural elements. Instead, it emphasizes the integration of cultural context and sensory experience, thereby fostering what may be termed placial naturalness—a sense of naturalness embedded in the unique cultural and spatial context of a place [26]. Therefore, biophilic interventions in urban alleys are significant as strategies to sustain place identity while enhancing cultural and ecological resilience.

Kellert [8] conceptualized biophilic design within three broad categories of experience, further specified into 25 design attributes (Figure 1). Direct experience of nature involves incorporating actual natural elements, such as street trees, potted plants, or architectural openings that allow natural light to enter alleys. Indirect experience of nature refers to the use of materials, patterns, and colors that mimic natural qualities, including patterned façades, the use of wood or stone, and soundscapes evocative of natural environments. Experience of space and place highlights spatial and cognitive experiences, exemplified by alley curvatures, courtyards, or vantage points that shape one's engagement with the environment.

Biophilic design has been increasingly applied in diverse urban pedestrian spaces, parks, and alleyways as a practical strategy to enhance psychological and emotional resilience. Common approaches include maximizing daylight penetration, increasing greenery, and applying natural materials or patterns. These interventions are associated with a range of positive outcomes, such as emotional stability, attentional restoration, and the facilitation of social interaction [9]. Depending on the spatial context, biophilic design operates in different ways, producing varied effects. Importantly, by introducing narrative, experiential, and memorable qualities, it enables spaces to be recognized as places, thereby mirroring the structural role of place-making factors [15].

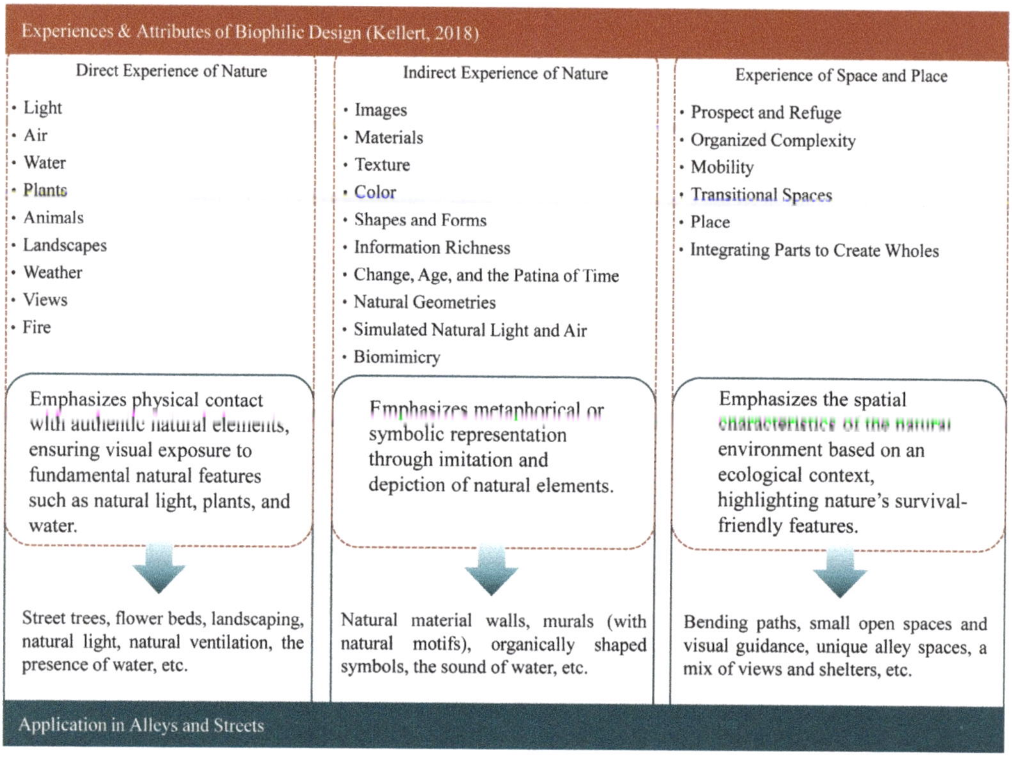

Figure 1. Biophilic design framework and application implications in urban landscape (adapted from Kellert, 2018 [8]).

This study links Kellert's three categories and their sub-attributes to the cognitive, emotional, and physical dimensions of place identity in historic-cultural landscapes. Through this conceptual mapping, it seeks to derive context-based strategies for biophilic design. These strategies are subsequently embedded into generative AI prompts, enabling experimental design scenarios that integrate place identity with naturalness and allow for simulation-based exploration.

2.3. LoRA Model and Prompt Framework for Landscape Visualization

Generative image AI has gained attention as a powerful medium that enables the exploration of emotional connections between users and spaces, as well as combinations of landscapes through scenario-based simulations. In particular, diffusion-based AI models [27] are capable of mapping textual prompts to image data, thereby producing high-resolution imagery that facilitates experimental inquiry into diverse design scenarios.

Building on this development, this study employs a Flux-based LoRA model. Flux, a state-of-the-art text-to-image model, performs better than conventional diffusion models due to its efficient memory structure and enhanced visual fidelity [28]. Notably, it excels in rendering contextual details such as texture, light, and spatial depth [28,29], making it particularly suitable for landscapes with layered meanings and embedded place identity, such as historic alleys.

Furthermore, advances in LoRA techniques have further enabled the use of relatively small datasets to train models according to specific user intentions or stylistic requirements [18]. LoRA adapts only a subset of low-rank parameters rather than retraining the entire weight set, thereby enabling efficient learning of specific styles or attributes from small datasets [28]. This advancement allows for the customization of models to reflect localized landscapes or place-specific identities.

Nevertheless, the quality of generative AI outputs is strongly influenced not only by model architecture but also by the systematic design of prompts. Lee and Park [17], for instance, proposed a biophilic design prompt framework to guide the application of generative AI in architectural and spatial contexts. Drawing on this approach, the present study adapts and structures prompt components for application within urban-landscape contexts, as illustrated in Figure 2.

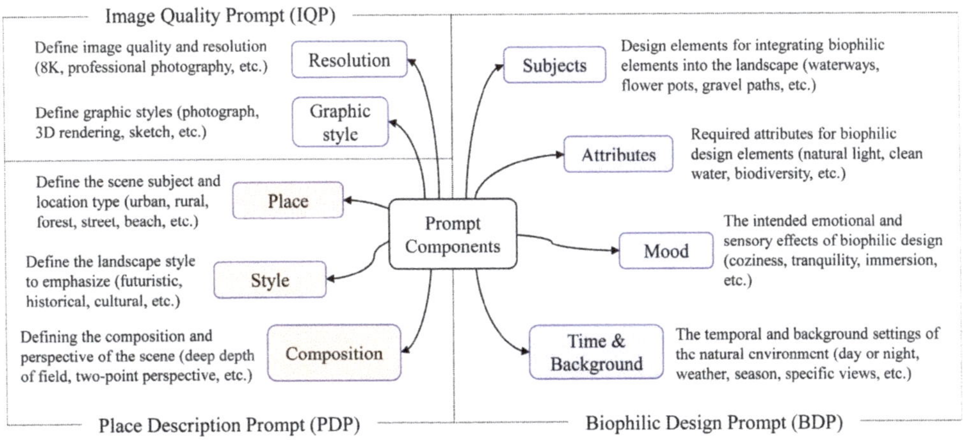

Figure 2. Structured prompt components for applying biophilic design in urban-landscape contexts.

The prompt framework is organized into three main categories. The image quality prompt (IQP) specifies the resolution and representational style of the generated output (e.g., drawing, isometric view, or photographic rendering). The place description prompt (PDP) defines the scene's fundamental structure by describing place type, spatial context, and the arrangement of objects. Finally, the biophilic design prompt (BDP) integrates natural elements and emotional experiences into the design process by employing key components such as subjects, attributes, mood, and temporal or background settings [17]. The BDP can be operationalized through interactive strategies that link place identity with biophilic design, thereby serving as a structured design tool for generative AI-based simulations.

2.4. Research Gap and Theoretical Linkage

Although place identity and biophilic design have been widely examined in architecture and urban studies, their integration through generative artificial intelligence remains largely unexplored. Previous studies have typically addressed cultural identity and biophilic enhancement separately, lacking methodological frameworks that connect these domains through data-driven visualization. In particular, the application of AI to visually simulate the coexistence of cultural authenticity and ecological resilience has not yet been systematically tested within urban landscape contexts.

To address this gap, the present study proposes an AI-driven framework that integrates place-identity-based analysis with biophilic design strategies through LoRA fine-tuning and structured prompts. This approach operationalizes cultural and ecological values within generative workflows, providing a methodological foundation for participatory and context-sensitive urban regeneration.

3. Methodology

3.1. Classification of Target and Alley Types

This study focuses on the historic alley landscapes of Jung-gu, Daegu, South Korea. Since the Japanese colonial period, Jung-gu has developed as the city's center of administration, commerce, and religion, resulting in a concentration of modernized urban functions. Today, it still retains numerous alleys where diverse layers of place identity and visual traces coexist.

To preserve and utilize these characteristics, the city of Daegu operates the Historical and Cultural Streets Tours, which connect major urban alleys through five designated routes. Among them, this study selects the "Historic Modern Culture Alley" route—an area where cultural and historical landmarks are most concentrated as the primary research site (Figure 3).

Figure 3. Site map of Jung-gu, Daegu, indicating the "Historic Modern Culture Alley" route and key landmarks (a–e) used to contextualize the four alley typologies. a: Daegu Modern History Museum, b: Yakjeon Jin Alley, c: Gyesan Cathedral, d: 3.1 Independence Movement Stairs, e: Dongsan Missionary Houses.

The study defines the site as a district unit, based on the premise that alleys function not merely as passages but as independent cognitive units with distinctive visual atmospheres and experiential qualities [25]. Within this district, physical and symbolic cues, such as landmarks, make visible the city's key character.

For visual analysis, alley landscapes were further categorized into four types: path, stairs, node, and edge. Stairs, while not included in Lynch's original framework, were added here due to their distinctive role in vertically connecting routes, reinforcing symbolic meaning, and ensuring functional and visual continuity in historic alleys. Each type is defined as follows:

- Path: The primary circulation axis and cognitive flow of the alley, shaped by its alignment, rhythm, scale, and walking experience. Features such as straightness or curvature, width, and boundary composition directly influence pedestrian perception and movement.
- Stairs: A vertical connector linking horizontal paths, introducing rhythm through gradients and level changes. In historic alleys, stairs often combine with topographic or defensive elements, reinforcing uniqueness while ensuring continuity between upper and lower spaces.
- Edge: Boundaries or transitional zones, typically formed by high walls, enclosed spaces, or dead ends. In historic alleys, materiality, height differences, and wall textures accentuate the perception of enclosure and openness, directly shaping pedestrian experience.
- Node: Points where paths intersect or visual attention converges, often associated with public spaces or clusters of urban functions. Nodes serve as focal areas for gathering and interaction, enhancing both orientation and place identity.

3.2. Dataset Construction and LoRA Training

To ensure consistency in the training dataset, fieldwork was conducted in May 2025, during the same time window (13:00–15:00), to collect images representing the four alley landscape types. All images were captured at human eye level using an iPhone 13, following three criteria:

- Inclusion of key landmarks and visual cues of the study site;
- Clear representation of the spatial configuration and characteristics of each alley type;
- Variation in viewpoints, orientations, and distances within each type.

Through this process, a final set of 40 refined images was secured for LoRA model training. All images were preprocessed by resizing to a resolution of 1152 × 712 pixels. Figure 4 illustrates the workflow of the LoRA fine-tuning procedure.

In this study, the preprocessed images were mapped to text captions. Each caption included both a general label (e.g., historic alley) and multiple trigger tokens (e.g., path, node) to distinguish alley landscape types. The use of multi-trigger tokens enhanced the precision and controllability of the fine-tuning process, enabling the trained model to recognize and reproduce subtle structural variations within the same category. This not only improved the model's sensitivity but also strengthened its ability to generate type-specific images during inference.

Key hyperparameters for fine-tuning were configured using the Pinokio web user interface. To maximize dataset representativeness, potential augmentation techniques were applied, while adopting a conservative configuration: rank = 8, alpha = 16, learning rate = 5×10^{-5}, and batch size = 1 with gradient accumulation = 8. Training was performed for 16 epochs, with each image repeated 10 times to ensure data efficiency. These settings effectively balanced variability introduced through augmentation with overfitting control, allowing for stable and reliable model training.

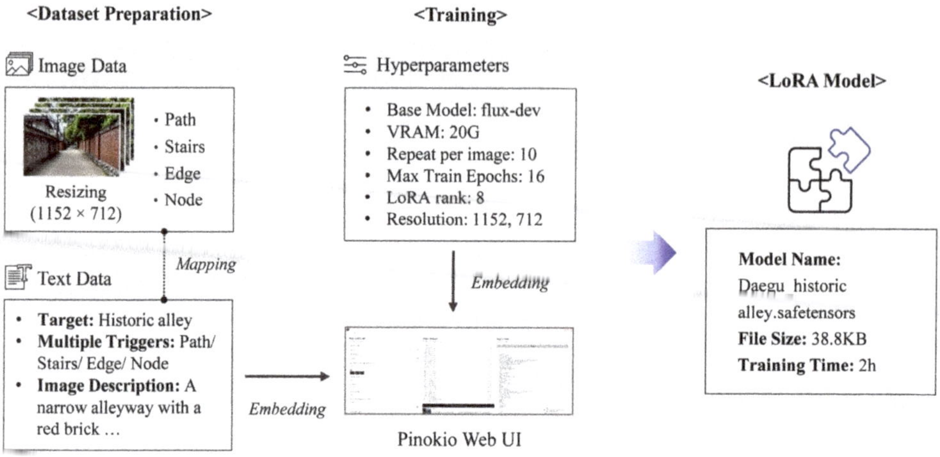

Figure 4. Workflow of LoRA fine-tuning with the Flux base model, including dataset preparation, captioning with trigger tokens, training stages, and example outputs.

3.3. Prompt Design and AI Image Generation

This study first conducted a field investigation and analysis to classify four alley types—path, stairs, edge, and node. Each spatial characteristic and visual indicator were categorized, and the elements of place identity were systematically organized into cognitive, emotional, and physical dimensions. These served as key considerations for place identity simulation and were incorporated into the construction of the PDP.

In the next step, a biophilic design strategy matrix was developed by aligning the place identity attributes of each alley type with biophilic experience and their sub-attributes. Previous research in architecture and urban planning was referenced to ensure contextual validity. The outcome was a set of keywords linking biophilic design strategies to each alley type. This process addressed a limitation noted in earlier studies [17], in which biophilic prompts often relied on intuition or vague references to natural imagery.

Subsequently, the biophilic design strategies were translated into prompts. Rather than simple keyword lists, these were hierarchically structured around the core components of the BDP (Figure 2). Figure 5 illustrates the workflow for prompt design and image generation.

PDPs and BDPs were combined with the LoRA model to generate alley landscape images that reflect both place identity and biophilic attributes. The image generation process involved two stages. First, PDPs were applied with the LoRA model to create place-identity-based alley images, which were compared with the training dataset to quantitatively evaluate the model's performance. Second, the validated LoRA model was combined with BDPs to generate place-identity-based images incorporating biophilic interventions. These outputs were then qualitatively assessed through expert evaluation. All image generation and computational processes were conducted on a local PC equipped with an RTX 4090 GPU and 64 GB of memory. Table 1 presents the key parameter settings used for image generation.

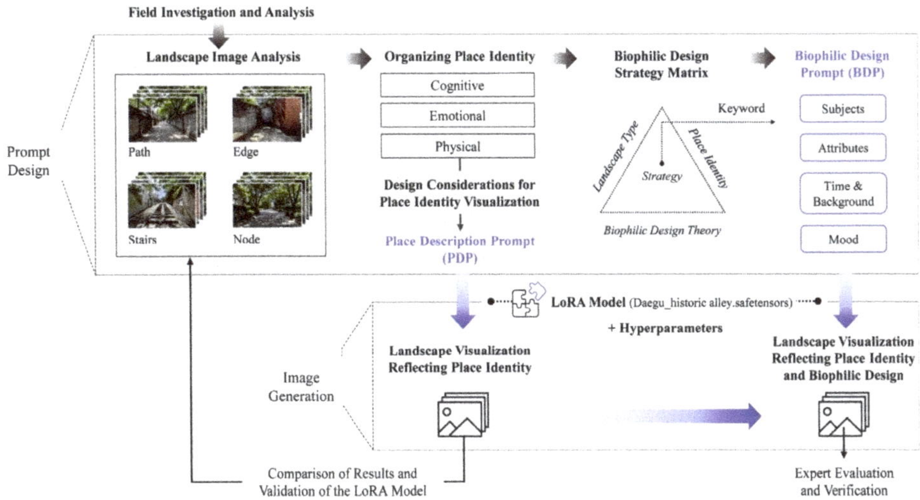

Figure 5. Workflow of prompt design and AI image generation.

Table 1. Flux configuration settings used for inference.

Item	Explanation	Value
Base model	The pre-trained generative model that serves as the foundation for LoRA fine-tuning; in this study, Flux1-dev, a text-to-image diffusion model optimized for contextual fidelity and efficient image synthesis, was used.	Flux1-dev
Sampling method	The algorithm used to control the diffusion trajectory during image generation; determines how the model iteratively refines noise into a coherent image.	Euler
Sampling steps	The number of iterative denoising steps performed in the diffusion process; a higher value generally yields more detailed and stable images.	25
Distilled CFG Scale	A distilled classifier-free guidance parameter that adjusts the balance between visual creativity and prompt adherence, maintaining realism while avoiding overfitting to textual prompts.	3.5
CFG scale	The base guidance intensity applied during image generation to regulate how strictly the model follows the input prompt.	1.5
Resolution	The output image size (in pixels), defining the width and height of generated images.	1152 × 712
LoRA weight	The blending coefficient determines how much the LoRA fine-tuned parameters influence the base model during inference.	0.7
Number of images	Total number of images generated per alley type during inference.	100 per type

3.4. Evaluation and Analysis

3.4.1. Image Similarity Evaluation Index

This study generated images using the PDPs derived from the place identity analysis of Daegu's alley landscapes. Two conditions were compared: (1) images generated using prompts only, and (2) images generated using prompts in combination with the LoRA model. For each alley type, 100 images were generated under each condition, resulting in a total of 800 AI-generated images (400 prompt-only and 400 prompt+LoRA) and 40 ground truth (GT) images collected from fieldwork.

Image quality and similarity were assessed using two complementary metrics: learned perceptual image patch similarity (LPIPS) [30] and fréchet inception distance (FID) [31] LPIPS measures perceptual distance between individual images by leveraging intermediate feature maps from deep neural networks [30]. In this study, a pre-trained VGG network was used to calculate the perceptual similarity between GT images and the generated images (prompt-only and prompt+LoRA). Lower LPIPS scores indicate smaller perceptual differences, meaning higher visual similarity. Thus, LPIPS served as a key indicator of how faithfully the generated outputs captured the characteristics of real alley landscapes.

FID evaluates differences in feature distributions between two sets of images, providing an index of overall quality and stylistic consistency [31]. Using features extracted from the final pooling layer of a pre-trained Inception-V3 network, FID scores were computed between the GT image set and each generated set. Lower FID scores indicate closer alignment in feature distributions, reflecting that the generated images more accurately reproduced the statistical and structural properties of real alley imagery.

3.4.2. Expert Evaluation

The expert evaluation involved eight professionals selected through purposive sampling based on three criteria: (1) specialization in architecture, landscape architecture, or urban design; (2) at least five years of professional or research experience; and (3) active engagement in spatial visualization or urban regeneration projects. The panel included three architects, three landscape architects, and two urban designers, representing both academic (n = 5) and professional practice (n = 3) backgrounds.

All experts were given a brief orientation on the study's aims, place identity, and biophilic design framework to establish a shared interpretive basis. They were instructed to assess contextual authenticity and ecological appropriateness rather than aesthetic preference, ensuring consistency across evaluations. We explicitly assumed that participants possessed sufficient professional literacy to evaluate design realism and contextual fit.

The evaluation materials included GT images and AI-generated outputs corresponding to the four alley types. For each type, one GT image and three LoRA-generated images (before/after intervention and one variation) were provided, resulting in a total of 16 images. These were arranged in paired and sequential comparisons to allow experts to intuitively perceive both the preservation and transformation of place identity.

Evaluation indicators were organized into three domains: place identity, biophilic attributes, and the suitability of AI-based simulations. Each item was rated on a five-point Likert scale (1 = very poor, 5 = excellent). In addition to completing itemized evaluations for each image set, the experts were invited to provide open-ended comments. Particular attention was given to collecting qualitative feedback related to the study's central question—"the balance between preserving place identity and enhancing restorative experiences." The detailed evaluation framework is presented in Table 2.

Table 2. Evaluation framework for expert review on AI-based biophilic alley images.

Evaluation Domain	Evaluation Items		N	Materials Provided
Place identity	Cognitive identity	Clarity of orientation, landmarks, and boundaries from a cognitive perspective	4	Paired comparison images (GT vs. AI) + explanation of place identity dimensions by alley type
	Emotional identity	Provision of stability, sense of openness, and fascination from an emotional perspective	4	
	Physical identity	Degree to which materials, scale, and historicity are represented from a physical perspective	4	
Biophilic design	Design integrity	Reflection of biophilic attributes such as vegetation, water, light, and seasonality	4	Sequential images (before–after–variation) + explanation of strategies and expected effects
	Design effectiveness	Quality and effectiveness of biophilic experiences	4	
Suitability of AI-based simulation	Applicability	Practical applicability of AI-generated results for design	1	Full image set + explanation of applicability
General feedback	Open comments	Suggestions for improvement, strengths, and limitations (free response)	1	-

4. Place-Identity-Based Biophilic Design Prompt

4.1. Place Identity Characteristics of Historic Alley Typologies

Table 3 summarizes the place identity of Daegu's historic alleys by typology. Each alley landscape type reveals distinctive cognitive, emotional, and physical characteristics, which serve as key cues for subsequent visualization and design interventions.

The path type combines linear pedestrian flows with tree canopies and traditional stone or brick walls, producing a clear sense of orientation and stability. Pedestrians experience a steady rhythm and protective enclosure, where consistent walking widths and boundary materials embed local identity and everyday memory.

The stairs type embodies the interplay of vertical transitions and historical context. The act of ascending and descending generates both anticipation and accomplishment, while rough stone textures and shifting shadows intensify embodied immersion. Stairs, often tied to Daegu's modern history, mediate collective memories associated with historical events.

The edge type represents transitional zones between openness and enclosure. High walls, textured façades, and contrasts of light and shadow accentuate spatial shifts, evoking salience and distinctiveness. Edges thus function as thresholds that transform the atmosphere and sensory perception of alleys.

The node type emerges where paths intersect and people gather, reinforcing both spatial orientation and communal exchange. Through openness and a sense of hospitality, nodes invite lingering, while central trees or sculptures provide symbolic meaning and enhance collective identity.

These typological distinctions serve as critical cues for visualization and design applications, offering a foundation to preserve historic place identity while accommodating new biophilic and regenerative interventions.

Table 3. Classification of alley landscape types and their place identity dimensions.

Type	Images	Spatial Characteristics	Place Identity Dimensions			Design Considerations for Visualization
			Cognitive	Emotional	Physical	
Path	 	Linear pedestrian axis with flanking walls, street tree canopy, and traditional architectural forms	Recognition of directionality through linear axis; continuity of walls and architecture	Calmness and security through steady walking rhythm	Stone/brick walls; consistent pedestrian width; traditional materials	Emphasize continuity of stone walls and pedestrian axis; reflect canopy effects; harmonize with historic architectural elements
Stairs	 	Combination of slopes and stairs; elevation change for visual and bodily experience; independence movement memorial stairs	Anticipation through stair rhythm and directional cues	Tension and accomplishment through ascent/descent; link to historical memory	Rough stone; shadow patterns; vertical elevation of steps	Highlight rhythmic stair patterns and elevation; reflect boundary walls; capture slope and shifting viewpoints
Edge	 	Dead-end alley; historical information guide and exhibition space	Perception of boundaries; thresholds defining spatial separation	Salience and distinctiveness evoked by contrast; memorable perception of spatial transition	Tall walls, light–shadow contrasts; textured surfaces	Smoothly connect spatial thresholds; integrate wall textures; utilize light–shadow contrast

The path type is characterized by linear pedestrian flows, tree canopies, and the rhythm of boundary walls. Introducing seasonal vegetation along tree lines and walls, together with permeable paving and groundcover, provides pedestrians with both clear orientation and a sense of protection. This strategy moves beyond simple wayfinding by fostering continuous interaction with nature and promoting restorative walking experiences [11,40].

The stairs type reflects the overlap of vertical transitions and historical context. Variations in stair height and wall elevation create sequential vantage points, while green walls and aquatic planting zones enhance visual and ecological diversity. These interventions allow pedestrians to experience both discovery and accomplishment during movement, generating immersive place experiences linked to historical meaning [15,36].

The edge type represents transitional zones where enclosure meets openness. Contrasts of textured walls, climbing vegetation, and dynamic light–shadow patterns emphasize spatial thresholds and produce a heightened sense of distinctiveness. At the same time, vertical greening and habitat provision strengthen ecological diversity and maintain continuity of the historic landscape [9,39].

The node type serves as a center of gathering and exchange. By integrating small resting areas, water features, permeable surfaces, and green infrastructure, nodes are transformed into restorative plazas. These spaces mediate social interaction and collective memory, reinforcing place attachment and enhancing community vitality [13,35].

Although the four types differ in their specific characteristics, several shared principles emerge: (1) visual integration of natural elements, (2) utilization of light and seasonality, (3) preservation of historic materials and textures, and (4) reinforcement of social place identity. Natural elements such as trees, vines, and water features provide orientation and restorative qualities; light and seasonal change reveal spatial dynamism; traditional materials sustain historical authenticity while combining with ecological interventions; and community-oriented devices enhance social attachment.

These strategies provide practical guidelines for urban landscape design that respect historical and cultural contexts while integrating human–nature experiences. Furthermore, the matrix informs generative AI–based biophilic simulation prompts, enabling exploration of restorative and ecological urban visions grounded in historic place identity.

4.3. Prompt Mapping for Biophilic Design Application

Structuring abstract design intentions into executable textual input components is a crucial step to ensure both reproducibility and contextual sensitivity in simulations. Based on the biophilic design strategy matrix, this study developed a structured prompt set (Table 5) to simulate alley landscapes that integrate place identity and biophilic attributes.

Trigger words serve as mediating prompts within the trained LoRA model, specifying alley landscape types and generating their corresponding visual characteristics. The BDP is systematized into four core components—subject, attributes, time & background, and mood—which explicitly embed the cognitive, emotional, and physical layers of place identity. This approach moves beyond a mere listing of keywords by establishing a hierarchical language system that enables generative AI to reflect spatial and cultural cues in a more nuanced manner. For instance, in the path type, "straight path" and "continuity of fence height" express cognitive clarity, while "tree canopy" and "dappled afternoon light" convey emotional protection. Each component is thus mapped to capture the biophilic design subject, attributes, spatiotemporal conditions, and sensory experiences.

Table 5. Biophilic design prompts (BDP) for generating alley images based on place identity, structured by subject, attributes, time & background, and mood per typology.

Type	Trigger	BDP			
		Subject	**Attributes**	**Time & Background**	**Mood**
Path	Historic alley in Daegu, a straight path	Tree canopy, textured stone and brick wall, continuity of fence height, tiled roof, seasonal planting, vine, and groundcover (green pavements)	Rhythmic paving, dynamic light patterns, natural texture variety, seasonal vegetation, tree shadows, and natural color	Season (e.g., spring, autumn), afternoon, dappled light through canopy, and historic alley context	Calm, contemplative, and sense of protection
Stairs	Historic alley in Daegu, stone-textured stairs	Vertical greenery, stone texture, middle landing with a small terrace, vertical cascading water fountain (water wall, stream), and waterway (drain) with aquatic plants	Step rhythm, natural (aged, mossy) texture, prospect and refuge, natural water features, and retaining wall planting	Season (e.g., spring, autumn), afternoon, hillside, city view beyond side wall, historic setting, and shade trees	Challenging, immersive, sense of achievement, peaceful, and awe
Edge	Historic alley in Daegu, courtyard with edge	High gray stone/red brick texture, vertical planting (garden), birds and insects, vine-covered upper boundary, and biophilic (organic, natural geometry, wavy) patterned wall (paving)	Weathered texture, softens the edge, textural richness, filtered light, creating a patina of time, light and shadow, and natural motifs (leaf patterns, fractal textural)	Season (e.g., spring, autumn), afternoon, semi-open views, transitional space between courtyard and alley (open plaza) and sky background	Blurred boundary, quiet immersion, softened tension, anticipation, extraordinary, and threshold experience
Node	Historic alley in Daegu, intersection points in the node	Central landmark (tunnel, tree, plaza), waterways (pond), curved paths, tree canopy with seating (bench, shelter), rain garden, small garden, and permeable pavements	Visible routes, soft paving, filtered shadows, clear (clear) water features, gentle (natural) water flow, native plants, and biophilic materials (shapes)	Season (e.g., spring, autumn), afternoon, urban greenery context, radiating paths, and central plaza	Sociable, communal, sense of place attachment, hospitality, openness, comfort, and community connection

Notes: Words in parentheses indicate alternative or supplementary prompt options.

Furthermore, the BDPs proposed in this study are not designed as isolated components but rather as interdependent elements that collectively reinforce the expression of place identity. Users can selectively emphasize particular biophilic attributes depending on context—for example, highlighting patterns of light and shadow in one scenario, or emphasizing vegetation networks and water features in another—allowing for comparative analysis across results. Such adaptability demonstrates that the BDP framework functions not only as a visualization tool but also as a platform for participatory design and experimental inquiry.

Ultimately, the structured prompt provides a direct operational linkage between the conceptual strategies of the biophilic design matrix and the generative execution of the AI workflow. By embedding layered attributes into prompt grammar, the framework supports image generation that is both consistent and flexible, thereby enabling subsequent quantitative and qualitative evaluation.

Detailed prompt templates and caption examples used in this process are summarized in Sections 5.1 and 5.3, illustrating how the structured prompts (PDP, BDP, and IQP) were applied within the AI workflow to generate contextually and visually coherent alley images.

5. AI-Generated Alley Images Results and Evaluation

5.1. Training Stability and Visual Comparison of LoRA-Generated Images

Figure 6 presents the loss curves of the LoRA model trained on images of Daegu's historic alleys. The average loss, current loss, and epoch-averaged loss all exhibit a steady decline as training progresses, indicating that the model learned the visual and structural characteristics of the alley dataset stably, without excessive oscillation or overfitting. In particular, the stabilization of loss values after several epochs suggests that the LoRA adaptation converged to an optimal state capable of effectively representing alley landscape types such as path, stairs, edge, and node.

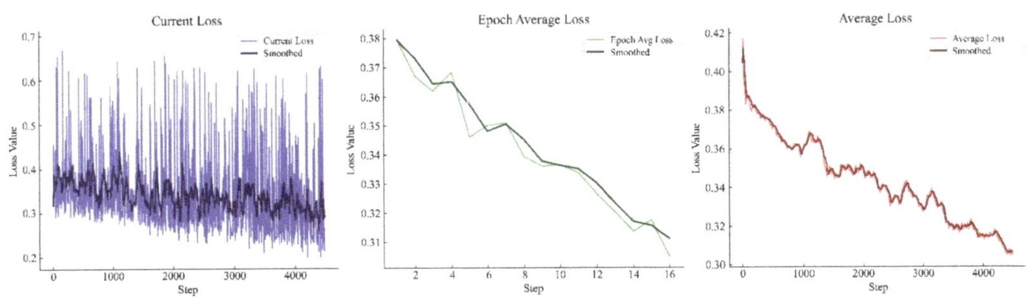

Figure 6. Training loss curves of the LoRA model. The figure shows the current loss, epoch average loss, and average loss across training steps, demonstrating convergence and stability of the model.

Table 6 compares image generation results with and without LoRA training. The PDPs were developed from keywords derived through the place identity analysis of alley types, incorporating simulation considerations and type-specific trigger words. The IQPs consisted of essential terms to ensure the production of high-quality landscape images.

Outputs generated by the prompt-only condition generally lacked contextual fidelity, failing to adequately reflect the distinctive visual characteristics of Daegu's historic alleys. In contrast, prompt+LoRA-based outputs demonstrated strong alignment with the intended alley types. For example, the path type faithfully preserved the rhythm of stone and brick boundary walls, while the stairs type captured not only the texture and rhythm of steps but also historical details such as the national flags and commemorative plaques. The edge type highlighted tall boundary walls and transitional effects, whereas the node type represented symbolic landmarks and open community spaces.

Table 6. Comparison of place-identity-based image generation results by alley landscape type.

Type	PDP/IQP	Prompt-Only	Prompt+LoRA
Path	PDP: Historic alley in Daegu, straight path with stone and red brick walls, traditional tiled roof, spatial continuity, boundary clarity, eye-level view, and deep perspective IQP: Full shot, high quality, 8K, and professional photography		
Stairs	PDP: Stairs of historic alley in Daegu, downward perspective view, stone wall with memorial plaques, other side wall topped with fencing and Korean flags, trees, mossy retaining wall, and deep perspective IQP: Full shot, high quality, 8K, and professional photography		
Edge	PDP: Edge of historic alley in Daegu, courtyard with brick and concrete wall, side bench, bushes and barbed wire on the walls, old texture, Korean historic sculpture in the center, spatial transition, enclosed to open, eye-level view, and mdi perspective IQP: Full shot, high quality, 8K, and professional photography		
Node	PDP: Historic alley with node in Daegu, intersection points, multiple curved paths converge, ivy-covered stone tunnel with mural depicting resistance, trees, open space, and wide-angle view IQP: Full shot, high quality, 8K, and professional photography		

These results provide a strong foundation for subsequent quantitative evaluations and for applying biophilic design prompts.

5.2. *Quantitative Evaluation of Image Fidelity and Structural Similarity*

To evaluate whether the LoRA model effectively captured the place identity of the four alley types, quantitative analyses were conducted using LPIPS and FID (Figure 7). Detailed numerical results for each alley type, including mean $\pm$ SD and effect sizes, are provided in Table A1 (Appendix A), which further confirms the substantial improvements achieved through LoRA fine-tuning.

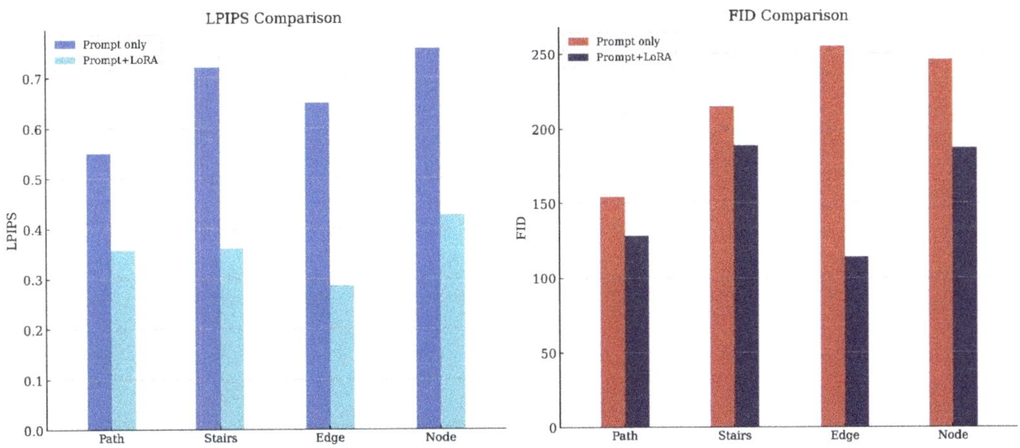

Figure 7. Comparison of LPIPS and FID between prompt-only and prompt+LoRA across four typologies; lower scores indicate better perceptual and distributional alignment with GT images.

The results consistently showed lower LPIPS scores for LoRA-based outputs across all types, indicating better perceptual alignment with the GT images. Since LPIPS reflects human-perceived visual similarity, this suggests that LoRA training did not merely refine textures and forms but also internalized subtle contextual elements such as patterns of light and shadow, temporal traces of materials, and symbolic cues. This finding is particularly meaningful given that place identity is rooted less in geometric accuracy and more in experiential and perceptual authenticity.

The FID scores further confirmed these results. Lower FID values for LoRA-based outputs demonstrated that the statistical feature distributions of generated images became more closely aligned with those of the real dataset. This indicates that LoRA captured not only individual attributes but also the overall diversity and contextual richness of alley scenes. Distributional fidelity is essential to prevent generative outputs from becoming homogenized while preserving contextual variability.

Taken together, these results show that the LoRA-based approach improved both image-level perceptual similarity and dataset-level statistical fidelity. This dual improvement underscores the reliability and validity of the proposed method, providing a solid foundation for the subsequent application of biophilic design prompts and expert evaluations.

5.3. *Generative Results Integrating Place Identity and Biophilic Design*

This section reports the outcomes of integrating the BDP with the validated LoRA model. The analysis investigates whether structured prompts can embed ecological attributes without undermining the historical and cultural identity of the four alley types. The generative results of place-identity-based biophilic design interventions are presented in Table 7.

Table 7. Examples of LoRA+BDP-driven generations by typology, showing prompts and resulting images that embed ecological attributes while preserving identity.

Type	BDP	Generated Images	
Path	<lora:daegu-historic-alley:0.7>, path PDP, textured stone and brick wall with vine, natural texture variety, continuity of fence height, tiled roof, tree canopy with seasonal vegetation, dappled light through canopy, spring (autumn) afternoon, calm, and IQP		
Stairs	<lora:daegu-historic-alley:0.7>, stairs PDP, vine fencing, middle landing with a small terrace, prospect and refuge, vertical cascading waterwall with natural stone texture, waterway with aquatic plants, afternoon, shade trees, city view beyond side wall, immersive, challenging, and IQP		
Edge	<lora:daegu-historic-alley:0.7>, edge PDP, high gray/red brick walls, weathered texture, vertical planting softens the edge, vine-covered upper boundary, biophilic patterned wall and paving, afternoon, semi-open views, softened tension, anticipation, and IQP		
Node	<lora:daegu-historic-alley:0.7>, node PDP, tree canopy with seating, waterways and rain garden, clear and gentle water, biophilic shapes, permeable pavements, summer afternoon, urban greenery context, communal, hospitality, IQP		

Notes: Each BDP was derived from the structured prompt components presented in Table 5, while the corresponding PDP and IQP were referenced from Table 6. Words in parentheses indicate alternative or supplementary prompt options used to explore generative variations. Together, these examples demonstrate the exact textual inputs applied within the AI workflow to generate contextually and visually coherent alley images.

The analysis shows that the BDP maintained contextual fidelity while enhancing the restorative dimensions of the generated images. For the path type, climbing vines along walls, seasonal vegetation, and light filtered through tree canopies enriched the sense of linear perspective. While preserving the historic rhythm of fences and tiled roofs, the images conveyed continuity and calmness, reinforcing both clarity and restorative qualities.

For the stairs type, vine planting, sound-buffering trees, and water features—such as cascades and channels along stone retaining walls—softened the rigid geometric form of stairs. These elements introduced visual and auditory stimuli derived from vegetation and water, creating a more immersive and dynamic atmosphere. Mid-landing spaces for outlook and rest further highlighted the prospect-and-refuge qualities of biophilic design, balancing anticipation and ecological richness.

The edge type was transformed through vertical greening, rain gardens, and patterned paving, softening its closed boundary into a semi-open transitional space. Serving as a distinctive place where historical traces coexist with ecological renewal, the edge type offered new emotional experiences by integrating heritage markers with environmental functions.

The node type emphasized symbolic and communal functions by integrating canopy shading, communal seating, water channels, and permeable paving. The central plaza evolved into a space of both cultural symbolism and biophilic coherence, where ecological forms and water features were combined to highlight collective identity and restorative capacity.

In summary, restorative experiences diverged between linear spaces (path/stairs) and nodal spaces (edge/node). Linear spaces primarily strengthened individual cognitive stability and bodily immersion, while nodal spaces reinforced emotional attachment and communal identity. These results demonstrate that AI-based simulation not only translates place identity into design cues but also explores how such cues interact to generate new forms of restorative experience.

5.4. Expert Evaluation of Place-Identity-Based Biophilic Design

An expert evaluation was conducted to assess the generated alley images. Internal consistency was examined using Cronbach's α [41], and inter-rater agreement was tested using the Intraclass Correlation Coefficient (ICC) based on a two-way random-effects model with absolute agreement [42].

The results showed good internal consistency (Cronbach's $\alpha = 0.83$) and high inter-rater reliability (ICC = 0.86), indicating satisfactory agreement among the eight experts. These findings confirm that the expert assessments were statistically consistent and dependable. A summary of the expert evaluation results for place identity and biophilic design is provided in Table 8.

Table 8. Summary of expert evaluation results.

Evaluation Items	Path	Stairs	Edge	Node	Average
Cognitive place identity	4.12 ± 0.99	4.00 ± 0.53	3.75 ± 0.89	4.12 ± 0.99	4.06 ± 0.93
Emotional place identity	4.38 ± 0.92	3.98 ± 0.64	3.58 ± 0.74	4.05 ± 0.89	3.98 ± 0.86
Physical place identity	3.88 ± 0.46	4.12 ± 0.83	4.00 ± 1.07	4.38 ± 0.74	4.10 ± 0.80
Biophilic design fidelity	4.12 ± 0.99	3.75 ± 0.99	4.25 ± 0.89	3.98 ± 0.99	4.06 ± 0.84
Biophilic design effectiveness	4.38 ± 0.74	4.25 ± 0.46	4.00 ± 0.93	4.25 ± 1.04	4.22 ± 0.79
AI applicability	4.06 ± 0.76				4.06 ± 0.76

Notes: Mean ± Standard deviation.

Overall, LoRA+BDP images received consistently positive assessments, with most items scoring above 4.0 on average. Particularly high ratings were observed for physical place identity (4.10 ± 0.80) and biophilic design effectiveness (4.22 ± 0.79), indicating that experts considered the outputs to be both materially authentic and ecologically enriched.

By type, path and node received the most favorable evaluations. These types effectively integrated vegetation, daylight, and communal features, thereby reinforcing both cognitive clarity and emotional bonding, while also highlighting symbolic meaning. The stairs type was positively evaluated for physical realism and rhythmic immersion, with mid-landing viewpoints and water features contributing to memorable spatial qualities. However, its biophilic fidelity was rated comparatively lower, likely due to structural constraints of stair spaces limiting design variability. This limitation, however, represents a potential area for refinement through more sophisticated prompt design and model improvement. In contrast, the edge type scored highest in biophilic fidelity (4.25 ± 0.89) but lowest in emotional place identity (3.58 ± 0.74). Experts noted that the closed boundary conditions tended to encourage overly dense application of biophilic elements, suggesting the need for additional training on appropriate proportionality between spatial scale and design attributes.

Across categories, emotional place identity received relatively lower scores (3.98 ± 0.86) but still remained above average. Experts emphasized that the emotional dimension is inherently the most challenging for AI to capture; nevertheless, the results demonstrated meaningful representation and significant potential for future enhancement. Iterative refinement of prompt design and model training was considered key to improving these outcomes further.

Experts also underscored the practical applicability of LoRA+BDP results. The "AI applicability" item was rated at 4.06 ± 0.76, suggesting that the generated images hold stable utility beyond specific alley types. Beyond quantitative scores, experts highlighted the usefulness of these images as visual negotiation tools in early-stage design exploration, participatory workshops, and educational settings. Although not a substitute for detailed design drawings, the outputs were recognized as valuable tools for stimulating creativity, visualizing alternatives, and facilitating stakeholder communication.

6. Discussion

6.1. Reproducing Place Identity Through LoRA Training

The LoRA fine-tuning in this study showed that generative AI can reproduce the cultural and spatial attributes that constitute the place identity of historic alleys. Unlike prompt-only outputs, which tended to produce generic images lacking contextual specificity, LoRA-based outputs consistently retained typological cues that reflect the multi-layered characteristics of the urban fabric. This indicates that fine-tuning functions as a methodological device capable of embedding local knowledge into the data-driven generative process.

The LoRA-generated images accurately represented both the physical and symbolic dimensions of place identity. Physical attributes—such as the rhythm of stone and brick walls, stair textures, and boundary forms—were faithfully reproduced, while cognitive and relational cues—such as transitional edges and communal landmarks—were also expressed. This suggests that LoRA internalized not only surface visual patterns but also the underlying structural logics sustaining each type. As a result, the model produced outputs aligned not only with external resemblance but also with the cognitive and cultural dimensions of place identity.

Quantitative metrics further supported these findings: lower LPIPS values and improved FID scores indicated greater perceptual and statistical consistency with GT images.

These outcomes are consistent with recent empirical studies reporting that context-aware fine-tuning enhances perceptual realism and the retention of typological cues in architectural or heritage imagery. For instance, LoRA-based adaptations have improved the cultural specificity and visual fidelity of place-dependent subjects [28], while identity-oriented generation studies show that tailoring models to local data reduces generic outputs and strengthens contextual alignment [20].

Taken together, LoRA fine-tuning provides a robust methodological foundation for adapting generative AI to urban-heritage contexts. By stably reproducing place-identity cues, LoRA establishes the conditions necessary for integrating biophilic design prompts and verifying applicability in design practice.

6.2. Integrating Biophilic Strategies Through Structured BDP

This study reconceptualized place identity not only as a set of physical features to be preserved but as design cues translatable into biophilic attributes. The structured BDP provided a systematic framework for embedding ecological qualities into alley typologies while reinforcing their cultural and spatial identities.

Results showed distinct effects: path emphasized continuity and clarity through vegetation and light; stairs deepened immersion with vertical greening and water features; edge transformed rigid boundaries into interpretive spaces with greenery and light gradients; and node highlighted symbolism and communal identity through green and water elements. These findings demonstrate that linear spaces (path, stairs) primarily strengthen individual clarity and immersion, while nodal spaces (edge, node) reinforce emotional attachment and collective identity.

These typology-specific patterns are consistent with empirical findings in biophilic and restorative urban design research. Previous studies have shown that vegetation, natural lighting, and visual openness enhance orientation, attentional focus, and perceived safety in linear environments such as streets and corridors [40,43,44]. In contrast, water features, tree canopies, and small plazas have been empirically linked to greater social interaction, emotional bonding, and collective memory formation at nodal urban spaces [45,46]. Furthermore, research on material texture and rhythmic light shadow variation indicates that such sensory attributes foster perceptual coherence and restorative experiences by mediating visual complexity and natural pattern recognition [15,47]. Collectively, these empirical insights support the present study's findings that biophilic attributes embedded in different alley typologies can simultaneously reinforce cognitive clarity and affective attachment within culturally significant environments.

Building upon these empirical implications, the proposed BDP framework extends such evidence into a generative design context, offering a practical means to simulate and evaluate biophilic interventions through AI. The methodological value of the BDP lies in its modularity and adaptability. Prompts can simulate seasonal variation or emphasize specific design variables, enabling generative AI to function not as a static visualization tool but as an experimental platform for testing biophilic interventions under varying contextual conditions.

6.3. Evaluating Reliability and Applicability of AI-Generated Images

In terms of effectiveness, consistent improvements in LPIPS and FID confirmed that LoRA-based images enhanced both perceptual authenticity and statistical fidelity. Experts also rated AI applicability positively (4.06 $\pm$ 0.76), suggesting that the proposed approach can serve as a practical tool in early design exploration. Notably, LoRA+BDP images not only reproduced historic alley features but also introduced biophilic attributes without undermining cultural specificity. Path and node types excelled in cognitive clarity and communal symbolism, stairs in physical realism and immersive experience, and edge type in biophilic fidelity. These results indicate that generative AI can strengthen historic and cultural identity by enriching spatial variation and experiential depth.

Nevertheless, several limitations were observed. Experts highlighted the restricted expression of emotional place identity, reflecting the inherent difficulty of reproducing affective resonance, symbolic associations, and personal memories through static imagery. Because emotional identity is closely tied to lived experience, future work should integrate 3D simulation and immersive technologies to better convey spatial presence. Furthermore, the reliance of generative AI on training data and prompt design introduces risks of visual oversimplification and cultural misrepresentation. If uncritically adopted as design output, such results may distort both research and practice. To address these issues, four measures are proposed:

(1) Expand multi-city and multicultural datasets to mitigate contextual bias.
(2) Present multiple alternative scenarios in parallel to avoid overreliance on single outcomes.
(3) Incorporate expert and citizen feedback loops to critically validate AI-generated results.
(4) Ensure transparency in data sources and prompt design, enhancing trust and accountability in interpretation.

In sum, generative AI demonstrates clear potential as an exploratory tool for biophilic alley design grounded in place identity, enabling typology-specific strategies. However, its application requires contextual validation and critical interpretation, particularly to overcome limitations in representing emotional and symbolic dimensions. Accordingly, LoRA+BDP simulation should be viewed as a mediating platform that balances identity preservation with restorative design interventions in urban regeneration, offering essential insights for the responsible integration of AI into architectural practice.

7. Conclusions

7.1. Contributions and Novelty

This study contributes to both theoretical and practical domains at the intersection of place identity, biophilic design, and generative AI.

Theoretically, three methodological innovations are proposed. First, LoRA fine-tuning was shown to effectively reproduce typological cues of place identity, capturing not only material realism but also symbolic resonance beyond mere visual similarity. Second, the study proposed a structured BDP framework that translates conceptual design strategies into modular prompt components, systematically embedding ecological attributes into AI simulations. Third, by integrating quantitative similarity indices with expert evaluation, a multi-layered verification framework was established, enhancing the reliability of generative outputs and offering a transferable methodology for future studies. Collectively, these contributions position generative AI as a third pathway for urban regeneration—beyond the dichotomy between physical preservation and modern intervention—by enabling a simulated balance between identity conservation and restorative design.

Practically, the study demonstrates the potential of AI-generated alley images as decision-support and participatory tools for design and planning. The framework provides a visual and analytical basis for testing design alternatives in heritage-sensitive contexts, helping practitioners communicate design intentions and evaluate ecological and cultural trade-offs. The findings are applicable not only to Daegu's historic alleys but also to similar neighborhoods undergoing regeneration, where cultural heritage must coexist with ecological enrichment. Typology-specific insights further reveal differentiated spatial values: path enhances legibility and continuity, stairs support immersive sequential experiences, edge transforms rigid boundaries into ecologically enriched transitional spaces, and node reinforces symbolism and communal identity.

Overall, this study presents a novel attempt to position generative AI as both a methodological and practical platform that harmonizes cultural specificity with biophilic strategies. By linking theoretical constructs with design experimentation, it establishes a foundation for applying generative AI in heritage-sensitive urban regeneration and for guiding future interdisciplinary approaches that connect artificial intelligence, ecological design, and cultural heritage conservation.

7.2. Practical Implications for Urban Planning and Design

This study provides practical implications for urban planners, architects, and designers seeking to integrate cultural heritage and ecological resilience in regeneration projects. The proposed LoRA+BDP framework provides a structured workflow that allows rapid generation of design alternatives prior to physical implementation, supporting visualization of options that preserve historic identity while introducing biophilic attributes such as vegetation, light, and water features.

First, AI-generated images can serve as decision-support tools in feasibility studies and early-stage design reviews, allowing planners to assess spatial coherence, environmental compatibility, and contextual continuity prior to detailed design.

Second, these visualizations facilitate participatory communication, helping citizens and non-experts intuitively understand complex spatial ideas during workshops and consultations, thereby improving transparency and inclusiveness in the decision-making process.

Finally, municipal authorities can employ the framework to support heritage-sensitive policies, using visual evidence to evaluate ecological and cultural impacts and prioritize sustainable interventions. Through these applications, generative AI can evolve from a visualization tool into a platform for collaborative and evidence-based urban regeneration.

7.3. Limitations and Future Research Directions

This study was exploratory in nature, with several limitations. First, there were constraints in both the dataset and the evaluation scope. LoRA fine-tuning was conducted using a limited dataset composed of Daegu's alleys image, and expert evaluation was based on a relatively small sample. These limitations inevitably reduced the diversity of visual patterns and professional perspectives represented in the results, which may have influenced the generalizability of the findings. Accordingly, the study should be regarded as an exploratory experiment rather than a fully generalizable application. Nevertheless, its primary contribution lies in demonstrating how generative AI can be structured to integrate place identity and biophilic design within historic-alley contexts, thereby providing a methodological foundation for larger-scale research. Future work should expand to multi-city datasets and involve broader, more diverse groups of experts to capture cross-context variability.

Second, the broader application of AI-based approaches to other heritage contexts presents additional challenges. Architectural heritage varies significantly across regions in spatial typology, materiality, symbolism, and cultural meaning. Consequently, a model fine-tuned on a specific urban or cultural dataset—such as Daegu's historic alleys—may not accurately represent sites shaped by different historical, climatic, or aesthetic conditions. Such contextual dependency could lead to representational bias in generative outputs, particularly in color, ornamentation, or symbolic interpretation. To ensure contextual validity, future research should incorporate region-specific datasets, collaborative validation with local experts, and comparative experiments across multiple heritage environments.

These efforts would strengthen both the adaptability and reliability of AI-driven heritage-design methodologies.

Third, this study primarily focused on visual fidelity and typological cues, while cognitive and emotional dimensions of place identity were interpreted mainly through visual representation. As experts noted, affective resonance and symbolic association remain domains that are difficult for AI to replicate. Future studies should employ multi-modal approaches—such as participant narratives, community-based feedback, and physiological indicators—to complement and capture these intangible layers.

Fourth, although the structured BDF framework was effective in embedding ecological attributes, the simulations were static and scenario-based. The absence of temporal or interactive variation constrained the analysis of how users might experience environmental changes over time. Future research should therefore integrate parametric variations—such as seasonal transitions, diurnal lighting differences, and gradual transformations of vegetation and water features—to enhance both robustness and design-decision support.

Finally, the potential risks and ethical issues associated with generative AI must remain a focus of continued attention. Dataset bias, intellectual property concerns over generated images, and the possibility of cultural misrepresentation highlight the need for interpretive safeguards. Establishing transparent protocols for data sourcing, prompt design, and stakeholder participation will be essential to the responsible integration of AI into heritage-sensitive design contexts.

Although the study was methodologically confined to image-based simulations rather than physical implementations, it nonetheless yields valuable insights into the digital transformation of urban design processes, particularly in the pre-assessment of design outcomes and the anticipation of community responses. The directions proposed herein can further advance both the methodological maturity of generative AI and its practical applicability to urban regeneration.

Author Contributions: Conceptualization, E.-J.L. and S.-J.P.; Methodology, E.-J.L. and S.-J.P.; Validation, E.-J.L.; Investigation, E.-J.L.; Data Curation, E.-J.L.; Writing—original draft preparation, E.-J.L.; Writing—review and editing, S.-J.P.; Visualization, E.-J.L.; Supervision, S.-J.P.; Project Administration, S.-J.P. All authors have read and agreed to the published version of the manuscript.

Funding: This research was supported by Basic Science Research Program through the National Research Foundation of Korea (NRF) funded by the Ministry of Education (Grant No. RS-2023-00275564) and the Korean Government (MSIT) (Grant No. RS-2021-NR058648).

Data Availability Statement: The data presented in this study are available on request from the corresponding author.

Conflicts of Interest: The authors declare no conflicts of interest.

Appendix A

Table A1. LPIPS and FID results for each alley type and effect size.

Alley Type	Condition	LPIPS	FID	Cohen's d (LPIPS)	Cohen's d (FID)
Path	Prompt-only	0.55 ± 0.15	154.17 ± 8.5	1.3	3.1
	Prompt+LoRA	0.36 ± 0.14	128.12 ± 7.4		
Stairs	Prompt-only	0.72 ± 0.16	214.72 ± 9.6	2.3	3.1
	Prompt+LoRA	0.36 ± 0.15	188.62 ± 8.9		

Table A1. *Cont.*

Alley Type	Condition	LPIPS	FID	Cohen's d (LPIPS)	Cohen's d (FID)
Edge	Prompt-only	0.65 ± 0.05	255.01 ± 10.2	3.4	12.6
	Prompt+LoRA	0.29 ± 0.14	114.02 ± 8.1		
Node	Prompt-only	0.76 ± 0.16	245.84 ± 9.8	2.2	5.9
	Prompt+LoRA	0.43 ± 0.14	186.89 ± 8.5		
Average	Prompt-only	0.67 ± 0.06	217.94 ± 9.5	2.3 (avg.)	6.2 (avg.)
	Prompt+LoRA	0.36 ± 0.04	154.91 ± 8.2		

Notes: Quantitative comparison of image similarity metrics (LPIPS and FID) between prompt-only and prompt+LoRA conditions for each alley type. Values are expressed as mean ± SD. Both metrics showed large effect sizes (Cohen's d > 1 in all cases), indicating that LoRA fine-tuning significantly improved perceptual and statistical similarity to ground-truth images.

References

1. Antrop, M. Why landscapes of the past are important for the future. *Landsc. Urban Plan.* **2005**, *70*, 21–34. [CrossRef]
2. Lowenthal, D. *The Past is a Foreign Country–Revisited*; Cambridge University Press: Cambridge, UK, 2015.
3. Yu, H. A Study on Urban Regeneration and Creating Place Assets in Daegu Modern Alleyway. Ph.D. Thesis, Kyungpook National University, Daegu, Republic of Korea, 2019.
4. Daegu Metropolitan City Jung-Gu: Historical and Cultural Street Tours. Available online: https://www.jung.daegu.kr/new/english/pages/tour/page.html?mc=5038 (accessed on 25 June 2025).
5. Ujang, N.; Zakariya, K. The notion of place, place meaning and identity in urban regeneration. *Procedia-Soc. Behav. Sci.* **2015**, *170*, 709–717. [CrossRef]
6. Ernawati, J. Dimensions underlying place identity for sustainable urban development. *MATTER Int. J. Sci. Technol.* **2018**, *3*, 271–285. [CrossRef]
7. Kellert, S.; Heerwagen, J.; Mador, P. *Biophilic Design: The Theory, Science, and Practice of Bringing Buildings to Life*; John Wiley & Sons: Hoboken, NJ, USA, 2008.
8. Kellert, S.R. *Nature by Design: The Practice of Biophilic Design*; Yale University Press: New Haven, CT, USA; London, UK, 2018.
9. Beatley, T. *Handbook of Biophilic City Planning & Design*; Island Press: Washington, DC, USA, 2017.
10. Li, Y.; Li, W.; Liu, Y. Remedies from nature: Exploring the moderating mechanisms of natural landscape features on emotions and perceived restoration in urban parks. *Front. Psychol.* **2025**, *15*, 1502240. [CrossRef]
11. Hartig, T.; Mitchell, R.; De Vries, S.; Frumkin, H. Nature and health. *Annu. Rev. Public Health* **2014**, *35*, 207–228. [CrossRef]
12. Relph, E. *Place and Placelessness*; Pion: London, UK, 1976; Volume 67.
13. Lewicka, M. Place attachment: How far have we come in the last 40 years? *J. Environ. Psychol.* **2011**, *31*, 207–230. [CrossRef]
14. Scannell, L.; Gifford, R. Defining place attachment: A tripartite organizing framework. *J. Environ. Psychol.* **2010**, *30*, 1–10. [CrossRef]
15. Browning, W.D.; Ryan, C.O.; Clancy, J.O. *14 Patterns of Biophilic Design: Improving Health & Well-Being in the Built Environment*; Terrapin Bright Green, LLC: New York, NY, USA, 2014.
16. Benedict, M.A.; McMahon, E.T. *Green Infrastructure: Linking Landscapes and Communities*; Island press: Washington, DC, USA, 2012.
17. Lee, E.J.; Park, S.J. A Structured Prompt Framework for AI-Generated Biophilic Architectural Spaces. *J. Build. Eng.* **2025**, *111*, 113326. [CrossRef]
18. Hu, E.J.; Shen, Y.; Wallis, P.; Allen-Zhu, Z.; Li, Y.; Wang, S.; Wang, L.; Chen, W. Lora: Low-rank adaptation of large language models. *ICLR* **2022**, *1*, 3. [CrossRef]
19. Campo-Ruiz, I. Artificial intelligence may affect diversity: Architecture and cultural context reflected through ChatGPT, Midjourney, and Google Maps. *Humanit. Soc. Sci. Commun.* **2025**, *12*, 24. [CrossRef]
20. Jang, K.M.; Chen, J.; Kang, Y.; Kim, J.; Lee, J.; Duarte, F.; Ratti, C. Place identity: A generative AI's perspective. *Humanit. Soc. Sci. Commun.* **2024**, *11*, 1156. [CrossRef]
21. Proshansky, H.M.; Fabian, A.K.; Kaminoff, R. Place-identity: Physical world socialization of the self. *J. Environ. Psychol.* **1983**, *3*, 57–83. [CrossRef]
22. Raagmaa, G. Regional identity in regional development and Planning1. *Eur. Plan. Stud.* **2002**, *10*, 55–76. [CrossRef]
23. Paasi, A. Region and place: Regional identity in question. *Prog. Hum. Geogr.* **2003**, *27*, 475–485. [CrossRef]
24. Groote, P.; Haartsen, T. *The Communication of Heritage: Creating Place Identities. The Routledge Research Companion to Heritage and Identity*; Routledge: London, UK, 2016; pp. 181–194.

25. Lynch, K. *The Image of the City*; MIT press: Cambridge, MA, USA, 1964.
26. Xue, F.; Gou, Z.; Lau, S.S.-Y.; Lau, S.-K.; Chung, K.-H.; Zhang, J. From biophilic design to biophilic urbanism: Stakeholders' perspectives. *J. Clean. Prod.* **2019**, *211*, 1444–1452. [CrossRef]
27. Rombach, R.; Blattmann, A.; Lorenz, D.; Esser, P.; Ommer, B. High-resolution image synthesis with latent diffusion models. In Proceedings of the IEEE/CVF Conference on Computer Vision and Pattern Recognition, New Orleans, LA, USA, 18–24 June 2022; pp. 10684–10695. [CrossRef]
28. Dai, M.; Feng, Y.; Wang, R.; Jung, J. Enhancing the digital inheritance and development of Chinese intangible cultural heritage paper-cutting through stable diffusion LoRA models. *Appl. Sci.* **2024**, *14*, 11032. [CrossRef]
29. Batifol, S.; Blattmann, A.; Boesel, F.; Consul, S.; Diagne, C.; Dockhorn, T.; English, J.; English, Z.; Esser, P.; Kulal, S. FLUX. 1 Kontext: Flow Matching for In-Context Image Generation and Editing in Latent Space. *arXiv* **2025**, arXiv:2506.15742.
30. Zhang, R.; Isola, P.; Efros, A.A.; Shechtman, E.; Wang, O. The unreasonable effectiveness of deep features as a perceptual metric. In Proceedings of the IEEE Conference on Computer Vision and Pattern Recognition, Salt Lake City, UT, USA, 18–22 June 2018; pp. 586–595.
31. Obukhov, A.; Krasnyanskiy, M. Quality assessment method for GAN based on modified metrics inception score and Fréchet inception distance. In Proceedings of the Computational Methods in Systems and Software, Konstanz, Germany, 23–25 September 2020; Springer: Berlin/Heidelberg, Germany, 2020; pp. 102–114.
32. Beatley, T. *Biophilic Cities: Integrating Nature into Urban Design and Planning*; Island Press: Washington, DC, USA, 2010.
33. Lee, E.-J.; Park, S.-J. Biophilic Experience-Based Residential Hybrid Framework. *Int. J. Environ. Res. Public Health* **2022**, *19*, 8512. [CrossRef]
34. Reeve, A.; Desha, C.; Hargroves, K.; Newman, P. Informing healthy building design with biophilic urbanism design principles: A review and synthesis of current knowledge and research. In Proceedings of the 10th International Conference on Healthy Buildings, Brisbane, Australia, 8–12 July 2012.
35. Lee, E.J.; Park, S.J. Toward the Biophilic Residential Regeneration for the Green New Deal. *Int. J. Environ. Res. Public Health* **2021**, *18*, 2523. [CrossRef]
36. Newman, P.; Beatley, T.; Boyer, H. *Resilient Cities: Responding to Peak Oil and Climate Change*; Island Press: Washington, DC, USA, 2009.
37. Ryan, C.O.; Browning, W.D.; Clancy, J.O.; Andrews, S.L.; Kallianpurkar, N.B. Biophilic design patterns: Emerging nature-based parameters for health and well-being in the built environment. *ArchNet-IJAR Int. J. Archit. Res.* **2014**, *8*, 62. [CrossRef]
38. Thomson, G.; Newman, P. Green infrastructure and biophilic urbanism as tools for integrating resource efficient and ecological cities. *Urban Plan.* **2021**, *6*, 75–88. [CrossRef]
39. Africa, J.; Heerwagen, J.; Loftness, V.; Ryan Balagtas, C. Biophilic design and climate change: Performance parameters for health. *Front. Built Environ.* **2019**, *5*, 28. [CrossRef]
40. Kaplan, R.; Kaplan, S. *The Experience of Nature: A Psychological Perspective*; Cambridge University Press: Cambridge, NY, USA, 1989.
41. Cronbach, L.J. Coefficient alpha and the internal structure of tests. *Psychometrika* **1951**, *16*, 297–334. [CrossRef]
42. Koo, T.K.; Li, M.Y. A guideline of selecting and reporting intraclass correlation coefficients for reliability research. *J. Chiropr. Med.* **2016**, *15*, 155–163. [CrossRef]
43. Tennessen, C.M.; Cimprich, B. Views to nature: Effects on attention. *J. Environ. Psychol.* **1995**, *15*, 77–85. [CrossRef]
44. Jiang, B.; Chang, C.-Y.; Sullivan, W.C. A dose of nature: Tree cover, stress reduction, and gender differences. *Landsc. Urban Plan.* **2014**, *132*, 26–36. [CrossRef]
45. Browning, M.H.; Saeidi-Rizi, F.; McAnirlin, O.; Yoon, H.; Pei, Y. The role of methodological choices in the effects of experimental exposure to simulated natural landscapes on human health and cognitive performance: A systematic review. *Environ. Behav.* **2021**, *53*, 687–731. [CrossRef]
46. White, M.; Smith, A.; Humphryes, K.; Pahl, S.; Snelling, D.; Depledge, M. Blue space: The importance of water for preference, affect, and restorativeness ratings of natural and built scenes. *J. Environ. Psychol.* **2010**, *30*, 482–493. [CrossRef]
47. Joye, Y.; Van den Berg, A. Is love for green in our genes? A critical analysis of evolutionary assumptions in restorative environments research. *Urban For. Urban Green.* **2011**, *10*, 261–268. [CrossRef]

land

MDPI

Article

The Dehesa as Landscape Heritage from the Perspective of the New Generation

Rebeca Guillén-Peñafiel [1], Ana-María Hernández-Carretero [1] and José-Manuel Sánchez-Martín [2,*]

1 Department of Social Sciences, Language and Literature, Faculty of Education, University of Extremadura, University Campus, Avenida de las Letras s/n, 10003 Cáceres, Spain; rebecagp@unex.es (R.G.-P.); ahernand@unex.es (A.-M.H.-C.)
2 Department of Art and Territorial Sciences, University of Extremadura, University Campus, Avenida de las Letras s/n, 10003 Cáceres, Spain
* Correspondence: jmsanche@unex.es

Abstract: The dehesa, as a socio-ecological system and cultural landscape, is a strategic resource for environmental education, territorial sustainability, and the intergenerational transmission of knowledge. This study analyzes the perception of primary school students in Extremadura regarding this environment, using a mixed methodology that combines statistical, semantic, and spatial analysis. The results show a generally positive assessment of the dehesa heritage, although accompanied by a disconnect between this symbolic assessment and direct experience of the territory, especially in urban contexts. It identifies significant differences between students from rural and urban environments in terms of their knowledge of trades, products, and dehesa spaces, as well as their preferred activities in the dehesa. While rural students show greater interest in operational activities and direct contact with the environment (such as feeding livestock and milking), urban students lean toward sensory or symbolic experiences (such as consuming products or occasional harvesting), reflecting different ways of connecting with the territory. Spatial analysis reveals that more than 80% of schools are located less than 5 km from well-preserved dehesa areas, which represents an opportunity to integrate these landscapes into formal education. However, inequalities in access from special education centers have been detected, posing challenges in terms of territorial and educational equity. This study concludes that the dehesa should be recognized as an open classroom, capable of fostering roots, ecological literacy, and cultural sustainability through contextualized and territory-sensitive pedagogical approaches.

Keywords: dehesa; environmental education; heritage education; cultural landscape; rural sustainability

1. Introduction

The *dehesa*—or *montado* in Portugal—is defined as a Mediterranean agro-silvo-pastoral system resulting from a long process of anthropization of the sclerophyllous forest, characterized by the scattered presence of species of the genus Quercus over grasslands and rotational crops [1]. In the Iberian context, this model is also recognized as a cultural landscape and as a complex socio-ecological system, as it integrates productive, ecological, and symbolic-identity functions [2]. Its contemporary relevance lies in the provision of ecosystem services, the conservation of Mediterranean biodiversity, and its contribution to territorial resilience in the face of climate change, while sustaining rural economies and identity values across approximately 3.5 million hectares between Spain and Portugal [3].

Moreover, the dehesa aligns with the concept of cultural landscape proposed by the European Landscape Convention (2000), as a symbolic and relational construction resulting from the prolonged interaction between human communities and their environment [4].

Its ecological dimension is crucial to understanding its uniqueness. With its mosaic structure—grasslands, cleared woodlands, dryland crops, and extensive livestock farming—the dehesa emerges as a resilient system adapted to poor soils and the seasonality and irregularity of Mediterranean rainfall. It is capable of co-producing agricultural, forestry, and livestock goods under extensive and sustainable management practices that maintain high levels of biodiversity, conserve soils, regulate the hydrological cycle, and contribute to carbon sequestration; indeed, it covers around four million hectares in the Iberian Peninsula [5,6]. This functioning has been possible thanks to a historical-environmental evolution in which traditional ecological knowledge has harmonized resource exploitation with the conservation of fundamental processes such as tree regeneration, soil fertility, and water balance [7–9]. Within this convergence of knowledge and practices lies its condition as a high-value biocultural landscape, where rural lifestyles, traditional crafts, and local knowledge configure an intangible heritage of notable ethnobiological richness [10–12].

However, the dehesa is currently subject to pressures that threaten its sustainability. The intensification of agricultural uses, the progressive abandonment of rural areas, landscape fragmentation, and the loss of profitability of traditional farms combine with the cumulative effects of climate change, weakening both the ecological functionality and the social structures that historically favored its resilience [13–16]. The globalization of agri-food markets has reinforced trends toward intensive production models disconnected from territorial logics, with impacts on cultural and economic homogenization in rural areas [17]. These transformations call for rethinking its management through approaches that recognize the system's multifunctionality, its associated cultural capital, and the opportunities arising from its comprehensive valorization.

At this point, the legal-political framework provides elements of recognition and, at the same time, limitations that must be considered. In Extremadura, Law 1/1986, of 2 May, on the Dehesa, establishes its uniqueness and promotes its conservation, regeneration, and valorization as natural and cultural heritage, although it emphasizes agronomic criteria (minimum area and extensive livestock use) that do not always capture the socio-ecological complexity of the dehesa [18]. In contrast, the Green Book of the Dehesa proposes an integrative vision that conceives it as a rational and sustainable resource exploitation system based on the coexistence of multiple uses and the predominance of Quercus species producing acorns, introducing variables such as tree density and the ratio between wooded and unwooded areas to understand its spatial and functional diversity [19]. This tension between normative definitions and more holistic ecosystem conceptualizations underscores the need for place-sensitive policies that integrate production, conservation, and culture [20,21].

The intangible heritage associated with the dehesa—traditional ecological knowledge, crafts, and cultural practices—has been explicitly recognized by the UNESCO Convention (2003), which highlights intergenerational transmission as a condition for territorial sustainability [22]. Research shows that such transmission is not only an essential component of cultural heritage but also a key vector for sustainable resource management and socio-ecological resilience [23–26]. When transmission is interrupted, communities' capacity to manage territory and conserve biodiversity is weakened [24,27]; conversely, its strengthening through support policies, education, and community participation increases adaptation to challenges such as climate change [23]. Incorporating this knowledge transmission into environmental governance frameworks and conservation strategies ensures greater equity, cultural relevance, and effectiveness of actions [26], while contributing to transforming cultural capital into territorial capital: a set of identities, practices, and knowledge that reinforce social cohesion, collective identity, and crisis response capacity [28–31].

Spatial inequalities typical of rural and peripheral territories, where the dehesa extends, justify the need to place spatial justice at the center of educational and territorial agendas. Distance, settlement dispersion, and lack of infrastructure limit access to educational and cultural resources. These deficits impact social cohesion and sustainable development, requiring public policies that address territorial specificities, improve accessibility, and promote equity [32–34]. In the educational sphere, equity is not limited to equal access: it involves recognizing and integrating local knowledge and practices into curricula, a condition that strengthens the cultural relevance of teaching in rural territories [35,36]. Hence, educational policies sensitive to place are recommended, with community participation in decision-making to ensure relevance and sustainability [37,38]. In regions such as Extremadura, the dehesa can and should operate as a proximity resource that compensates for the lower density of educational infrastructure and brings learning opportunities closer to students.

Education for Sustainable Development (ESD) precisely promotes this logic of contextualization. UNESCO (2014) advocates integrating local contexts into educational processes, understanding that sustainability is learned through direct contact with territories and their problems [39]. Evidence indicates that incorporating local and indigenous knowledge into sustainability education fosters more meaningful learning, strengthens identity and cultural relevance, and improves academic outcomes by connecting content with students' realities [40–42]. Furthermore, territorial contextualization enhances community participation, real problem-solving, and the development of key competencies for sustainability, such as critical thinking, collaboration, and transformative action [43,44]. All this gives the dehesa a central role as an open classroom where environmental and heritage education can be articulated in a situated and effective manner.

Nevertheless, recent research reveals a growing disconnection between younger generations and the rural environment due to increasing urbanization processes, so much so that the perception of the landscape by children living in urban areas differs from that of children living in rural areas [45]. Furthermore, Özdemir (2018) determined that learning verbal information through experience with visual data increased students' levels of perception and awareness [46]. Alongside this, it is worth noting that children are aware of environmental problems evident in the places where they live and of possible future issues. They show clear environmental awareness [47]. In Primary Education, limited knowledge of the ecosystem is observed, with greater deficiencies in cultural and historical dimensions [48–50]. This situation is aggravated by the scarce presence of the dehesa in curricula and the lack of didactic proposals that integrate cultural and natural components in a transversal and meaningful way [51]. Although heritage education appears as a curricular axis, an excessively material and monumental orientation persists, relegating cultural landscapes and intangible knowledge linked to everyday life [52]. The result is a missed opportunity to develop critical, inclusive, and sustainability-oriented heritage education, as required by international ESD frameworks [39].

Incorporating the dehesa into Primary Education processes should not be interpreted as a mere expansion of content but as an opportunity to rethink methodologies, approaches, and objectives from territorial, ecological, and cultural perspectives. Specialized literature emphasizes that situated learning, based on direct experience and student participation, favors the development of competencies such as critical thinking, empathy, collaboration, and transformative action capacity [53]. In practice, this translates into proposals such as sensitive cartographies, school biodiversity laboratories, interviews with elders, and field trips, which promote interdisciplinary learning, a critical view of the environment, the identification of local problems, and the formulation of creative and contextualized solutions [54–56]. This experiential pedagogy—learning in, from, and with the territory—activates cognitive, emotional, and social processes that are difficult to reproduce in conventional classrooms and promotes inclusion, equity, and engaged citizenship [57–61].

Collaboration between schools, universities, and local actors—livestock farmers, artisans, cultural associations, environmental technicians—reinforces this orientation through participatory governance schemes that enrich educational processes, generate territorial synergies, and strengthen the social fabric. Immersive experiences, service-learning, and inter-institutional projects have proven effective in promoting meaningful learning and real commitments to sustainability, while favoring intergenerational knowledge transmission and the revaluation of crafts and practices [62–64]. In parallel, responsible tourism modalities—agrotourism, ecotourism, and educational tourism—help diversify the local economy, generate employment, strengthen the social fabric, and revalue heritage by facilitating direct interaction with the landscape and participation in traditional activities under sustainable management criteria [65,66]. In this framework, education and tourism act as complementary levers for sustainable rural development, provided they are designed with respect for ecosystem rhythms and capacities and with informed community participation [67–69].

On these bases, theoretical approaches to cultural landscape and cultural capital provide fertile analytical frameworks for understanding the dehesa and guiding educational interventions. The cultural landscape approach allows conceiving the dehesa as a symbolic and relational construction, an interweaving of materialities, practices, and meanings that are historically co-produced [70,71]. Bourdieu's (1997) theory of cultural capital helps interpret the knowledge, skills, and dispositions linked to dehesa crafts as symbolic capital that can be mobilized to generate social recognition, collective self-esteem, and economic opportunities [72]. Integrating these frameworks into curriculum design and educational actions means recognizing and legitimizing local knowledge as a first-order pedagogical resource, in line with ESD recommendations and the need to strengthen community resilience [73,74].

Consistent with the above, studying the dehesa in Primary Education is justified for ecological, cultural, pedagogical, and territorial reasons. Ecologically, it offers the possibility of understanding the complexity of socio-ecological systems and the interdependence between society and nature, as well as experimenting with observation, analysis, and environmental care practices that translate into learning about biodiversity, soils, and the water cycle [5–9]. Culturally, it activates processes of symbolic appropriation of the territory and strengthening of the sense of belonging, essential for building resilient local identities, and enables the transmission and updating of traditional knowledge and crafts (transhumance, acorn collection, artisanal cheese-making, extensive livestock management), with impacts on collective self-esteem and continuity of intangible heritage [6,22,75,76]. Pedagogically, it articulates a critical and situated heritage education that aligns curricular content with students' experience, improves motivation, and facilitates the development of sustainability competencies [44]. Finally, territorially, it contributes to spatial justice by bringing high-quality educational resources closer to rural contexts, mitigating inequalities in access and relevance [32,34].

It should be acknowledged that rural youth today face a double process of disengagement: on the one hand, limited job opportunities and social mobility in their home territories; on the other, the symbolic devaluation of the rural environment, often perceived as a space without a future [77]. This disconnection translates into a loss of cultural capital linked to the dehesa and, consequently, a reduction in communities' adaptive capacity. The response requires integrated, interdisciplinary, and territorially adapted approaches that recognize the system's multifunctionality and its value as a strategic resource for sustainable rural development [78–80]. In this horizon, the intersection between education, heritage, and territorial planning—supported by participatory governance, social innovation, and multi-actor alliances—emerges as a promising path to ensure the continuity of the dehesa as a resilient socio-ecological system and shared cultural legacy [81,82].

From this applied orientation, the present study focuses on childhood as a key agent for rearticulating links between territory and community. Direct contact of students with the dehesa and its traditional practices—visits to livestock farms, dairies, artisanal workshops—favors the revaluation of TEK, the strengthening of local identity, and the intergenerational transmission of knowledge, while promoting a critical, empathetic, and committed citizenship in the face of sustainability challenges [6,75,83]. Integrating these learnings into coherent curricular proposals—from classroom projects to didactic itineraries—implies recognizing that the dehesa is not only a productive ecosystem but a complex socio-ecological construction in which ecological, economic, cultural, and symbolic dimensions converge [10–12,70–72].

The continuity of this discourse therefore demands a dual movement: on the one hand, ensuring public policies and regulatory frameworks that recognize and support the diversity of dehesa configurations and management practices, incorporating ecosystemic and cultural criteria alongside agronomic ones [20,21]; on the other, deploying situated educational practices that integrate local heritage into the curriculum, reduce territorial gaps, and promote school-community-university alliances, as well as educational tourism and agrotourism modalities compatible with carrying capacity and heritage safeguarding [32–38]. Only in this way can the dehesa continue to offer a privileged environment for integral learning, value acquisition, and the construction of life projects linked to the territory, in line with sustainability and territorial cohesion agendas [73,74].

To contribute to this agenda, this study presented here is structured around a series of objectives and empirical contrasts that take as their starting point the perceptions and knowledge of Primary Education students about the dehesa. Specifically, the following objectives are proposed: (i) to explore the degree of knowledge and perception that students have about the dehesa; (ii) to identify the heritage elements that are most attractive to them; (iii) to analyze differences in heritage valuation according to variables such as place of residence or frequency of visits to the natural environment; (iv) to determine the activities preferred by students to carry out in the dehesa; and (v) to locate educational centers close to well-preserved dehesa areas, in order to promote their use as educational spaces. These objectives derive from the premise that the dehesa, as a cultural landscape, constitutes a strategic educational resource to align heritage and environmental education with territorial planning in a sustainability key [8,10,11,42,44].

Based on this justification and the reviewed literature, the following starting hypotheses are formulated: (H1) Primary Education students have limited knowledge about the dehesa, especially regarding its cultural components, such as traditional crafts, tree species, and derived products; (H2) their perception tends to focus on natural values and biodiversity, while cultural aspects go relatively unnoticed; (H3) there is a greater appreciation of natural, landscape, and gastronomic resources compared to elements of intangible heritage; (H4) urban or rural origin does not significantly influence heritage valuation, although it does condition the frequency of visits and contact with the environment; and (H5) place of residence modulates the activities that are most attractive to students, which can guide the design of more contextualized educational proposals. These hypotheses guide the empirical analysis and, in turn, aim to provide evidence for the design of policies and educational practices that strengthen the link between childhood, territory, and heritage, an essential condition to guarantee the continuity of the dehesa as a resilient socio-ecological system and as a shared cultural legacy [72–74].

2. Materials and Methods

2.1. The Study Area

The region selected for this research is Extremadura, an autonomous community located in southwestern Spain, comprising the provinces of Cáceres and Badajoz. With an area of 41,634 km^2, it is one of the largest regions in the country, although also one of the least populated. According to official data from the National Institute of Statistics, in January 2025 the total population was 1,052,998, of which 108,250 were under 12 years of age, representing approximately 10.3% of the total population [84]. These students are distributed among 572 nursery and primary schools, in addition to 18 special education centers [85].

In the field of education, during the 2024–2025 school year, Extremadura had a total of 1047 non-university educational centers, distributed among early childhood education, primary education, secondary education, high school, vocational training, and special education 1. In terms of school enrollment, there were 22,940 students enrolled in the first and second cycles of early childhood education, in addition to 55,888 students in primary education and 779 in special education. In addition, 47,312 students were enrolled in compulsory secondary education (ESO), 15,204 were enrolled in upper secondary education, and 17,890 were enrolled in vocational training courses, both intermediate and advanced [85].

Furthermore, more than 80% of Extremadura's students attend public schools, placing the region among the autonomous communities with the highest proportion of public education in Spain. This characteristic reflects the strong influence of the public education system in the region, especially in rural and sparsely populated areas, where the network of subsidized and private schools is more limited.

In addition, the area of dehesa landscapes in this autonomous community covers more than 1,450,000 hectares. This makes it one of the areas with the highest representation of this agrosystem [75] in the Iberian context and makes it an ideal setting for implementing specific policies to enhance its value (Figure 1). In addition, it is necessary to highlight their good ecological condition, possibly due to the low demographic pressure they experience. To understand this, it is necessary to consider that the region has a population density of barely 25 inhabitants per km^2. This figure is even lower in the dehesa areas, where the dispersion of the population and rurality are particularly pronounced.

Beyond its productive function, the dehesa is also a space steeped in cultural and heritage values. A significant percentage of Extremadura's dehesas are located within protected areas, such as the Natura 2000 Network, natural parks, and biosphere reserves, which reinforces their role in biodiversity conservation and ecosystem services [86]. They also preserve material elements such as livestock trails, farmhouses, huts, dry stone walls, fountains, and hermitages, many of which are protected or listed as immovable cultural heritage. Unfortunately, however, many others are abandoned and almost in ruins. In addition, there are monumental sites of great historical value, such as Trujillo, and natural areas of international importance, such as Monfragüe and Villuercas-Ibores-Jara. Likewise, the associated intangible heritage—traditional crafts, transhumance, knowledge linked to land management—constitutes a living legacy, although one that is increasingly fragile due to the lack of generational renewal [77,83]. In this sense, rural tourism, hunting tourism [87], and agrotourism [82], together with initiatives to enhance heritage, are emerging as ways to diversify the economy and strengthen territorial roots. Although their demographic impact is still limited, these activities contribute to revaluing the rural environment and strengthening the link between the population and the territory [88,89].

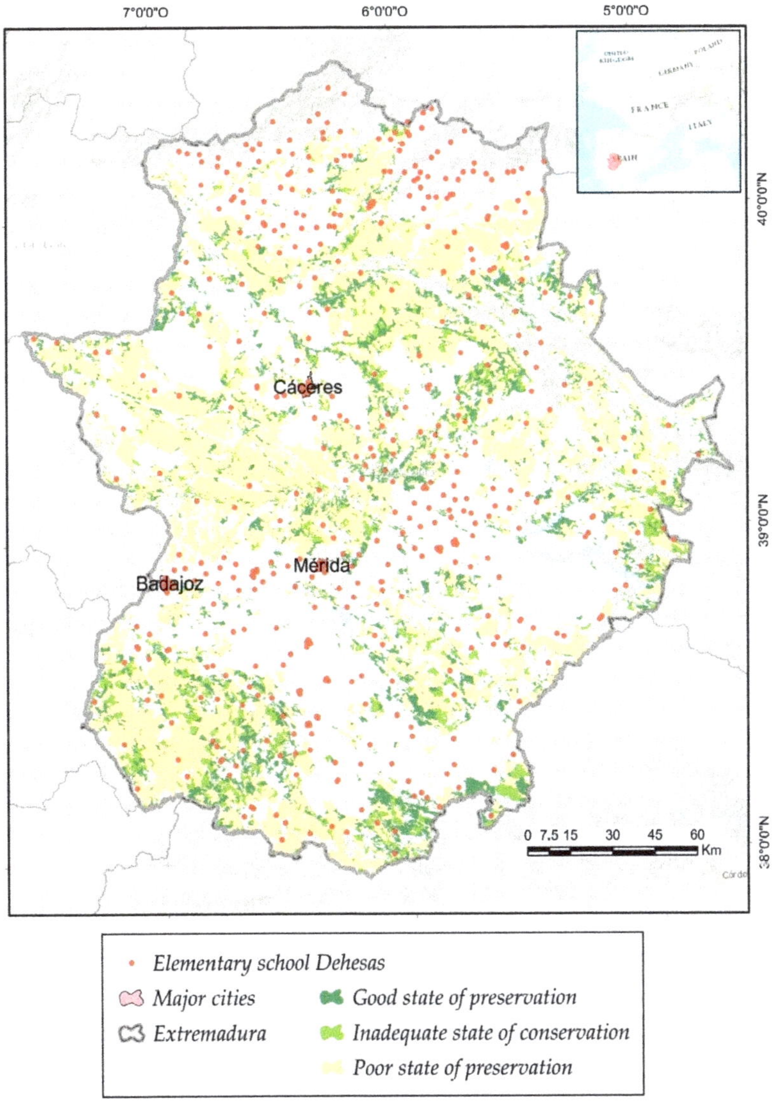

Figure 1. Study area.

Therefore, the integration of heritage education and the promotion of sustainable tourism experiences are key strategies for ensuring the active conservation of the dehesa and addressing the demographic challenges that threaten its continuity as a cultural landscape and way of life.

2.2. Data and Processing

The data were obtained through a survey conducted among third-cycle primary education students during May and June 2025, as detailed in the survey technical data sheet (Table 1).

Table 1. Survey technical data sheet.

Study title:	Image of the dehesa among third-cycle primary education students
General objective:	To ascertain the image of the dehesa among third-cycle primary school, as well as their preferences in relation to different activities that take place in this landscape students.
Geographical scope:	Area where this study was conducted: Extremadura
Study population:	55,888 (primary school students)
Sample size:	233 people
Type of sampling:	Intentional and convenience
Margin of error and confidence level:	±3.5% for a 95% confidence level
Data collection technique:	Face-to-face survey
Method of application:	CAPI (Computer-Assisted Personal Interviewing)
Average interview duration:	15–20 min
Fieldwork period:	May and June 2025
Ethical Considerations: The survey, conducted within the framework of the research project "Dehesa Heritage: A Comparison Between Primary School Students in Rural and Urban Areas", received approval from the Ethics Committee of the University of Extremadura on 12 May 2025. Given that the participants were minors, explicit authorization was also obtained from the principals of the participating schools, as well as informed consent from the students' parents or legal guardians.	

The survey consists of 11 items, 10 closed-ended and 1 open-ended, designed to capture different dimensions of students' relationship with the dehesa as a cultural landscape (Table 2).

Table 2. Survey questions.

Questions	Hypotheses	Objectives
Q1. Do you usually go to the countryside? Q2. Do you usually have contact with farm animals other than pets: pigs, cows, sheep…? Q3. Do you like being in nature? Q4. Who do you usually go to the countryside with? Q5. Do you live in a village or a city?	H4: Urban or rural origin does not significantly influence heritage valuation, although it does condition the frequency of visits and contact with the environment.	(iii) Analyze differences in heritage valuation according to variables such as place of residence or frequency of visits to the natural environment.
Q6. What traditional trades do you know? Q7. What products do the dehesa provide? Q8. What vegetation is typical of the dehesa?	H1: Limited knowledge about the dehesa, especially regarding its cultural components (traditional crafts, tree species, derived products).	(i) Explore the degree of knowledge and perception that students have about the dehesa.
Q9. How much do you like it? a. Dehesa with oak trees and a pond b. Shearer c. Pigs in the montanera d. Shepherd e. Traditional shepherd's huts f. Beekeeper g. Production of traditional products (cheeses, sausages, etc.) h. Cork extractor i. Pigsties/pens j. Charcoal burner	H2: Perception tends to focus on natural values and biodiversity; cultural aspects go relatively unnoticed. H3: Greater appreciation of natural, landscape, and gastronomic resources compared to elements of intangible heritage.	(ii) Identify the heritage elements that are most attractive to students.
Q10. Write down the first three words that you associate with the dehesa.	Related to H2 and H3: Predominance of natural terms over cultural ones.	(i) Explore perceptions about the dehesa.
Q11. Which of the following activities would you prefer?	H5: Place of residence modulates the activities that are most attractive to students.	(iv) Determine the activities preferred by students in the dehesa.

The first group of questions (1–5) explores the degree of interaction with rural environments and the place of residence, whether urban or rural. These variables provide indicators of exposure to nature, frequency of visits to the countryside, and social context, such as whether visits occur alone or accompanied by family or peers. Understanding these aspects is essential for interpreting how direct contact with natural settings influences heritage perception and environmental attitudes. The second group (Questions 6–8) focuses on cultural knowledge of the dehesa, addressing recognition of traditional trades, characteristic products, and typical vegetation. These items are critical for assessing awareness of tangible heritage and for determining whether cultural dimensions—often less visible than ecological ones—are integrated into students' conceptual frameworks. Identifying gaps in this knowledge is fundamental for designing educational interventions that reinforce cultural heritage awareness.

Question 9 introduces a perceptual dimension through AI-generated images depicting natural and cultural elements of the dehesa, allowing the evaluation of attractiveness and preference for specific resources. Question 11 complements this by identifying activities students would most like to engage in within the dehesa, providing insights for the development of contextualized educational proposals aligned with students' interests and territorial realities.

The open-ended question (Question 10) captures spontaneous associations with the dehesa, offering qualitative data on mental representations and symbolic interpretations. This item is particularly relevant for detecting implicit values and emotional connections that structured questions may not reveal. Additionally, this study includes the identification of educational centers located near well-preserved dehesa areas, supporting the integration of cultural landscapes into formal education and reinforcing the link between heritage, territory, and sustainability. The complete questionnaire is available in the Supplementary Material.

The responses were used to create a database that underwent two types of processing. On the one hand, statistical analysis was used, and on the other, semantic analysis was applied.

The database generated has been subjected to a basic statistical analysis, but one that is highly relevant to achieving the objectives proposed in this research. It focuses on the joint analysis of frequency distribution tables, contingency tables, the chi-square test, and Cramer's V coefficient. Despite its simplicity, it constitutes a robust methodological strategy for exploring and quantifying relationships between categorical variables, widely used in disciplines such as epidemiology, psychology, sociology, and behavioral sciences. This approach allows not only the identification of patterns of association but also the evaluation of their statistical significance and intensity, providing a solid empirical basis for the interpretation of complex phenomena.

Frequency distribution tables are the starting point for organizing and visualizing categorical data, facilitating the identification of trends, asymmetries, and possible biases in the distribution of categories [90]. When the analytical objective is oriented toward exploring relationships between two qualitative variables, contingency tables are used, which allow us to observe the joint frequency of occurrence of the categories and detect possible structural dependencies between them [91,92].

The chi-square test of independence is applied to these tables, which contrasts the observed frequencies with those expected under the null hypothesis of independence between variables. The resulting statistics allow us to determine whether the observed association is statistically significant, that is, whether it is unlikely to have occurred by chance [93]. However, this contrast does not provide information on the magnitude of the effect, which limits its interpretive capacity in practical terms. It is therefore essential to consider that the validity of the chi-square test depends on certain assumptions, such as the sample size and the magnitude of the expected frequencies in each cell [94]. To overcome this limitation, Cramer's V coefficient is incorporated, a normalized measure of association that quantifies the strength of the relationship between categorical variables. This coefficient is especially useful in non-square contingency tables or tables larger than 2 × 2, and its value ranges from 0 (no association) to 1 (perfect association), which facilitates its comparative interpretation [95].

Taken together, this integrated methodological approach not only allows us to detect the existence of associations between categorical variables, but also to estimate their empirical relevance and understand the underlying structure of the data. Its rigorous application is essential to ensure the internal validity of studies and to support solid inferences in quantitative research of an explanatory or exploratory nature.

The present analysis is complemented by the application of computer-assisted qualitative data analysis (CAQDAS) techniques, which allow for the systematic, rigorous, and traceable management of unstructured data—such as texts, interview transcripts, images, audio and video files—in the context of scientific research. This methodological approach is based on the segmentation of the corpus into units of meaning, the analytical coding of relevant fragments, and the construction of conceptual networks that enable the identification of patterns, semantic relationships, and emerging categories. This study used Atlas.ti v.25 software, a specialized tool that offers advanced features for code management, analytical memo creation, co-occurrence visualization, and conceptual map generation. These capabilities not only optimize the organization of qualitative material but also reinforce the transparency of the analytical process by allowing for the traceability of interpretive decisions and the replicability of the procedure [96].

Specifically, semantic analyses based on word frequencies were carried out to identify the terms most frequently repeated by students. In this process, empty terms that are too general or do not provide any excluded information (much, quite, no, be, aspect, etc.). Concepts that refer to the same reality (tree and grove; flora and have also been unified vegetation; livestock farmer and livestock farming; tranquility and tranquil, etc.), choosing the word that is repeated the most times. After this selection process, a total of 610 terms were recognized. Finally, this analysis has made it possible to synthesize and project the students' perceptions in an illustrative way using word clouds. To do this were, terms repeated more than 10 times considered.

About spatial analysis, the geographical location of the educational centers, as well as information on the distribution and conservation status of the dehesas in the autonomous community of Extremadura, was obtained from the Extremadura Territorial Information System (SITEX). This platform, managed by the Regional Government of Extremadura, provides geospatial data under a Creative Commons Attribution 4.0 International (CC-BY 4.0) license, which guarantees free access, reuse, and scientific traceability [97].

Based on these layers of georeferenced information, the minimum distance between each educational center and the dehesas classified as best preserved has been calculated. To do this, ArcGIS version 10.8 software was used, employing spatial analysis tools that allow proximity relationships between vector entities to be determined. This procedure has made it possible to generate a key quantitative variable for this study, which serves as an indicator of the potential accessibility of schools to the dehesa do environment in optimal conservation conditions.

2.3. Workflow

This research is structured around six distinct phases, ranging from a review of the literature to a presentation of the main conclusions of this study (Figure 2).

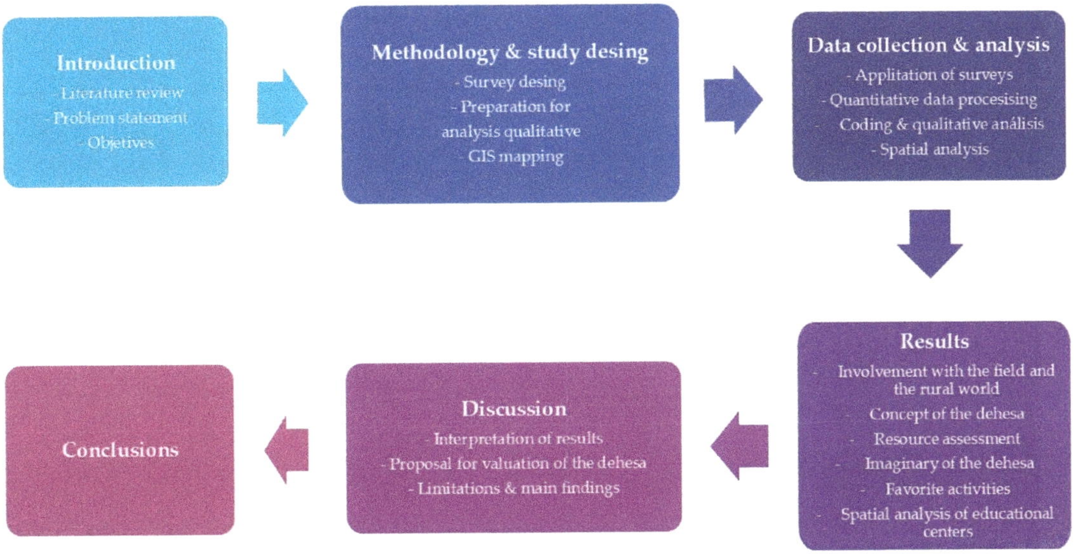

Figure 2. Workflow.

The first phase corresponds to the introduction, conceived as the space for theoretical foundation and delimitation of the object of study. At this stage, a systematic and critical review of the specialized literature is carried out, aimed at identifying gaps in knowledge, conceptual tensions, and relevant methodological approaches. Based on this analysis, the research problem is formulated with analytical precision, and the general and specific objectives are defined, which serve as the backbone of the methodological design and empirical scope of this study.

The second phase focuses on the description of the methodology and design of this study, where the epistemological position is made explicit and the most appropriate methodological strategies (quantitative, qualitative, or mixed) are selected. This section details the data collection instruments, sampling criteria, units of analysis, and validation procedures. Specific techniques are incorporated, such as the design of structured surveys, the development of protocols for qualitative analysis, and the application of geographic information systems (GIS), which allow for an accurate spatial representation of the phenomena observed.

The third phase, corresponding to data collection and analysis, constitutes the empirical core of the research process. It includes the implementation of field instruments (surveys and participant observation), the statistical processing of quantitative data, and the thematic coding of qualitative information.

The results section presents the findings in a systematic, objective manner, without interpretation. This section will include thematic maps, comparative analyses between population subgroups, and descriptions of territorial patterns.

The discussion constitutes the space for critical interpretation, where the results are contextualized in relation to the theoretical framework, contrasted with previous studies, and theoretical, methodological, and practical implications are derived. Likewise, the limitations of this study are identified, and its original contributions are highlighted. In applied research, this section may include proposals for intervention or territorial management strategies.

Finally, the conclusion summarizes the main findings, evaluates the degree to which the objectives have been met, and proposes future lines of research. This phase closes the research cycle, highlighting this study's contribution to the advancement of scientific knowledge and its applicability in real and complex contexts.

3. Results

3.1. Involvement with the Field and the Rural World

The results obtained through the questionnaire reveal a complex and nuanced reality: the relationship between students and the rural and natural environment is neither homogeneous nor linear, but is constructed from multiple interacting factors—geographical, cultural, emotional, and educational. From both a quantitative and qualitative perspective, the data reveal patterns of behavior and perception that, while showing a frequent presence in natural environments, also hint at a certain disconnect with the dynamics of the rural world (Table 3).

Table 3. Main characteristics of respondents and their relationship with the countryside.

Question	Frequency		Percentage	
	No	Yes	No	Yes
1. Do you usually go to the countryside?	89	144	38.2	61.8
2. Do you usually have contact with farm animals other than pets: pigs, cows, sheep...?	134	99	57.5	42.5
3. Do you like being in nature?				
Very much	107		45.9	
Quite	77		33.0	
Somewhat	40		17.25	
Very little	7		3.0	
None	2		0.9	
4. Who do you usually go to the countryside with?				
With friends	25		10.7	
With school or on field trips	4		1.7	
With my family	187		80.3	
I don't usually go to the countryside	17		7.3%	
5. Do you live in a village or a city?				
Village	134		57.5%	
City	99		42.5	

A first reading reveals that 61.8% of students regularly visit the countryside. This figure suggests good accessibility to the natural environment and a certain continuity in contact with it, due to geographical proximity and established family habits. However, when we look more closely at the quality of this experience, the picture becomes more nuanced: only 42.5% say they have contact with farm animals such as cows, pigs, or sheep. This data points to a more superficial experience of the countryside, focused on walking or enjoying the scenery, but far removed from the productive and cultural processes that have historically defined life in rural areas.

In emotional terms, the relationship with nature is stronger. Almost 79% of students say they are "very much" or "quite" like being in nature, which constitutes an important emotional basis on which to build meaningful educational proposals. This affinity, however, is not universal: a small but significant 3.9% express little or no interest in the natural environment. This indifference could be related to a lack of previous experiences, the urbanization of lifestyles, or the absence of cultural references that value the rural environment as a space for learning and belonging.

The context in which visits to the countryside take place also offers valuable insights. Most students (80.3%) visit the countryside accompanied by their families, while only 10.7% do so with friends and a mere 1.7% as part of school activities. This data reinforces the role of the family environment as the main mediator in the relationship with nature, but it also highlights a structural deficiency: the lack of trips to the rural environment in formal educational programming. Schools, as socializing institutions, have not yet systematically integrated the territory as a pedagogical resource, which represents a strategic opportunity for the development of active methodologies such as experiential learning, fieldwork, and heritage education, as well as for educational tourism experiences.

Finally, the place of residence of the students—57.5% in villages and 42.5% in cities—introduces a contextual variable that helps to understand part of the results. Although one might expect that living in a rural environment would foster a closer relationship with nature, the data show that geographical rurality does not guarantee experiential or symbolic rurality. In other words, living in a village does not necessarily imply being connected to its practices, values, or landscapes.

Overall, the results paint a picture of a frequent but functionally weak relationship with the rural environment. Students are physically present in the countryside, but they do not always experience it in an emotional, transformative, or critical way. This finding invites us to rethink the role of territory in teaching and learning processes. It is not just a matter of taking students to the countryside, but of helping them to see, feel, understand, and appreciate what happens there. To this end, it is necessary to design educational strategies that integrate ecological knowledge, rural issues, rural memory, and sustainability as the backbone of the curriculum, promoting an emotional and cultural appropriation of the environment that contributes to forming citizens who are more aware, sensitive, and committed to their territory.

Statistical analysis reveals differences in the relationship with the rural environment between people living in urban and rural contexts. In line with this, the results obtained allow us to identify different patterns that reflect not only physical access to the natural environment, but also the ways in which it is experienced, valued, and shared by different population groups (Table 4).

Table 4. Main characteristics of respondents and their relationship with the countryside according to place of residence.

	Q1. Do you normally go to the countryside?						
	No	Yes				Chi	Cramer's V
City	44.8	55.2				0.016	0.016
Village	29.3	70.7					
	Q2. Do you usually have contact with farm animals?						
	No	Yes				Chi	Cramer's V
City	68.7%	31.3				0.000	0.26
Village	42.4	57.6					
	Q3. Do you like being in nature?						
	Very much	Quite	Somewhat	Not very	Not at all	Chi	Cramer's V
City	48.5	30.6	17.9	1.5	1.5	0.286	0.15
Village	42.4	36.4	16.2	5.1	0.0		
	Q4. Who do you usually go to the countryside with?						
	With friends	With school	With my family	I don't usually go to the countryside		Chi	V de Cramer
City	36.0	100.0	58.8	64.7		0.042	0.19
Village	64.0	0.0	41.2	35.3			

First, it was observed that the frequency with which people visit the countryside varies significantly depending on their place of residence. Seventy-seven percent of rural residents said they go to the countryside regularly, compared to 55.2% of urban residents. Although this difference was statistically significant, $\chi^2 = 0.016$, the effect size was very low (Cramer's V = 0.016), suggesting a weak association between the two variables.

In relation to direct contact with farm animals, the data revealed a more pronounced difference. Fifty-seven-point six percent of rural residents reported having regular contact with animals, compared to 31.3% of urban residents. This difference was highly significant, $\chi^2 = 0.000$, and had a moderate effect size (Cramer's V = 0.26), indicating a substantial relationship between the residential environment and interaction with rural wildlife.

Regarding personal preferences for nature, both groups showed a positive assessment, although with nuances. Forty-eight-point five percent of respondents in urban schools said they "very much" like being in nature, compared to 42.4% of rural respondents. The latter, however, showed a more balanced distribution between the categories "very much" and "quite a lot." The difference was statistically significant, $\chi^2 = 0.286$, although with a low effect size (Cramer's V = 0.15), suggesting that, while there are differences in the intensity of enjoyment, appreciation for nature is a widely shared value.

Finally, when analyzing who participants usually go to the countryside with, relevant differences were identified in the social dynamics associated with this activity. Rural residents tend to go more with friends (64%), while urban residents do so more frequently with family (58.8%) and, notably, with school (100%). This difference was significant, $\chi^2 = 0.042$, and had a low-moderate effect size (Cramer's V = 0.19), suggesting that the context of residence influences the forms of socialization linked to the natural environment.

Taken together, these results emphasize that the residential environment conditions not only access the natural environment, but also the way in which it is experienced, valued, and shared. The differences observed, although subtle in some cases, offer important clues for understanding contemporary relationships with nature and can inform educational, environmental, and territorial policies aimed at strengthening the link with the rural environment from an inclusive and contextualized perspective.

3.2. *Concept of the Dehesa*

3.2.1. Traditional Trades

A descriptive analysis was carried out of the declared knowledge of various traditional trades in the rural environment, with the aim of exploring possible differences between people living in urban and rural contexts. This exploration is part of a broader interest in understanding how knowledge about the working heritage linked to the natural environment is preserved, transmitted, and transformed depending on the living environment (Table 5).

Table 5. Traditional trades associated with the dehesa.

	Total		City		Village	
	n	%	n	%	n	%
Q6a. Shepherd	219	94.0	125	57.1	94	42.9
Q6b. Beekeeper	127	54.5	77	60.6	50	39.4
Q6c. Shearer	144	61.8	81	56.3	63	43.8
Q6d. Carbonero	87	37.3	53	60.9	34	39.1
Q6e. Cork extractor	104	44.6	71	68.3	33	31.7
Q6f. Fisherman	2	0.9	1	50.0	1	50.0
Q6g. Farmer	21	9.0	9	42.9	12	57.1
Q6h. Livestock farmer	27	11.6	9	33.3	18	66.7
Q6i. Farmer	1	0.4	0	0.0	1	100.0
Q6j. Hunter	3	1.3	3	100.0	0	0.0
Q6k. I don't know	6	2.6	4	66.7	2	33.3

The results show that the profession of shepherd is by far the most widely recognized by the sample, with 94.0% of mentions. This high frequency suggests that shepherding maintains a strong symbolic and cultural presence, possibly due to its representation in educational, cultural, and tourist media. The distribution between urban (57.1%) and rural (42.9%) residents is relatively balanced, indicating that this knowledge transcends the immediate geographical context and is part of the collective imagination.

Other occupations also enjoy high levels of recognition, such as shearer (61.8%), beekeeper (54.5%), and cork harvester (44.6%). In these cases, although most mentions come from urban dwellers, the difference is not extreme, suggesting a certain cultural dissemination of these trades, possibly through school programs, environmental education activities, or initiatives to promote rural heritage. The case of beekeeping is particularly interesting, as 60.6% of those who mention it live in cities, which could reflect a growing interest in urban beekeeping or its role in environmental sustainability, given the increasing visibility of bees in contemporary ecological discourse.

In contrast, some occupations show a clear association with rural areas. 66.7% of those who mention livestock farming as a known occupation live in villages, compared to 33.3% in cities. Similarly, farming is more widely recognized by people in rural areas (57.1%) than in urban areas (42.9%). These data suggest that, although these occupations remain fundamental to the rural economy and culture, their visibility in urban contexts is more limited, which could be related to processes of territorial disconnection and the loss of traditional productive references in cities.

Other occupations, such as charcoal burner (37.3%) or cork harvester (44.6%), although less well known in absolute terms, maintain a relatively balanced distribution between the two groups. This could be interpreted as an indication that, although their active practice has declined, their presence in the collective memory is maintained thanks to their heritage value or their inclusion in historical and cultural narratives.

The least associated occupations were fisherman (0.9%), farmer (0.4%), and hunter (1.3%), with very low frequencies and unrepresentative distributions. It should be noted that hunters were mentioned exclusively by urban respondents, while farmers were only recognized by one rural respondent. Although these data do not allow for firm conclusions due to the low number of cases, they do invite reflection on the invisibility of certain occupations or activities in contemporary discourse, as well as on the possible stigmas or transformations associated with them.

Finally, a small percentage of participants (2.6%) stated that they did not know any of the occupations listed. This response, although a minority, can be interpreted as an indicator of cultural disconnection from the rural world, especially considering that most of the occupations mentioned have historically been fundamental to life in the countryside.

Overall, the results show that knowledge of traditional rural occupations is not distributed evenly among the population, and that the place of residence influences familiarity with certain professions. While some trades maintain a strong symbolic presence in both contexts, others seem to be more rooted in direct experience of the rural environment. These findings reinforce the need to promote educational and cultural strategies that reconnect the urban population with rural knowledge, not only as a way of preserving intangible heritage, but also as a means of fostering a deeper and more respectful understanding of the territory and its knowledge.

3.2.2. Products of Dehesa Landscapes

A descriptive analysis of declared knowledge about rural products was carried out, with the aim of identifying differential patterns between people living in urban and rural contexts. This distribution suggests that, although knowledge about rural products is present in both contexts, their nature and depth may vary depending on the life experience and everyday environment of the participants (Table 6).

Table 6. Products associated with the dehesa landscapes.

	Total		City		Village	
	n	%	n	%	n	%
Q7a. Iberian pork	150	64.4	83	55.3	67	44.7
Q7b. Beef	101	43.3	48	47.5	53	52.5
Q7c. Sheep	101	43.3	48	47.5	53	52.5
Q7d. Goat	101	43.3	48	47.5	53	52.5
Q7e. Firewood and charcoal	92	39.5	43	46.7	49	53.3
Q7f. Cork	89	38.2	53	59.6	36	40.4
Q7g. Honey	103	44.2	62	60.2	41	39.8
Q7h. Mushrooms	124	53.2	63	50.8	61	49.2
Q7i. Asparagus	124	53.2	63	50.8	61	49.2
Q7j. Acorns	4	1.7	2	50.0	2	50.0
Q7k. I don't know	50	21.5	32	64.0	18	36.0

The most widely recognized product was Iberian pork, with 64.4% of mentions. This result is consistent with the symbolic, economic, and cultural centrality of this animal in regions such as Extremadura, where its extensive breeding in dehesas is not only a productive activity but also a deeply rooted element of identity. The distribution of mentions between cities (55.3%) and villages (44.7%) indicates that awareness of this product transcends the rural sphere, probably due to its presence in gastronomy, tourism, and the media. This phenomenon can be interpreted as an example of how certain rural products manage to enter the urban collective imagination through processes of cultural and commercial valorization.

At a second level of recognition are livestock products such as beef, sheep, and goat, all with 43.3% of mentions. Unlike Iberian pork, these products have a slight majority of mentions in rural areas (52.5%), suggesting greater familiarity derived from direct experience or the local productive environment. This difference, although not extreme, points to a persistent connection between rural life and knowledge of traditional livestock farming, which continues to be a source of identity, livelihood, and intergenerational transmission of knowledge.

Other products, such as firewood and charcoal (39.5%) and cork (38.2%), show different patterns that reflect different ways of relating to natural resources. Firewood and charcoal are more widely recognized by rural people (53.3%), which may be linked to their continued use in homes or agricultural activities, especially in areas where traditional energy sources remain relevant. In contrast, cork is mentioned more often by urban students (59.6%), which could be due to its growing value as an ecological and renewable material in urban contexts, especially in sectors such as design, sustainable architecture, and fashion. In addition, it is also possible that this is related to the information included in the textbooks they use. This contrast illustrates how the same resource can take on different meanings depending on the context: functional and every day in rural areas, symbolic and aesthetic in urban areas.

Products such as honey (44.2%), mushrooms (53.2%), and asparagus (53.2%) show a more balanced distribution between the two groups, although with a slight urban majority in the case of honey (60.2%). These products, associated with both harvesting and artisanal production, seem to have gained visibility in both contexts, possibly due to their link to healthy eating practices, ecotourism, or local cuisine. Their presence in local markets, agri-food fairs, and responsible consumption campaigns may have contributed to their spread beyond rural areas.

The case of acorns is marginal, with only 1.7% of mentions and an equal distribution between cities and villages. This low level of recognition may be since, although they are essential in the diet of Iberian pigs, they are not usually perceived as a product for direct human consumption, which limits their visibility. However, their low mention rate can also be interpreted as a symptom of the disconnect between end products and the ecological processes that sustain them.

Finally, 21.5% of participants stated that they did not know how to identify products from the rural environment. This response was more frequent among urban dwellers (64.0%), which could be interpreted as an indicator of disconnection from the natural environment or less exposure to rural contexts. This finding is particularly relevant from an educational perspective, as it highlights the need to strengthen knowledge about the rural environment in urban contexts, not only as part of the curriculum, but also as part of a broader territorial literacy. Lack of familiarity with farm products not only limits understanding of food and ecological systems but also impoverishes the symbolic and emotional relationship with the territory.

In summary, the results show that knowledge about agricultural products is conditioned by the environment in which people live, although there are significant areas of convergence. While rural inhabitants tend to be more familiar with products linked to livestock farming and the direct use of natural resources, urban dwellers are more familiar with products that have been revalued in contemporary environmental, gastronomic, or cultural discourse. These findings reinforce the importance of promoting territorially sensitive education that recognizes and values rural knowledge as part of our common heritage and as the basis for a more sustainable, informed, and respectful relationship with the environment. Furthermore, they invite us to rethink cultural and educational policies from a perspective that integrates the rural world not as a fading past, but as a living present that is necessary for imagining more balanced and resilient futures [20].

3.2.3. Dehesa Vegetation

The dehesa is defined precisely by its vegetation structure, consisting of a mosaic of scattered trees, Mediterranean scrub, and grasslands, the result of centuries of human interaction in balance with nature. The analysis of the responses allows us to identify not only which species are most recognized, but also how the vegetation that sustains this cultural landscape is perceived and valued. To achieve this, a descriptive analysis of the declared knowledge about different types of vegetation and rural landscapes was carried out, with the aim of identifying differential patterns between people living in urban and rural contexts.

The total sample collected 747 mentions, of which 55.2% came from students living in cities and 44.8% from residents of villages. This relatively balanced distribution suggests that knowledge about the rural landscape is not restricted exclusively to the environment in which one lives, although significant differences are observed depending on the type of vegetation recognized (Table 7).

Table 7. Vegetation associated with the dehesa.

	Total		City		Village	
	n	%	n	%	n	%
Q8a. Holm oaks and cork oaks	162	69.5	85	52.5	77	47.5
Q8b. Rockroses	87	37.3	36	41.4	51	58.6
Q8c. Robles	74	31.8	43	58.1	31	41.9
Q8d. Olive trees	141	60.5	82	58.2	59	41.8
Q8e. Eucalyptus trees	49	21.0	20	40.8	29	59.2
Q8f. Gall oaks	74	31.8	43	58.1	31	41.9
Q8g. Fruit trees	70	30.0	46	65.7	24	34.3
Q8h. Dehesas without trees	43	18.5	24	55.8	19	44.2
Q8i. I don't know	47	20.2	33	70.2	14	29.8

The most widely identified type of landscape was formed by holm oaks and cork oaks, with 69.5% of mentions. This result reflects the centrality of the dehesa as an emblematic ecosystem of the southwestern peninsula, especially in regions such as Extremadura. Its widespread recognition in both urban (52.5%) and rural (47.5%) contexts indicate that this landscape has transcended its local dimension to become a widely recognized cultural and ecological symbol. In addition to its productive value, the dehesa represents a model of sustainable land management, which may explain its presence in educational, tourism, and environmental discourse.

Olive trees were the second most mentioned type of vegetation (60.5%), with many urban mentions (58.2%). This data can be interpreted because of the strong presence of olive trees in Mediterranean food culture, as well as their visibility in everyday consumer products such as olive oil. Unlike other elements of the rural landscape, the olive tree has been widely integrated into the urban imagination, not only as a crop, but also as a symbol of tradition, health, and sustainability. However, its inclusion reveals a significant lack of knowledge about the trees that make up the dehesa.

At an intermediate level of recognition are species such as oak and gall oak (both with 31.8% of mentions), whose distribution is slightly higher in urban environments (58.1%). This trend may be because, although these species are not predominant in all territories, their presence in schoolbooks, urban parks, and reforestation campaigns has contributed to their cultural dissemination. In this sense, knowledge does not always derive from direct experience, but also from educational and symbolic mediation.

In contrast, rock roses (37.3%) and eucalyptus trees (21.0%) were more widely recognized by people in rural areas (58.6% and 59.2%, respectively). In the case of rock roses, their abundance in scrubland areas and their traditional use in activities such as beekeeping or the production of essences may explain their greater familiarity among those who live in direct contact with the natural environment. The case of eucalyptus trees, an exotic species that has nothing to do with the dehesa, is more complex, although they form part of the forest landscape in many rural areas due to their use in the timber industry, which also explains their greater recognition in these contexts. However, their lesser presence in urban environmental discourse, where they are often associated with negative impacts on biodiversity, could explain their scarcity of mention among urban residents.

Fruit trees (30.0%) were mentioned mainly by urban dwellers (65.7%), which could be interpreted as an idealization of the rural landscape from the city. In many cases, fruit trees are associated with family or community, gardens, or recreational activities, rather than intensive agricultural operations. This perception, although positive, may be more linked to a symbolic image of the countryside than to a deep understanding of its productive reality.

Treeless dehesas were the least mentioned type of landscape (18.5%), with a slightly higher distribution in urban areas (55.8%). This low frequency may be because open spaces without tree vegetation tend to be less valued or recognized as part of the rural landscape, despite their ecological and livestock importance. Their lack of visibility may be related to an aesthetic perception that favors wooded landscapes or to a lack of information about their ecological function.

Finally, 20.2% of participants stated that they did not know how to identify types of vegetation or rural landscapes. This response was significantly more frequent among urban dwellers (70.2%), which could be interpreted as an indicator of disconnection from the natural environment or less exposure to rural settings. This finding is relevant from an educational perspective, as it highlights the need to reinforce knowledge about elements of the natural landscape as part of our shared cultural and ecological heritage. Lack of familiarity with the elements that make up the natural environment not only limits understanding of ecosystems but also improves the symbolic and emotional relationship with the territory.

Therefore, the results reflect that knowledge about types of vegetation and rural landscapes is conditioned by the environment of residence, although with significant areas of convergence. While rural residents tend to recognize elements of Mediterranean scrubland and forest use more frequently, urban residents show greater familiarity with more symbolic or idealized landscapes. This highlights the importance of promoting ecological and territorial literacy that connects the population with their environment, recognizing the diversity of rural landscapes and their productive and cultural value. Promoting this type of knowledge not only contributes to the conservation of the natural environment but also strengthens the sense of belonging and shared responsibility towards the territories we inhabit and on which we depend.

3.3. Resource Assessment

Table 8 presents the results of a Likert-type rating scale (1 = not at all, 5 = very much) applied to a series of representative elements of the landscape, traditional crafts, and products linked to the dehesa. Measures of central tendency (mean, median, mode) and dispersion (standard deviation) are included, allowing for an accurate quantitative reading of the degree of social appreciation of each resource.

Table 8. Resource assessment (5 = very much/1 = not at all).

	Mean	SD	Md	Mo	City	Village	% City	% Village	Chi	V Cramer
Q9a. Dehesa with holm oaks and pond	4.3	0.8	4	5	581	421	58.0	42.0	0.140	0.170
Q9b. Shearer	3.6	1.1	4	3	467	375	55.5	44.5	0.076	0.190
Q9c. Pigs in free-range farming	3.8	1.1	4	4	507	369	57.9	42.1	0.113	0.180
Q9d. Shepherd	4.1	1.1	4	5	541	415	56.6	43.4	0.151	0.170
Q9e. Traditional shepherd's huts	3.4	1.3	4	4	440	346	56.0	44.0	0.070	0.190
Q9f. Beekeeper	3.2	1.4	3	3	416	320	56.5	43.5	0.800	0.080
Q9g. Production of traditional products (cheese, cured meats, etc.)	4.4	1	5	5	592	427	58.1	41.9	0.170	0.170
Q9h. Cork extractor	3.3	1.2	3	3	423	335	55.8	44.2	0.258	0.150
Q9i. Pigsties/pens	3.4	1.2	4	4 & 3	450	353	56.0	44.0	0.643	0.100
Q9j. Carbonero	3	1.3	3	4	395	309	56.1	43.9	0.043	0.210

The results of the assessment of a series of elements representative of the landscape and traditional rural trades show that the elements most highly valued by the participants as a whole were the production of traditional products (M = 4.4, SD = 1.0), the dehesa landscapes with holm oaks and ponds (M = 4.3, SD = 0.8), and the figure of the shepherd (M = 4.1, SD = 1.1). These scores reflect a highly positive assessment of practices and landscapes that are not only part of the rural heritage but are also deeply rooted in the cultural identity of many regions. The mode in these three cases was 5, indicating that most participants gave the highest possible score, reinforcing the idea that these elements are perceived as valuable, authentic, and representative of a sustainable and meaningful way of life.

The case of traditional product manufacturing (cheeses, sausages, etc.) stands out in particular, not only for its high average score, but also for its low dispersion (SD = 1.0), suggesting a broad consensus around its value. This result can be interpreted as an expression of appreciation for the quality, craftsmanship, and connection to the local area that these products represent, both in rural and urban contexts. In this sense, food acts as a symbolic bridge between the countryside and the city, facilitating the transmission of cultural and emotional values.

The dehesa landscapes with holm oaks and ponds, as an emblematic landscape of the southwestern peninsula, also received a very high rating (M = 4.3), with a low standard deviation (SD = 0.8), indicating a shared perception of its ecological, aesthetic, and cultural value. This type of landscape, in addition to its productive function, represents a model of balance between human activity and biodiversity conservation, which may explain its high rating in both environments.

The profession of shepherd (M = 4.1) was equally well rated, with a mode of 5 and a median of 4, suggesting a positive perception of this profession as a symbol of ancestral wisdom, connection with nature, and resilience. Although in many contexts this profession has lost visibility, its representation continues to evoke respect and admiration, especially at a time in history when traditional knowledge is being revalued in the face of contemporary ecological challenges.

At an intermediate level of appreciation are elements such as free-range pigs (M = 3.8), the shearer (M = 3.6), traditional shepherd's huts (M = 3.4), pigsties or corrals (M = 3.4), and the cork harvester (M = 3.3). These scores reflect a positive perception, although somewhat more nuanced, possibly due to less direct familiarity or more limited visibility in public discourse. In particular, the shearer and the cork extractor, although fundamental to the rural economy, may be less well known to those who have not had direct contact with these activities.

The figure of the beekeeper (M = 3.2) showed the greatest dispersion (SD = 1.4), suggesting wide variability in responses. This variability may be due to differences in the degree of knowledge, the perception of risk associated with the activity, or the lack of direct contact with beekeeping, especially in urban contexts. However, its average rating remains positive, indicating a general recognition of its ecological and productive importance.

The occupation with the lowest rating was charcoal burner (M = 3.0, SD = 1.3), despite its historical relevance in many rural communities. This score may reflect a perception of obsolescence or a disconnect with occupations that have disappeared from the contemporary landscape. However, it is important to note that this item had the highest Cramer's V value (0.21), indicating a moderate association between the rating and the residential environment, with it possibly being more highly valued in rural contexts where its memory remains alive.

In terms of distribution by residence, most items were rated slightly higher by urban dwellers (between 55% and 58% of mentions), suggesting a symbolic appropriation of the rural world from the city. This trend can be interpreted as a form of idealization or a search for authenticity in contrast to urban life. However, the differences were not statistically significant in most cases, as indicated by the chi-square values (all >0.05, except in the case of the charcoal burner, $\chi^2 = 0.043$), suggesting that the positive assessment of rural heritage is widely shared between both groups.

Overall, the results show a high appreciation of the elements of the landscape and traditional rural crafts, both in urban and rural contexts. This widespread appreciation can be interpreted as an indicator of the cultural, identity, and emotional value that these elements continue to have in today's society. Likewise, the data invite reflection on the need to preserve and raise awareness of this knowledge and these practices, not only as part of tangible and intangible heritage, but also as educational, community, and territorial resources that can contribute to strengthening the bond between people and their environment. In a context of increasing urbanization and disconnection from nature, these elements can play a key role in building a more conscious, rooted, and committed citizenry.

3.4. Imaginary of the Dehesa

Semantic analysis based on word frequency reveals the main terms with which students associate the dehesa (Figure 3). This analysis provides a conceptual corpus that defines the image that young people have of this landscape.

Figure 3. Imaginary of the dehesa.

The results show that students associate the dehesa with the terms countryside, animals, nature, trees, and trades. This reveals that students are aware of the natural value and biodiversity of this landscape. Likewise, the term "holm oak" is also one of the most repeated, suggesting an accurate identification of the characteristic elements of the dehesa.

However, it is particularly relevant that 34 students say they do not know what is a dehesa. This may be associated with a lack of knowledge of the explicit name of this landscape, even though many of them do recognize its characteristic elements (such as "pig," "cattle," "sheep," "acorn," "cork oak," or "vegetation") or show familiarity with its appearance and functions. This lack of knowledge highlights the need to give greater prominence to local ecosystems in the curriculum. This lack of knowledge is also evident in the high representation of the term "olive tree," a tree that is not associated with this landscape, which reinforces the importance of promoting greater education about it.

Finally, it is worth noting the presence of concepts such as "beautiful" or "freedom," which suggest a positive emotional connection with the dehesa, associating it with feelings of well-being and beauty.

3.5. Favorite Activities

Several studies have shown that the residential environment—urban or rural—significantly influences children's preferences and participation in activities related to nature and the rural environment. The specialized literature indicates that schoolchildren living in rural areas tend to engage in a wider range of activities and more frequently than their urban peers, especially those of physical, recreational, and direct interaction with the natural and productive environment. This greater diversity and frequency are associated with an operational and experiential familiarity with the rural environment, derived from daily exposure to agricultural practices and outdoor life.

In line with this evidence, the results of this study—based on a sample of primary school students who selected up to three activities related to the dehesa—show a general convergence in interest in the rural environment, with an average selection per activity of 32.67% among urban and r students and 32.10% among rural students. However, analysis of the percentage differences reveals clearly differentiated patterns of preference between the two groups (Table 9).

Table 9. Preferred activities for participation.

	Total		City		Village	
	n	%	n	%	n	%
Q11a. Accompanying the herder in moving livestock	87	37.3	41	30.6	46	46.5
Q11b. Feeding livestock	131	56.2	71	53.0	60	60.6
Q11c. Helping to strip the bark from cork trees and extract firewood	35	15.0	18	13.4	17	17.2
Q11d. Helping with sheep shearing	49	21.0	27	20.1	22	22.2
Q11e. Eating products from the dehesa such as meat, honey, etc.	107	45.9	74	55.2	33	33.3
Q11f. Going for walks in the countryside	95	40.8	50	37.3	45	45.5
Q11g. Milking	63	27.0	34	25.4	29	29.3
Q11h. Doing wool-related work	46	19.7	27	20.1	19	19.2
Q11i. Gathering wild products (mushrooms, asparagus, etc.)	97	41.6	60	44.8	37	37.4
Q11j. Seeing the old buildings in the dehesa, such as huts, corrals, or pigsties...	57	24.5	33	24.6	24	24.2

The activities with the greatest positive difference in favor of students living in villages were walking in the dehesa (45.5% in villages compared to 37.3% in cities), feeding livestock (60.6% vs. 52.9%), and milking (29.3% vs. 25.4%). These preferences reflect a functional and everyday relationship with the landscape, based on direct experience with agricultural and livestock practices. In addition, a stronger connection with nature and more consolidated pro-environmental behavior is observed among rural schoolchildren, mediated by regular contact with the natural environment.

In contrast, urban schoolchildren showed a marked preference for sensory, symbolic, or recreational activities, such as eating products from the dehesa (55.2% in cities compared to 33.3% in villages), gathering wild products such as mushrooms or asparagus (44.8% vs. 37.4%), and working with wool (20.1% vs. 19.2%). These choices suggest a more idealized and representational approach to the rural environment, focused on consumption and aesthetic experience rather than participation in productive processes.

The differences observed are not solely due to physical access to the territory, but also to social and cultural factors and attachment to the place, which influence the value and meaning attributed to the landscape. In this sense, rural schoolchildren show a greater degree of territorial attachment, while urban schoolchildren tend to build a more recreational and episodic relationship with the natural environment.

These findings have important implications for the design of educational programs and heritage interpretation strategies, which must consider contextual differences in children's perception of the territory. Promoting cultural sustainability and the intergenerational transmission of territorial knowledge requires pedagogical approaches that are sensitive to the residential environment and capable of integrating both direct experience and symbolic representation of the landscape.

3.6. Spatial Analysis of Educational Centers

Analysis of the spatial distribution of educational centers in Extremadura reveals a notable proximity between preschool and primary schools and the best-preserved dehesas in the region. This proximity represents a strategic opportunity to integrate these ecosystems as outdoor learning environments, facilitating educational experiences linked to traditional knowledge, biodiversity, and sustainability. The dehesas offer an ideal context for the development of key skills in the students, especially in relation to environmental education and knowledge of natural and cultural heritage.

Based on the analysis of geospatial data, the results of which can be seen in the Supplementary Material, four distance intervals between the dehesas and the educational centers were established: less than 1 km, between 1 and 2.499 km, between 2.5 and 4.999 km, and 5 km or more. These ranges allow for the evaluation of different levels of accessibility, from walking to those requiring motorized transport. The choice of these thresholds therefore responds to criteria of sustainable mobility and logistical feasibility for carrying out educational activities in the immediate natural environment (Table 10).

Table 10. Distance from nursery, primary, and special education schools to well-preserved dehesas.

Distance	Nursery and Primary Schools
<1 Km	129
1 < 2.5 Km	182
2.5 < 5 Km	170
≥5 Km	91

The data analyzed shows that 129 schools are located less than 1 km from dehesa landscapes, 182 between 1 and 2.5 km, 170 between 2.5 and 5 km, and 91 at a distance equal to or greater than 5 km. This distribution reveals that 54.4% of educational centers are located less than 2.5 km from these areas, which represents immediate and highly favorable accessibility for the development of regular educational activities in direct contact with the environment. In many cases, this proximity allows for travel on foot, reducing dependence on motorized transport and minimizing logistical costs, while strengthening the daily link between students and their cultural landscape.

In the intermediate range, corresponding to distances between 2.5 and 5 km, 29.7% of schools are located. Although this range requires more careful planning, it is still viable for regular activities, especially if school transportation or collaboration agreements with local entities are available. In this case, distance is not an insurmountable barrier, but it does require greater institutional coordination and an explicit willingness to integrate environmental education into the teaching program.

On the other hand, 15.9% of schools are located 5 km or more from dehesa landscapes areas, which can be a structural limitation for the regular implementation of activities in these environments. In these cases, it is necessary to explore complementary strategies, such as the creation of contextualized teaching materials, the use of digital and immersive technologies [98] for virtual visits, or the organization of occasional outings with institutional support. Likewise, the possibility of establishing agreements with private farms or closer public spaces that reproduce the ecological and cultural characteristics of the dehesa landscapes could be considered.

From a territorial perspective, these data reflect a relatively favorable distribution for the development of environmental education programs linked to the dehesa, especially considering that more than 80% of the centers are located less than 5 km from these spaces (Figure 4). This proximity is a strategic asset for promoting territorial roots, ecological literacy, and the intergenerational transmission of traditional knowledge, in line with the objectives of educational and heritage sustainability. The planning of educational activities in these environments must consider not only physical distance, but also mobility conditions, the involvement of the educational community, and the design of contextualized and culturally relevant pedagogical proposals.

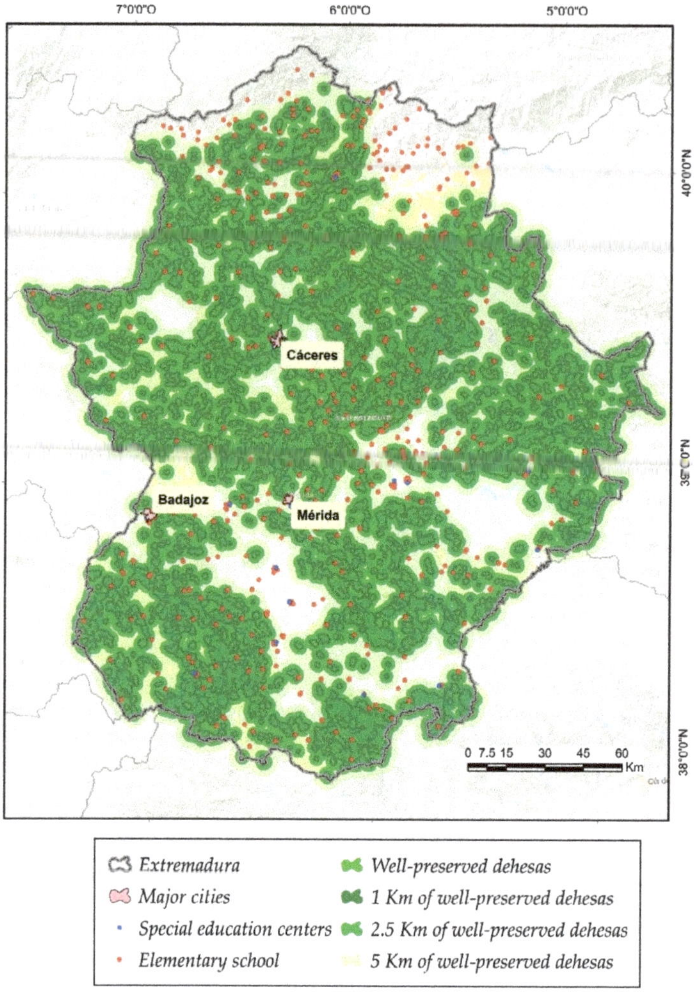

Figure 4. Distances from dehesas to preschool, primary school, and special education centers.

In this sense, environmental education takes on a central role as a tool for reconnecting new generations with their immediate environment. Numerous studies have shown that learning in direct contact with nature improves students' attention, intrinsic motivation, emotional self-regulation, and overall well-being [58]. In addition, these open and sensory-rich environments promote educational inclusion, as they allow for greater methodological flexibility and a more natural adaptation to the diversity of learning rhythms, styles, and needs [57]. In this context, the dehesa should not only be understood as an ecological or productive resource, but also as a privileged pedagogical space for the comprehensive education of students.

In short, the territorial accessibility of the dehesa to educational centers is a key facilitating factor for its integration into teaching-learning processes. Its use as an educational resource requires strategic planning that articulates public policies, institutional commitment, and community participation. Only in this way will it be possible to consolidate an environmental education rooted in the territory, capable of contributing to the ecological, cultural, and social sustainability of rural landscapes.

4. Discussion

The results show that, although young people value the dehesa positively and recognize elements such as the Iberian pig and holm oaks, there is a disconnect between this symbolic value and practical knowledge of the territory. Direct experience is key to building strong territorial ties [99], and ecological memory plays an essential role in the sustainability of systems such as the dehesa [100]. However, being physically present in the rural environment does not guarantee meaningful appropriation or the transmission of knowledge [101,102].

This phenomenon, known as "appreciation without roots" [103], reflects how young people can value the rural environment without feeling connected to it. Although many visit the countryside, few interact with its productive elements, such as animals or traditional crafts. This indicates that the rural environment is not always experienced as a space for learning or belonging.

Education reinforces the need to use the dehesa as a living classroom. Active outdoor methodologies improve student motivation, attention, and well-being [104] and foster respect for the environment, especially if they take place in landscapes with cultural significance such as the dehesa [105].

The literature on educational inclusion emphasizes the importance of designing learning environments that are accessible, flexible, and culturally relevant to all students [57]. In this context, the dehesa can become a privileged resource for inclusive education, provided that physical, pedagogical, and symbolic access to its values and resources is guaranteed. This involves not only adapting infrastructure and materials but also training teachers in methodologies that integrate diversity as a value rather than an obstacle.

Another relevant aspect is the intergenerational transmission of knowledge. The results show that traditional trades, although valued, are little known to young people, especially in urban contexts. This loss of practical and symbolic knowledge threatens the continuity of the intangible heritage associated with the dehesa. The sustainability of agroforestry systems depends not only on their economic or ecology, but also on their ability to be understood, valued, and culturally reproduced [106,107].

In addition, rural students show a greater preference for activities involving direct interaction with the environment, such as walking in the dehesa, feeding livestock, and milking. This reflects a functional and everyday relationship with the landscape, based on direct experience with agricultural and livestock practices [108], which is common in areas where agriculture and livestock farming are the livelihood of many families. In contrast, urban schoolchildren tend to prefer sensory or symbolic activities, such as eating products from the dehesa, gathering wild products, and working with wool. This trend suggests a more idealized approach to the rural environment, focused on consumption and aesthetic experience rather than participation in productive processes [109]. These results are consistent with previous research indicating that urban children tend to construct a more symbolic and recreational image of the natural environment, while rural children develop a more operational and experiential relationship [76,83]. In line with this, it can be deduced that the differences observed are not only due to physical access to the territory, but also to sociocultural factors and attachment to the place, which influence the value and meaning attributed to the landscape. In this sense, rural schoolchildren show a greater degree of territorial attachment, while urban schoolchildren tend to construct a more episodic and fragmented relationship with the natural environment. This divergence in the symbolic construction of the territory has direct implications for environmental education, territorial planning, and heritage management.

In addition, educational tourism and agrotourism are emerging as complementary tools for strengthening the connection between young people and the territory. In this regard, it should be noted that rural environments are becoming recreational areas, although they are subject to paradoxes such as ignorance of what is closest to them [109]. Immersive experiences, when developed under the principles of sustainability, authenticity, and community participation, generate meaningful and transformative learning. These practices not only raise awareness of the importance of conserving natural and cultural heritage but also strengthen the emotional bond with the territory and promote responsible attitudes towards its management.

The triangulation between the empirical results obtained and the hypotheses initially formulated allows for an accurate assessment of the degree of achievement of the objectives set.

Firstly, hypothesis H1, which anticipated significant gaps in primary school students' knowledge of the dehesa and its associated heritage, is corroborated by the data. The limited identification of traditional trades, derivative products, and plant species characteristic of the dehesa ecosystem confirms the existence of a significant cognitive gap, especially in urban contexts.

Hypothesis H2, which posited a predominantly naturalistic perception of the dehesa to the detriment of its cultural dimension, is also confirmed. Semantic analysis and resource assessment show a greater association of the landscape with ecological and sensory elements, while components linked to intangible heritage—such as traditional crafts—are less recognized and valued.

In relation to H3, the results show a clear preference for natural, landscape, and gastronomic resources over cultural elements, which validates the hypothesis. The production of traditional products and the contemplation of the landscape receive significantly higher scores than trades such as shearer, charcoal burner, or cork extractor, suggesting a symbolic hierarchy of rural heritage.

Hypothesis H4, which proposed that place of residence had no significant influence on the valuation of heritage, is partially rejected. Although the overall valuation is positive in both groups, statistically significant differences are identified in the frequency of visits to the countryside, contact with animals, and preferred activities, indicating that the residential environment conditions the experience and connection with the dehesa.

Finally, hypothesis H5, regarding the influence of place of residence on preferred activities, is fully confirmed. Rural students show a preference for operational activities and direct interaction with the environment (such as feeding livestock and milking), while urban students opt for symbolic or sensory experiences (such as consuming products or occasional harvesting), reflecting different ways of connecting with the land.

In terms of meeting the specific objectives, there is a direct correlation between the findings and the goals set. The following has been achieved:

SO1: Explore students' level of knowledge and perception of the dehesa landscapes, highlighting relevant educational gaps.

SO2: Identify the most attractive heritage elements, highlighting traditional products and the natural landscape.

SO3: To determine significant differences in the valuation of heritage based on origin and direct experience with the environment.

SO4: Assess the activities preferred by students, which allows for the development of contextualized educational proposals.

SO5: Establish the proximity of educational centers to well-preserved dehesa areas, opening possibilities for their integration into the curriculum.

This interrelationship between hypotheses, objectives, and results reinforces the internal validity of this study and provides a solid basis for the design of territorially sensitive educational strategies aimed at the cultural, ecological, and pedagogical sustainability of the dehesa landscape.

Despite the results obtained, this study has some limitations that should be considered. First, the use of intentional and convenience sampling limits the generalizability of the findings. In addition, the application of the survey at a specific time of year may have influenced responses related to the seasonality of rural activities. The qualitative analysis, based on self-reports, could be affected by social desirability bias. Likewise, the measurement of accessibility to dehesas has been carried out solely in terms of geographical distance, without considering factors such as the availability of transport, terrain conditions, or institutional barriers. Finally, unequal access has been detected between early childhood and primary education centers and special education centers, which poses challenges in terms of territorial and educational equity.

These limitations open new possibilities for future research. It would be relevant to extend this study to other regions with similar agroforestry systems, as well as to incorporate ethnographic methodologies that delve deeper into the cultural practices linked to the dehesa landscapes. It is also recommended that longitudinal studies be carried out to assess the evolution of young people's perceptions and the impact of specific educational programs. In the area of inclusion, it is necessary to explore the barriers faced by special education centers in accessing natural environments and to design strategies that guarantee equitable environmental education. Finally, the use of immersive technologies could be a complementary way to bring the dehesa closer to those groups with physical access difficulties, thus promoting more inclusive and transformative territorial literacy.

5. Conclusions

The dehesa, as a cultural landscape and socio-ecological system, represents much more than a productive space: it is a living setting of memory, identity, and sustainability. This study has shown that although the younger generations value its most emblematic elements—such as Iberian pigs, holm oaks, and traditional products—there remains a worrying disconnect between this symbolic value and a deep, experiential, and everyday knowledge of the territory.

The data reveal that contacts with the rural environment do not always translate into a meaningful experience. Many young people visit the countryside, but they do so for leisure or contemplation, without a clear understanding of the ecological, cultural, and economic processes that sustain it. This gap between physical presence and cultural appropriation directly challenges educational systems, which have not yet systematically integrated the territory as a pedagogical resource.

Despite this, clear opportunities can be identified. The high value placed on the resources of the dehesa by students, regardless of their place of residence, suggests fertile ground for the design of educational proposals that connect emotion, knowledge, and action. Outdoor education, heritage education, and immersive experiences in rural areas are emerging as effective strategies for strengthening the link between young people and the territory. In addition, the analysis of the activities preferred by students provides a key dimension for understanding how relationships with the rural environment are built. In this sense, schoolchildren living in villages show a greater inclination towards operational and experiential activities—such as feeding livestock, milking, or accompanying the shepherd—while urban children prefer sensory or symbolic activities—such as eating products from the dehesa or occasional gathering.

Likewise, spatial analysis has revealed inequality in access to dehesas between preschool and elementary schools and special education centers. This difference poses a challenge in terms of educational equity and requires public policies that guarantee inclusive access to natural spaces, especially for those groups that face greater barriers.

It is concluded that the dehesa landscape represents a valuable educational, cultural, and socio-ecological resource. Its continuity depends on the effective integration of academic knowledge, experiential learning, and collective memory, as well as educational and territorial policies that are sensitive to the diversity of contexts. Only by building bridges between knowledge, experiences, and values will it be possible to guarantee the transmission of this cultural landscape as a living, diverse, and shared legacy for future generations.

In short, the dehesa can and should be an open classroom, a space for comprehensive learning where ancestral knowledge and contemporary challenges intertwine. To this end, it is necessary to build bridges between schools, communities, and the territory and to recognize that sustainability is not taught only through content but also through experiences that transform the perspective and commitment of those who inhabit and inherit these landscapes.

Supplementary Materials: The following supporting information can be downloaded at: https://www.mdpi.com/article/10.3390/land14112111/s1, Table S1: Survey; Table S2: Distance from schools to well-preserved dehesas.

Author Contributions: Conceptualization, R.G.-P., A.-M.H.-C. and J.-M.S.-M.; methodology, R.G.-P., A.-M.H.-C. and J.-M.S.-M.; validation, R.G.-P., A.-M.H.-C. and J.-M.S.-M.; formal analysis, R.G.-P., A.-M.H.-C. and J.-M.S.-M.; investigation, R.G.-P., A.-M.H.-C. and J.-M.S.-M.; research, R.G.-P., A.-M.H.-C. and J.-M.S.-M.; research, R.G.-P., A.-M.H.-C. and J.-M.S.-M.; resources, R.G.-P.; data curation, R.G.-P.; writing—original draft preparation, R.G.-P. and J.-M.S.-M.; writing—review and editing, R.G.-P.; writing—review and editing, R.G.-P., A.-M.H.-C. and J.-M.S.-M.; research, R.G.-P., A.-M.H.-C. and J.-M.S.-M.; supervision, A.-M.H.-C. and J.-M.S.-M.; project administration, J.-M.S.-M. All authors have read and agreed to the published version of the manuscript.

Funding: This research received no external funding.

Institutional Review Board Statement: This study was conducted in accordance with the Declaration of Helsinki and approved by the Ethics Committee of University of Extremadura (15 May 2025) for studies involving humans.

Data Availability Statement: The original contributions presented in the study are included in the article, further inquiries can be directed to the corresponding author.

Conflicts of Interest: The authors declare no conflicts of interest.

References

1. Schnabel, S.; Pulido-Fernández, M.; Lavado-Contador, J.F. La dehesa: Un paisaje cultural afrontando los desafíos del futuro. In *Atlas de la Desertificación de España*; Instituto de Investigación INTERRA, Universidad de Extremadura: Badajoz, Spain, 2019. Available online: https://atlas-desertificacion.ua.es/es/9.-la-dehesa-un-paisaje-cultural-afrontando-los-desafios-del-futuro.html (accessed on 15 October 2025).
2. Silva Pérez, R.; Fernández Salinas, V. Claves para el reconocimiento de la dehesa como paisaje cultural de UNESCO. *An. Geogr. Univ. Complut.* **2015**, *35*, 121–142. [CrossRef]
3. Silva Pérez, R. La dehesa vista como paisaje cultural: Fisonomías, funcionalidades y dinámicas históricas. *Ería* **2010**, *82*, 143–157.
4. Consejo de Europa. Convenio Europeo del Paisaje. Available online: https://www.mapa.gob.es/dam/mapa/contenido/desarrollo-rural/temas/desarrollo-territorial/paisaje/convenio-europeo-del-paisaje/090471228005d489.pdf (accessed on 16 October 2025).
5. Parra-López, C.; Sayadi, S.; Garcia-Garcia, G.; Ben Abdallah, S.; Carmona-Torres, C. Prioritising conservation actions towards the sustainability of the dehesa by integrating the demands of society. *Agric. Syst.* **2023**, *206*, 103613. [CrossRef]
6. Ramirez-Hernandez, A.; Mico, E.; Marcos-García, M.D.L.Á.; Brustel, H.; Galante, E. The "dehesa", a key ecosystem in maintaining the diversity of *Mediterranean saproxylic* insects (*Coleoptera* and *Diptera*: *Syrphidae*). *Biodivers Conserv.* **2014**, *23*, 2069–2086. [CrossRef]

7. Hearn, K.; Prada Llorente, E. Spanish dehesas as heritage: Using participatory mapping and archival data to characterise the temporal evolution of an agrosilvoshepherdal cultural landscape. *Landsc. Hist.* **2024**, *45*, 5–32. [CrossRef]
8. Vicente, Á.M.; Alés, R.F. Long Term Persistence of Dehesas. Evidences from History. *Agrofor. Syst.* **2006**, *67*, 19–28. [CrossRef]
9. Huntsinger, L.; Oviedo, J.L. Ecosystem services are social–ecological services in a traditional shepherdal system: The case of California's Mediterranean rangelands. *Ecol. Soc.* **2014**, *19*, 8. [CrossRef]
10. Plieninger, T.; Bieling, C. Resilience-based perspectives to guiding high-nature-value farmland through socioeconomic change. *Ecol. Soc.* **2013**, *18*, 20. [CrossRef]
11. Pinto-Correia, T.; Mascarenhas, J. Contribution to the extensification/intensification debate: New trends in the Portuguese montado. *Landsc. Urban Plan.* **1999**, *46*, 125–131. [CrossRef]
12. González Santa Cruz, F.; Pemberthy Gallo, L.S.; López-Guzmán, T.; Pérez Gálvez, J.C. Tourist segmentation in an intangible heritage setting: The Holy Week processions in the city of Popayán, Colombia. *J. Herit. Tour.* **2019**, *15*, 485–499. [CrossRef]
13. González, E.P.H.; Picazo, M.T.M. Retro-innovación, sostenibilidad y viabilidad de la dehesa española y el montado portugués. *UCJC Bus. Soc. Rev.* **2024**, *21*, 60–95. [CrossRef]
14. Lozano-Parra, J.; Sánchez-Martín, J.M. Exploring the Impact of Climate Change on Water Resources for Vegetation Covers in Extremadura (Spain). *Water* **2024**, *16*, 1418. [CrossRef]
15. Plieninger, T.; Bieling, C. *Resilience and the Cultural Landscape: Connecting Cultural Landscapes to Resilience*; Cambridge University Press: Cambridge, UK, 2012. [CrossRef]
16. Plieninger, T.; Flinzberger, L.; Hetman, M.; Horstmannshoff, I.; Reinhard-Kolempas, M.; Topp, E.; Moreno, G.; Huntsinger, L. Dehesas as high nature value farming systems: A social-ecological synthesis of drivers, pressures, state, impacts, and responses. *Ecol. Soc.* **2021**, *26*, 23. [CrossRef]
17. Rodrigues Ferreira, D.I.; Sánchez Martín, J.M. La función de las áreas agrícolas en el debate epistemológico sobre el turismo rural, el agroturismo y el agroecoturismo. *Rev. Geogr. Norte Gd.* **2022**, *81*, 235–261. [CrossRef]
18. Gobierno de España. Ley January 1986, de 2 de Mayo, Sobre la Dehesa en Extremadura. «DOE» Núm. 40, de 15 May 1986, «BOE» Núm. 174, de 22 July 1986. Available online: https://www.boe.es/eli/es-ex/l/1986/05/02/1/con (accessed on 23 July 2025).
19. Pulido, F.; Picardo, A. (Coord). Libro Verde de la Dehesa. *Junta de Extremadura, Mérida: Consejería de Agricultura y Comercio.* 2010. Available online: https://studylib.es/doc/4559173/libro-verde-de-la-dehesa{-}{-}-universidad-de-extremadura (accessed on 23 July 2025).
20. Olea, L.; Miguel-Ayanz, A.S.; The Spanish dehesa. A traditional Mediterranean Silvoshepherdal System Linking Production and Nature Conservation. 2006. Available online: https://www.fao.org/fileadmin/templates/agphome/images/iclsd/documents/wk1_c5_radomski.pdf (accessed on 23 July 2025).
21. Campos, P.; Huntsinger, L.; Oviedo, J.L.; Starrs, P.F.; Díaz, M.; Standiford, R.B.; Montero, G. (Eds.) *Mediterranean Oak Woodland Working Landscapes: Dehesas of Spain and Ranchlands of California*; Springer: Berlin/Heidelberg, Germany, 2013. [CrossRef]
22. UNESCO Convention for the Safeguarding of the Intangible Cultural Heritage. Adopted by the General Conference of UNESCO at Its 32nd Session, Paris, 17 October 2003. Available online: https://ich.unesco.org/en/convention (accessed on 23 July 2025).
23. Hosen, N.; Nakamura, H.; Hamzah, A. Adaptation to Climate Change: Does Traditional Ecological Knowledge Hold the Key? *Sustainability* **2020**, *12*, 676. [CrossRef]
24. Mistry, J.; Berardi, A. Bridging indigenous and scientific knowledge. *Science* **2016**, *352*, 1274–1275. [CrossRef]
25. Ruiz-Mallén, I.; Corbera, E. Community-Based Conservation and Traditional Ecological Knowledge: Implications for Social-Ecological Resilience. *Ecol. Soc.* **2013**, *18*, 12. [CrossRef]
26. Gómez-Baggethun, E.; Corbera, E.; Reyes-García, V. Traditional Ecological Knowledge and Global Environmental Change: Research findings and policy implications. *Ecol. Soc.* **2013**, *18*, 72. [CrossRef] [PubMed]
27. Aswani, S.; Lemahieu, A.; Sauer, W. Global trends of local ecological knowledge and future implications. *PLoS ONE* **2018**, *13*, e0195440. [CrossRef] [PubMed]
28. Fabbricatti, K.; Boissenin, L.; Citoni, M. Heritage Community Resilience: Towards new approaches for urban resilience and sustainability. *City Territ. Arch.* **2020**, *7*, 17. [CrossRef]
29. Wardekker, A.; Nath, S.; Handayaningsih, T.U. The interaction between cultural heritage and community resilience in disaster-affected volcanic regions. *Environ. Sci. Policy* **2023**, *145*, 116–128. [CrossRef]
30. Musialik, W.; Łukaniszyn-Domaszewska, K.; Karaś, E.; Malik, K. Cultural heritage capital as a resilience factor in sustainable development policy of the region. *Econ. Environ.* **2025**, *91*, 964. [CrossRef]
31. Hale, J.; Irish, A.; Carolan, M.; Clark, J.K.; Inwood, S.M.; Jablonski, B.; Johnson, T. A systematic review of cultural capital in U.S. community development research. *J. Rural. Stud.* **2023**, *103*, 103113. [CrossRef]
32. Pitarch-Garrido, M.D. Social Sustainability in Metropolitan Areas: Accessibility and Equity in the Case of the Metropolitan Area of Valencia (Spain). *Sustainability* **2018**, *10*, 371. [CrossRef]
33. Madanipour, A.; Shucksmith, M.; Brooks, E. The concept of spatial justice and the European Union's territorial cohesion. *Eur. Plan. Stud.* **2021**, *29*, 1457–1475. [CrossRef]

34. Chamusca, P. Measuring and Addressing Territorial Cohesion: A Framework for Regional Development in Portugal. *Sustainability* **2025**, *17*, 3061. [CrossRef]
35. Edwards, D.B., Jr.; Sustarsic, M.; Chiba, M.; McCormick, M.; Goo, M.; Perriton, S. Achieving and Monitoring Education for Sustainable Development and Global Citizenship: A Systematic Review of the Literature. *Sustainability* **2020**, *12*, 1383. [CrossRef]
36. Mahon, M.; Woods, M.; Farrell, M.; Jones, R.; Goodwin-Hawkins, B. A spatial justice perspective on EU rural sustainability as territorial cohesion. *Sociol. Rural.* **2023**, *63*, 683–702. [CrossRef]
37. Cairney, P.; Kippin, S.; St Denny, E.; Mitchell, H. Policy design for territorial equity in multi-level and multi-sectoral political systems: Comparing health and education strategies. *Reg. Sci. Policy Pract.* **2022**, *14*, 1051–1061. [CrossRef] [PubMed]
38. Medeiros, E.; Rauhut, D. Territorial Cohesion Cities: A policy recipe for achieving Territorial Cohesion? *Reg. Stud.* **2018**, *54*, 120–128. [CrossRef]
39. UNESCO. *Roadmap for Implementing the Global Action Programme on Education for Sustainable Development*; UNESCO: París, France, 2014. Available online: https://unesdoc.unesco.org/ark:/48223/pf0000230514 (accessed on 23 July 2025).
40. Druker-Ibáñez, S.; Cáceres-Jensen, L. Integration of indigenous and local knowledge into sustainability education: A systematic literature review. *Environ. Educ. Res.* **2022**, *28*, 1209–1236. [CrossRef]
41. Bascopé, M.; Perasso, P.; Reiss, K. Systematic Review of Education for Sustainable Development at an Early Stage: Cornerstones and Pedagogical Approaches for Teacher Professional Development. *Sustainability* **2019**, *11*, 719. [CrossRef]
42. [illegible] Science Learning in the Context of 'Indigenous Knowledge' for Sustainable Development. *Int. J. Ethnoscience Technol. Educ.* **2024**, *1*, 28–41. [CrossRef]
43. Yu, Y.; Appiah, D.; Zulu, B.; Adu-Poku, K.A. Integrating Rural Development, Education, and Management: Challenges and Strategies. *Sustainability* **2024**, *16*, 6474. [CrossRef]
44. Nordén, B. Learning and teaching sustainable development in global–local contexts. *Environ. Educ. Res.* **2016**, *24*, 772–773. [CrossRef]
45. Januškienė, E.; Kamičaitytė, J. Landscape Perception of Children Under the Age of 12 Living in Different Settings: A Systematic Review. *Archit. Urban Plan.* **2024**, *20*, 2024. [CrossRef]
46. Özdemir, A. Cultural heritage readability: Children's perception of cultural landscape, Laodikeia Ancient City, Denizli. *Kastamonu Univ. J. For. Fac.* **2018**, *18*, 32–45. [CrossRef]
47. Ural, E. Exploring nature perceptions of children aged between 9–12 through their drawings in the context of today, living place, and future. *J. Outdoor Environ. Educ.* **2025**. [CrossRef]
48. Corraliza, J.A.; Collado, S. Ecological awareness and children's environmental experience. *Papeles Del Psicólogo* **2019**, *40*, 137–144. [CrossRef]
49. López Fernández, J.A.; Medina, S.; López, M.J.; García Morís, R. Perceptions of heritage among students of early childhood and primary education. *Sustainability* **2021**, *13*, 10636. [CrossRef]
50. Casonato, C.; Bonfantini, B. Cultural heritage education in the everyday landscape: School, citizenship, space, and representation. In *Digital Innovations in Architecture, Engineering and Construction*; Springer: Berlin/Heidelberg, Germany, 2022; pp. 1–285. [CrossRef]
51. Martínez Rodríguez, M.; Fontal Merillas, O. Dealing with heritage as curricular content in Spain's Primary Education. *Curric. J.* **2020**, *31*, 77–96. [CrossRef]
52. García Monteagudo, D.; Brotons Samper, S. La representación del paisaje: Pensamiento y conciencia geográfica en educación infantil. *Didáctica Geográfica* **2025**, *26*, 35–53. [CrossRef]
53. Kudryavtsev, A.; Stedman, R.C.; Krasny, M.E. Sense of place in environmental education. *Environ. Educ. Res.* **2011**, *18*, 229–250. [CrossRef]
54. Sánchez Fuster, M.C.; Miralles Martínez, P.; Serrano Pastor, F.J. School trips and local heritage as a resource in primary education: Teachers' perceptions. *Sustainability* **2023**, *15*, 7964. [CrossRef]
55. I'Anson Gutiérrez, S.; Suárez Suárez, M.Á.; Calaf Masachs, R. Local History and the Development of Heritage Bonds: A Primary Education Intervention. *Heritage* **2023**, *6*, 7215–7229. [CrossRef]
56. Becker, C.; Lauterbach, G.; Spengler, S.; Dettweiler, U.; Mess, F. Effects of regular classes in outdoor education settings: A systematic review on students' learning, social and health dimensions. *Int. J. Environ. Res. Public Health* **2017**, *14*, 485. [CrossRef]
57. Sanger, C.S. Inclusive Pedagogy and Universal Design Approaches for Diverse Learning Environments. In *Diversity and Inclusion in Global Higher Education*; Sanger, C., Gleason, N., Eds.; Palgrave Macmillan: Singapore, 2020. [CrossRef]
58. Mann, J.; Gray, T.; Truong, S.; Brymer, E.; Passy, R.; Ho, S.; Sahlberg, P.; Ward, K.; Bentsen, P.; Curry, C.; et al. Getting out of the classroom and into nature: A systematic review of nature-specific outdoor learning on school children's learning and development. *Front. Public Health* **2022**, *10*, 877058. [CrossRef] [PubMed]
59. Espada-Chavarria, R.; González-Montesino, R.H.; López-Bastías, J.L.; Díaz-Vega, M. Universal Design for Learning and Instruction: Effective strategies for inclusive higher education. *Educ. Sci.* **2023**, *13*, 620. [CrossRef]

60. Kiviranta, L.; Lindfors, E.; Rönkkö, M.-L.; Luukka, E. Outdoor learning in early childhood education: Exploring benefits and challenges. *Educ. Res.* **2024**, *66*, 102–119. [CrossRef]
61. Anyanwu, L.; Olaitan, D.A. Culturally responsive pedagogy and Universal Design for Learning: For building inclusive and accessible curriculum for diverse learners. *Int. J. Acad. Multidiscip. Res.* **2025**, *9*, 18–25.
62. Lane, B.; Kastenholz, E. Rural tourism: The evolution of practice and research approaches—Towards a new generation concept? *J. Sustain. Tour.* **2015**, *23*, 1133–1156. [CrossRef]
63. Garrod, B.; Wornell, R.; Youell, R. Re-conceptualising rural resources as countryside capital: The case of rural tourism. *J. Rural. Stud.* **2006**, *22*, 117–128. [CrossRef]
64. Ballantyne, R.; Packer, J. Using tourism free-choice learning experiences to promote environmentally sustainable behaviour: The role of post-visit 'action resources'. *Environ. Educ. Res.* **2011**, *17*, 201–215. [CrossRef]
65. Martín-Delgado, L.-M.; Jiménez-Barrado, V.; Sánchez-Martín, J.-M. Sustainable Hunting as a Tourism Product in Dehesa Areas in Extremadura (Spain). *Sustainability* **2022**, *14*, 10288. [CrossRef]
66. Ferreira, D.I.R.; Sánchez-Martín, J.-M. Agricultural Landscapes as a Basis for Promoting Agritourism in Cross-Border Iberian Regions. *Agriculture* **2022**, *12*, 716. [CrossRef]
67. Organización Mundial del Turismo (OMT). *Recomendaciones Sobre Turismo y Desarrollo Rural: Una Guía Para Convertir el Turismo en una Herramienta de Desarrollo Rural Efectiva*; Organización Mundial del Turismo (OMT): Madrid, Spain, 2020. [CrossRef]
68. Acosta-Naranjo, R.; Guzmán-Troncoso, A.J.; Gómez-Melara, J. The persistence of wild edible plants in agroforestry systems: The case of wild asparagus in southern Extremadura (Spain). *Agrofor. Syst.* **2020**, *94*, 2391–2400. [CrossRef]
69. Torralba, M.; Oteros-Rozas, E.; Moreno, G.; Plieninger, T. Exploring the Role of Management in the Coproduction of Ecosystem Services from Spanish Wooded Rangelands. *Rangel. Ecol. Manag.* **2017**, *71*, 549–559. [CrossRef]
70. Bidegain, Í.; López-Santiago, C.A.; González, J.A.; Martínez-Sastre, R.; Ravera, F.; Cerda, C. Social Valuation of Mediterranean Cultural Landscapes: Exploring Landscape Preferences and Ecosystem Services Perceptions through a Visual Approach. *Land* **2020**, *9*, 390. [CrossRef]
71. Maldonado, A.D.; Ramos-López, D.; Aguilera, P.A. The Role of Cultural Landscapes in the Delivery of Provisioning Ecosystem Services in Protected Areas. *Sustainability* **2019**, *11*, 2471. [CrossRef]
72. Bourdieu, P. *Capital Cultural, Escuela y Espacio Social*; Siglo XXI: Buenos Aires, Argentina, 2022.
73. Paulino, I.; Burgos-Tartera, C.; Aulet, S. Participatory governance of intangible heritage to develop sustainable rural tourism: The timber-raftsmen of La Pobla de Segur, Spain. *J. Herit. Tour.* **2023**, *18*, 710–729. [CrossRef]
74. Guillén-Peñafiel, R.; Hernández-Carretero, A.M.; Sánchez-Martín, J.M. Role-Playing as a Didactic Strategy for Teaching Intangible Heritage: An Intervention with Trainee Teachers. *Heritage* **2025**, *8*, 90. [CrossRef]
75. Gaspar, P.; Mesías, F.J.; Escribano, M.; Rodríguez de Ledesma, A.; Pulido, F. Economic and management characterization of dehesa farms: Implications for their sustainability. *Agrofor. Syst.* **2007**, *71*, 151–162. [CrossRef]
76. Guillén-Peñafiel, R.; Hernández-Carretero, A.M.; Sánchez-Martín, J.M. University Students' Perception of the Dehesa and the Associated Traditional Trades. *Sustainability* **2024**, *16*, 3843. [CrossRef]
77. Guillén-Peñafiel, R.; Hernández-Carretero, A.M.; Sánchez-Martín, J.M. Intangible Heritage, Education and Tourism as a Strategy for the Enhancement of the Value of the Dehesa and the Traditional Shepherd's Trade. *Sustainability* **2024**, *16*, 10997. [CrossRef]
78. Moreno, G.; Pulido, F.J. The Functioning, Management and Persistence of Dehesas. In *Agroforestry in Europe*; Advances in Agroforestry; Rigueiro-Rodróguez, A., McAdam, J., Mosquera-Losada, M.R., Eds.; Springer: Dordrecht, The Netherlands, 2009; Volume 6. [CrossRef]
79. Gómez-Campos, N.E.; Pérez-Martínez, L.; Cabrera-Hernández, J.A.; Nogueira-Rivera, D.; León, A.M. Integración de servicios ecosistémicos en la planificación territorial. Procedimiento y aplicación. *Ing. Ind.* **2023**, *44*, 59–76.
80. Ministerio de Agricultura, Pesca y Alimentación. La Conservación de la Dehesa y Las Intervenciones de Desarrollo Rural en el PEPAC 2023–2027. 2023. Available online: https://www.mapa.gob.es/dam/mapa/contenido/reforma-de-la-pac/plan-estrategico-pac-post-2020/desplegable/la-conservacion-de-la-dehesa-y-las-intervenciones-de-desarrollo-rural.pdf (accessed on 15 October 2025).
81. Cuenca-López, J.M.; Estepa-Giménez, J. Educación patrimonial para la inteligencia territorial y emocional de la ciudadanía. *MIDAS* **2017**, *8*, 1173. [CrossRef]
82. Rodríguez Alonso, G. El agroturismo, una visión desde el desarrollo sostenible. *Cent. Agrícola* **2019**, *46*, 62–70.
83. Guillén-Peñafiel, R.; Hernández-Carretero, A.M.; Sánchez-Martín, J.M. Intangible Heritage of the Dehesa: The Educational and Tourist Potential of Traditional Trades. *Heritage* **2023**, *6*, 5347–5373. [CrossRef]
84. Instituto Nacional de Estadística (INE). Población Residente Por Fecha, Sexo y Edad. 2024. Available online: https://ine.es/jaxiT3/Tabla.htm?t=56940&L=0 (accessed on 23 July 2025).
85. Gobierno de España. Ministerio de Educación, Formación Profesional y Deportes. *Alumnado Matriculado (Curso 2024–2025)*. 2025. Available online: https://www.educacionfpydeportes.gob.es/servicios-al-ciudadano/estadisticas/no-universitaria.html (accessed on 23 July 2025).

86. Sánchez-Martín, J.-M.; Blas-Morato, R.; Rengifo-Gallego, J.-I. The Dehesas of Extremadura, Spain: A Potential for Socio-Economic Development Based on Agritourism Activities. *Forests* **2019**, *10*, 620. [CrossRef]
87. Martín-Delgado, L.-M.; Rengifo-Gallego, J.-I.; Sánchez-Martín, J.-M. Hunting Tourism as a Possible Development Tool in Protected Areas of Extremadura, Spain. *Land* **2020**, *9*, 86. [CrossRef]
88. Ruiz-Ballesteros, E.; Gálvez-García, C.; Jaraíz-Arroyo, G. Community-based tourism and rural demographic decline. Reflections from Extremadura (Spain). *Curr. Issues Tour.* **2024**, *28*, 2329–2342. [CrossRef]
89. Ruiz-Labrador, E.-E.; Sánchez-Martín, J.-M.; Gurría-Gascón, J.-L. The Agritourism Value Chain: An Application to the Dehesa Areas of Extremadura. *Agriculture* **2023**, *13*, 2078. [CrossRef]
90. Aiken, L.R. A General Questionnaire Analysis Program. *Educ. Psychol. Meas.* **1978**, *38*, 167–169. [CrossRef]
91. Wooditch, A.; Johnson, N.J.; Solymosi, R.; Medina Ariza, J.; Langton, S. Chi-Square and Contingency Tables. In *A Beginner's Guide to Statistics for Criminology and Criminal Justice Using R*; Springer: Cham, Switzerland, 2021. [CrossRef]
92. Armstrong, R.A.; Hilton, A. Statnote 7: Chi-square contingency tables. *Microbiologist* **2006**, *2006*, 36–37. [CrossRef]
93. Lombardo, R.; Takane, Y.; Beh, E.J. Familywise decompositions of Pearson's chi-square statistic in the analysis of contingency tables. *Adv. Data Anal. Classif.* **2020**, *14*, 629–649. [CrossRef]
94. Goodman, L.A. Partitioning of Chi-Square, Analysis of Marginal Contingency Tables, and Estimation of Expected Frequencies in Multidimensional Contingency Tables. *J. Am. Stat. Assoc.* **1971**, *66*, 339–344. [CrossRef]
95. Camilli, G.; Hopkins, K.D. Applicability of chi-square to 2 × 2 contingency tables with small expected cell frequencies. *Psychol. Bull.* **1978**, *85*, 163–167. [CrossRef]
96. Scales, B.J. Qualitative analysis of student assignments: A practical look at ATLAS.ti. *Ref. Serv. Rev.* **2013**, *41*, 134–147. [CrossRef]
97. Junta de Extremadura, S.I.T.E.X. Centro de Descargas. Available online: https://sitex.juntaex.es/SITEX/centrodescargas (accessed on 23 July 2025).
98. Apaza-Panca, C.; Epiquen, A.; Nole, I.; Noriega, C.; Hidalgo, S.; Canova, J.; Ramos, E.; Taboada, M.; Alvarado, E. Immersive Technologies as Promotion Tools for Tourism. *J. Posthumanism* **2025**, *5*, 2417–2433. [CrossRef]
99. López-Santiago, C.A.; Oteros-Rozas, E.; Martín-López, B.; Plieninger, T.; Martín, E.G.; A González, J. Using visual stimuli to explore the social perceptions of ecosystem services in cultural landscapes: The case of transhumance in Mediterranean Spain. *Ecol. Soc.* **2014**, *19*, 27. [CrossRef]
100. Barthel, S.; Folke, C.; Colding, J. Social–ecological memory in urban gardens—Retaining the capacity for management of ecosystem services. *Glob. Environ. Change* **2010**, *20*, 255–265. [CrossRef]
101. Surová, D.; Ravera, F.; Guiomar, N.; Martínez Sastre, R.; Pinto-Correia, T. Contributions of Iberian Silvo-Shepherdal Landscapes to the Well-Being of Contemporary Society. *Rangel. Ecol. Manag.* **2018**, *71*, 560–570. [CrossRef]
102. Plieninger, T.; Torralba, M.; Hartel, T.; Fagerholm, N. Perceived ecosystem services synergies, trade-offs, and bundles in European high nature value farming landscapes. *Landsc. Ecol.* **2018**, *34*, 1565–1581. [CrossRef]
103. Buijs, A.E. Lay People's Images of Nature: Comprehensive Frameworks of Values, Beliefs, and Value Orientations. *Soc. Nat. Resour.* **2009**, *22*, 417–432. [CrossRef]
104. Beavington, L. A walk to the river: Nature and student mental health. *J. Adventure Educ. Outdoor Learn.* **2024**, *25*, 45–51. [CrossRef]
105. Vasilaki, M.-M.; Zafeiroudi, A.; Tsartsapakis, I.; Grivas, G.V.; Chatzipanteli, A.; Aphamis, G.; Giannaki, C.; Kouthouris, C. Learning in Nature: A Systematic Review and Meta-Analysis of Outdoor Recreation's Role in Youth Development. *Educ. Sci.* **2025**, *15*, 332. [CrossRef]
106. Pinto-Correia, T.; Azeda, C. Public policies creating tensions in Montado management models: Insights from farmers' representations. *Land Use Policy* **2017**, *64*, 76–82. [CrossRef]
107. Joffre, R.; Vacher, J.; de los Llanos, C.; Long, G. The dehesa: An agrosilvoshepherdal system of the Mediterranean region with special reference to the Sierra Morena area of Spain. *Agrofor. Syst.* **1988**, *6*, 71–96. [CrossRef]
108. Korpershoek, H.; Canrinus, E.T.; Fokkens-Bruinsma, M.; de Boer, H. The relationships between school belonging and students' motivational, social-emotional, behavioural, and academic outcomes in secondary education: A meta-analytic review. *Res. Pap. Educ.* **2019**, *35*, 641–680. [CrossRef]
109. Paniagua, A. The rural as a site of recreation: Evidence and contradictions in Spain from a geographical perspective. *J. Tour. Cult. Change* **2012**, *10*, 264–275. [CrossRef]

land

MDPI

Article

"Probably Much More Boring": Future Projections of Christmas Market Heritage Landscapes

Murray Parker [1,*], Dirk H. R. Spennemann [2] and Jennifer Bond [2]

[1] Faculty of Science and Health, Charles Sturt University, P.O. Box 789, Albury, NSW 2640, Australia
[2] Gulbali Institute of Agriculture, Water & Environment, Charles Sturt University, P.O. Box 789, Albury, NSW 2640, Australia; dspennemann@csu.edu.au (D.H.R.S.); jebond@csu.edu.au (J.B.)
* Correspondence: muparker@csu.edu.au

Abstract: Christmas markets are centres of place and culture, having heritage value through impressive histories and long-running traditions. With their array of congregation spaces, huts, and icons (Christmas trees, nativity scenes, carousels), they represent a cultural landscape en miniature, the structural elements being overlain by multisensorial elements (sounds, smells, tastes). Each of these factors are market-distinctive, as are space and place connections, with each heritage landscape being unique. Collectively, these markets represent a cross-national cultural landscape temporarily unifying communities with different cultural traditions. However, transition examination is limited, and missed experiences or future projections have not yet been addressed. Utilising on-site interviews at five Christmas markets in Germany and UK (Dresden, Mainz, Frankfurt, Offenbach, Birmingham) in 2023, we examine changes over remembered time—what was missed, what has emerged, and what future projections were envisaged. While each market is distinctive, overall trends included a decrease in artisanal products and a concomitant increase in gastronomy, an increase in tourism/commerciality, and diminished traditional experience—with factors including policy/administration control and shifting social demographics. Future projections included diminished ambiance, lessened authenticity, and more commercialisation. Our findings highlight potential conflicts considering transition/development across these cultural landscapes, and issues with respect to heritage and cultural integrity.

Keywords: authenticity; cultural landscape; intangible heritage; multisensory; sensory experience

1. Introduction

Cultural landscapes, as spatially circumscribed interactions of people with their environment, can have numerous manifestations. While the focus of most research tends to be on tangible landscapes in the form of observable and tangible traces of events and activities on a larger micro- or macro-landscape scale (e.g., structural evidence, field systems) [1,2], there are also intangible landscapes. They can be cultural constructs, manifesting themselves as toponymic landscapes, where settler colonialist societies supplanted the First Nations appellations of localities [3], indigenous landscapes, where locales are imbued with meaning and spiritual features [4], or they can have a form that is perceivable by a person's senses, such as smellscapes (e.g., spice bazaars) [5]. Indeed, the term 'soundscape' or 'smellscape' alludes to a sensory experience within a circumscribed physical space, being considered culturally important upon conveyance of social values.

Cultural landscapes can be permanent, such as landform modifications, or they can be ephemeral, such as election landscapes [6]. The majority of cultural landscapes literature examines past landscapes, most of which are a palimpsest of successive layers of cultural interaction on the same circumscribed geographical space (e.g., Refs. [7–9]. Given that most cultural landscapes research is intertwined with cultural heritage management which relies on hindsight [10,11], conceptualisations of future cultural landscapes are rare [8], although methodologies for futurist hindsight in heritage exist [12].

The intangible sensory experience of cultural landscapes has been previously acknowledged, evidenced by research into aural heritage pertaining to Zimbabwean soundscapes [13] and Chinese classical garden landscapes [14], visuality of industrial landscapes [15], tactility of rock climbing landscapes [16], and multisensory engagement (visual, olfactory, aural, tactile) of designed garden landscapes [17] and historic Indian water system cultural landscapes (tactile and visuality) [18].

Furthermore, marketplaces are recognised as dynamic cultural landscapes, typically being spaces of discernible sensory and multisensory focus and, therefore, intangibly representative. Prior research has investigated intangibility of social markets in part, such as olfactory characteristics of Indonesia's Chinese markets and Türkiye's Spice Markets [5,19], gastronomic experiences of Spanish food markets [20], and multisensory (total sensory) experience of German and Austrian Christmas markets [21].

Indeed, Christmas markets are recognised as having historical and cultural importance [22] with the sensory experience (visual, smell, aural, gustatory, tactile) combining to form an ephemeral annual cultural landscape within a given space [21]. Previous research has investigated visitor characteristics [23], visitor motivation [24], crowding issues [25], and use value of Christmas markets [26], but discussion of cultural heritage connections, cultural value, and intangible experience is an emerging topic of research [21,22,27,28].

As ephemeral cultural event landscapes deeply rooted in German history and tradition [22,29], Christmas markets straddle the traditional dichotomy of tangible and intangible heritage assets (Figure 1). They possess heritage value through the impressive histories associated with some markets (such as the Christmas markets in Dresden, Nuremberg, or Frankfurt) and a parallel long-running tradition (e.g., early fifteenth century), the settings in which they occur (primarily historic market squares), and the intergenerationally transmitted cultural practices of participation [22,30]. Recent research has uncovered other aspects of heritage at both long-established and contemporary Christmas markets through contributing factors such as portable material culture (Glühweinbecher, tokens) [31], tangible infrastructure (carousels, nativity scenes), and intangible sensory components (Christmas music sounds, Kartoffelpuffer odours, Glühwein tastes) [32]. Although they would qualify under the definitions of the UNESCO Convention for the Safeguarding of Intangible Cultural Heritage [33], they so far have not been afforded any formal recognition at the national level by the German Commission for UNESCO [34] nor by the various Ministers for Culture at the state level [35–41].

Using Christmas markets as a focus, this paper will examine the trajectories of change and development of cultural landscapes, and attempt to project what future cultural landscapes of such markets will look like. Christmas markets are an eminently suitable case study as they represent an annually reoccurring ephemeral cultural landscape that by virtue of its stalls and focal attractions (e.g., Christmas tree, nativity scene) has a tangible form, but that also possesses intangible manifestations in the form of soundscapes and smellscapes. In further exploring the heritage of Christmas markets as both tangible (place-centred, iconic infrastructure) and intangible (multisensory interaction) cultural landscapes, this study therefore sought to find how a sample of Christmas markets have changed over remembered time—what was missed, what has emerged, and what future projections of these cultural landscapes were envisaged.

Figure 1. Christmas market on the Römerberg, Frankfurt, 1936 (contemporary postcard, photographer unknown; Collection DHRS).

2. Materials and Methods

This study extends upon previous work by the authors investigating heritage perceptions of Christmas markets, using a parallel methodology [32]. For inclusion in this multiple case study [42], five Christmas markets were selected; four in Germany (Dresden Strietzelmarkt, Mainz Historischer Weihnachtsmarkt, Frankfurter Weihnachtsmarkt, and Weihnachtsmarkt Offenbach) and the fifth in the United Kingdom (Birmingham's Frankfurt Christmas Market, an annual market officially exported from Frankfurt). These markets were chosen due to the strong historical association of Christmas markets in Germany, with several German Christmas markets having been continually operating since the fifteenth century (e.g., Dresden's Striezelmarkt since 1434). Furthermore, German Christmas markets can be considered archetypal of a 'traditional' Christmas market, offering an experiential and social event which is recognised as culturally representative by both locals and tourists, often set in historic settings and encompassing rich sensory landscapes. Site selection included both historic and modern renditions of German Christmas markets, being geospatially and socially diverse, to incorporate a wide variety of site variables and thus a potentially more diverse range of perceptions (Table 1). The extension to include Birmingham's exported German-styled market further expanded site variables.

Data were collected through interviews across November/December 2023, targeting visitors (local, non-local), stallholders, and key informants—including representatives of market organisers and church figures of adjacent churches. Potential participants were approached on-site with the objective of exploring a large perspective range, including those from local and non-local visitors, and from across a range of demographics (e.g., age and gender). Thus, the sampling strategy objective was to collect a broad range of perspectives rather than to attain a fully representative sample. Semi-structured interviews were undertaken on location, or offsite as requested due to inclement weather. Key informant interviews were offsite (one online). Interview questions solicited perceptions of temporal change in markets (e.g., changes in market layout, stall composition, organisation, sensory experience, and visitor composition) alongside future projections, as well as missed experiences (if any) from past renditions. Interviews were undertaken in English and German, with the research team including a native German speaker for facilitation of multi-lingual communication. Length of interviews averaged twenty minutes, with a range from five to sixty minutes.

Table 1. Christmas market data collection specifics of those selected for inclusion in this study.

Market Site Location	Country	Established	Data Collection Dates	# Interviews (# of People)	Site Variables
Birmingham	England	2001	Nov 18–22	11 (n = 13)	Non-historic setting, large town, very recent market
Frankfurt	Germany	1401	Nov 27–29, Dec 10	9 (n = 15)	Historic setting, large town, historic market
Offenbach	Germany	1978	Nov 25–26, 30, Dec 1	5 (n = 10)	Non-historic setting, medium town, recent market (post-WWII)
Mainz	Germany	1975	Dec 1, 8–9, 11	9 (n = 21)	Historic setting, medium town, recent market
Dresden	Germany	1434	Dec 3–6	12 (n = 15)	Historic setting, large town, icon historic market

Fifty-eight interviews were employed in the original study. For the purposes of this study investigating temporal changes and projections, only repeat visitors from previous years were selected for inclusion, resulting in a total of forty-six interviews (Table 1). Informed consent was provided by prospective participants, and localness was self-identified. Procedures followed Charles Sturt University Human Research Ethics Committee protocols (Protocol Number H23548).

Coding and Analysis

Interviews were either audio-recorded or, in cases where participants declined consent for recording (n = 5), detailed notes were taken instead. All recordings were digitally transcribed and manually reviewed for accuracy. Each interview was coded according to its site location (e.g., Dresden = D1, D2; Mainz = M1, M2), with multiple participants at a given site labelled sequentially (e.g., Dresden = D1-1, D1-2). Survey data were also location coded (e.g., Dresden = DS1, DS2) (see Supplementary File). Textual data were thematically analysed in NVivo 12 Plus across two coding rounds, with nodes generated iteratively as key themes emerged [43]. The initial coding framework was informed by findings from an earlier pilot study [44].

3. Results

3.1. Birmingham

The Birmingham Frankfurt Christmas Market is a comparatively recent market, consisting of approximately sixty stalls since its formal establishment in 2001. It runs from the western core area of the architecturally distinctive Victoria Square eastwards across New Street—a narrow street lined by commercial buildings. The market is visually attractive with high levels of stall illumination and iconic infrastructure, and becomes a hive of social activity for locals—especially on evenings and weekends.

For this market, local visitors commented on negatively perceived recent changes, such as the increased price of offered products (notably Glühwein) being a major catalyst for reduced visitation "it's got progressively more expensive over the last few years to have drinks, so we now prefer to go somewhere else" [B1]. Other observations included positively perceived changes, such as the addition of LCD screens and the more noticeable presence of security staff alongside car barriers for pedestrian protection. However, some non-local visitors commented that little had changed recently, and others simply that numbers of seated eating places had reduced.

Other changes were also evident as a result of management adjustments reacting to the COVID-19 pandemic. There was a clear decline and/or change in stall numbers and offerings post-pandemic, with an organiser revealing that the market footprint was lessened in 2021 to reduce congestion, subsequent to the total cancellation of the market in 2020. A post-COVID one-way pedestrian system was reported to have significantly improved visitor flow and lessened apparent crowding issues.

Stallholders also commented on market size and composition change since the introduction of Brexit, resulting in fewer stalls operating, with a proportional increase in those offering foods and a concomitant decrease in crafts (e.g., Christmas decorations and nutcrackers). An account of overall composition evolution (including Brexit) was explained by a long-term market organiser. After an initial pilot market in 1997, the Birmingham Christmas market has run annually since 2001 (COVID year 2020 excepted), utilising German-sourced products and labour. Original stalls included Glühwein, bratwurst, and marzipan vendors, alongside a number of craft stalls such as those selling glass baubles and other Christmas ornaments. With market expansion from only ten stalls as a pilot, alcohol stall numbers increased concomitantly with the lifting of alcohol restriction zones, effectively forming a licenced premises over the duration and footprint of the market. Children's programmes were correspondingly developed, as were adjacent events on nearby Centenary Square, which included a Ferris wheel, an ice rink, and an English craft market. Of significance, however, was the implementation of Brexit in 2020, due to complications and increased expense regarding products and staff—with the notable consequence being that vendors have since dropped out and certain offerings (such as craft and gift stalls) are now considerably limited. Recognition of compositional change enabled management's attempts to rebuild what was lost, with actions including the targeted introduction of a craft stall in 2023 and a planned introduction for two more in 2024, and the re-supply of some iconic 'non-returner' products (such as paper stars)—enacted through other larger operating trader's entities. Stallholders commented that established stalls such as those selling crystals, ceramics, and gifts were recently offered additional stalls to manage by the market organisers, with the aim to increase the footprint of the market through 2024 and beyond.

Stemming from the above, future projections of the Birmingham market were centrally connected to immigration policies associated with Brexit. An organiser advised that product authenticity could remain to an extent, with German beer, crafts, and foods being sourced and imported directly, although at increased cost through customs agents. However, items not considered as having 'authenticity issues' would be sourced locally from around the UK, such as serviettes and incidental foodstuffs (lettuce for example).

The central issue moving forward related to staff and complicated immigration rules; notably, that new staff who own, run, or manage markets are explicitly excluded from a suitable visa route. Whilst a 'frontier worker permit' (costs covered by the organiser) allows for 180 days of work annually in the UK for those previously working pre-December 2020, and descendant children aged 18 to 21 may apply for pre-settled status through parental frontier worker permits, such permits expire after five years. Whilst management accepted renewal costs, it was revealed that around 100 individuals needed renewed permits in 2026. The issue was highlighted that as individuals retire, have health issues, or become more family-focussed, core German staff and stallholders are unable to be replaced as no immigration route is available. The age profile of current staff suggested that an eight to nine-year timeframe (~2031–2032) results in a large tranche of people reaching retirement age, and whilst local staff can be readily hired to undertake 'back of house' duties, the core team of supervisor and managers will be profoundly affected, being not sufficient in number to run the market in its present rendition. Reasonable future projects therefore entail that, although some kind of market would be put into the space—"for Birmingham without a Christmas market now would be unimaginable, just unimaginable" [B11], it could not be an authentic German Christmas market. Being managed by a UK operator, "they might do German style, but it will all have to be built and sourced and staffed from the UK because without a change to the law, we just can't get the staff in to do the work" [B11]. Critically, resultant authenticity issues were seen as the most significant challenge to the market moving forward, and significant to the social landscape of Birmingham as the current rendition remains as "the jewel in the crown of the Christmas package" [B11].

3.2. Frankfurt

Germany's Frankfurt Weihnachtsmarkt comprises over 200 stalls, with the market traversing a space north from the river Maim to Hauptwache, a distance of approximately 500 m. Whilst the core area has been centred on Römerberg since 1401 (a square surrounded by fifteenth century buildings) (Figure 2), the market now includes other distinctive spaces including Paulsplatz (with plane trees and commercial buildings), and the northern Roßmarkt (an open square bordered by modern skyscrapers).

Figure 2. Christmas market on the Römerberg, Frankfurt, 2023 (photograph DHRS).

Changes to this Christmas market were reported over both temporal and spatial scales. A regularly returning non-local commented on product price increases together with a substantially higher number of people now visiting the market, but that actual stall composition had not changed over the last ten years. Another local commented on the market's size expansion, being initially based on the Römerberg and since expanding into other districts, being negatively perceived and considered "just become too much, I would say" [F9-2].

Market organisers commented that security concerns and responses to the COVID pandemic were catalysts for market expansion, with very tight rows now making way for more space, based on safe visitor distancing requirements. This development resulted in an altered feel of the market, as closer lanes engendered more jostling between visitors. One local in her seventies commented that they had become accustomed to lower numbers of people in the pandemic time generally, and that current large numbers of visitors post-pandemic were a major consideration to constrain market visitation to quieter midweek times, with weekends being actively avoided. Interestingly, an organiser noted that the spatial increase now offered on the Römerberg currently adheres to fire department regulations, but that the more compact previous rendition of the market also adhered to regulations ten years prior—suggesting that future regulations will require further layout and design modification.

Footprint expansion was explained to be a reaction to contemporary visitor demands and expectations, such as the inclusion of the Pink Christmas market in 2012 (an LGBT+ friendly addition) and the more modern market on Rosmarkt in 2021. As a necessary post-Corona step, the newest market area of Rosmarkt was designed to generate more space for visitors in a modern style. With a skyscraper setting and located in the heart of Frankfurt city, it was described to perfectly contrast the traditional Römerberg Christmas market with a contemporary 'American' rendition, much like "the North Pole and the South Pole" [F2-2], shifting away from the traditional style across the market's entire footprint.

Frankfurt locals commented on other aspects of increased modernity of the market, such as the addition of a modern carousel, and comments were made by an 89-year-old that past market renditions were much simpler and modest, with modern iterations being glitzier and more ostentatious. Some locals commented on the preference of past smaller, more intimate markets, "Das war früher schöner [That used to be nicer]" [F6-1], generating disinterest of attending current renditions. Similarly to Birmingham, numerous visitors commented on the increased prices of products (notably Glühwein), and corresponding agreement of a shift away from specialist handicrafts like Hessian pottery to general commercialisation.

Other changes which were displeasing included the evolution to a modern container-style vendor stall (being easier to erect) from the more charming imperfect stalls of the past, and the shift from traditional incandescent bulb lights to LEDs, with an organiser stating "The light is warm also, but I don't know what, but I'm missing something. Maybe the defect one" [F2-2]. Of interest here is the recognition of change being important to locals and those deeply connected to the market: "I guess our normal visitor won't see the difference between the lights (. . .) It's like always experts will see it and newbies don't see it" [F2-2]. Such comments highlight that market familiarity engenders higher levels of connection and recognition. Other disliked changes include the shift away from traditional Christmas music to pop/disco music, as remarked by a 66-year-old local.

Future projections of Frankfurt's Christmas market were largely linked to local policy and management, being considered as notable concerns for the overall market experience for future generations. Organisers raised the probability of the market being "much more boring in 50 years" [F2-2], with the potential for many things being forbidden in the future through implementation of stricter rules to make the market safer, such as no open fires, lower alcohol limits on glühwein, and tighter control of foodstuffs. Additionally, negative future projections were also linked to climate change. An organiser considered that significant changing local demographics would not result in an impacted Christmas market, but that climate-change-induced increased temperatures would fundamentally change the market's essence.

3.3. Dresden

One of the oldest markets in Germany (proudly since 1434), Dresden Strietzelmarkt, continues to be located in the Altmarkt, a square bordered by architecturally distinctive buildings and the adjacent Kreuzkirche (Figure 3). It is considered a 'must-visit' traditional German Christmas market. Over two hundred thematically decorated and illuminated stalls are squeezed into the space, with several visual icons adorning the market, including an historic double-decker carousel and an impressive candle arch at the primary entrance.

Figure 3. Christmas market on the Altmarkt, Dresden, 2023 (photograph DHRS).

The Strietzelmarkt was perceived to have undergone several changes over remembered time, both temporally short and long. In the short term, the market was reported as not running for two years due to the COVID pandemic (2020–2021). On resumption, younger Dresden locals commented how pleasing it was to see the market return to a size comparable to pre-COVID times, as the first rendition post-COVID hiatus was limited to around 100 stalls. Compared to the 2022 rendition—stated as being "so small and you didn't get this feeling like today" [D1-1]—the 2023 version was significantly larger, with one local highlighting the positives "I'm happy that Christmas markets can be so big [as] before Corona because I missed it" [D1-1].

Over the long-term, older locals explained how the Striezelmarkt used to be limited locationally to the Altmarkt, but has since spread to the side streets, towards the Elbe, and continually expanded all the way to Albertplatz. Comparatively new additional market elements were the Schwibbogen (candle arch) and the pyramid (from around 2007). Another major change included the titled narrow rows of "Soldatenmäßig [solider-like]" [D5] stalls from the GDR period being developed into a more organic arrangement around 2008. One long-established stallholder explained that the preceding rows created equal opportunities for stallholders to purvey their wares without the recent problem of "really dead corners" which "looks good, but once you are here, it's very confusing" [D10]. This alteration, combined with the addition of the subterranean garage, fountains, and round stalls, was explained to have taken much of the old 'flair' away, and the previous market design was missed by some locals. However, local preference of design was mixed overall, with some preferring the less dense market as one aged, and others preferring the original organic rendition.

The sensory experience of the market was also reported to have changed temporally. Some locals drew attention to the improved visual amenity of the space over time, as during the GDR times (former East Germany) the surrounding sandstone buildings were black because they were covered in soot. Mixed feelings were raised—the black buildings made the lights more intense as a child in the past due to the visual contrast, and that modern market renditions look more colourful now due to the absence of black. Others drew attention to the diminished aural amenity, with recorded sounds being noticeably absent as a result of licencing issues and substantially increased costs, with national music no longer played. This loss of traditional sound was highly missed by older locals, with the additional feeling or unique flair that was engendered through the aural realm being notably absent.

Comments highlighted the deficiency of previous children's evening programmes, apart from the recent inclusion of a children's bakery over the last 15 years. The dearth of stalls offering interactive craft opportunities for children (painting, Christmas ball decoration, singing of German Christmas songs) was highly missed by younger locals, being considered a reminisced childhood activity, engagement of which had significant personal importance.

Older locals compared the current market to previous renditions operating during the GDR, which focused more on crafts and handicrafts (knitting, stonework) before a gradual change in composition to less handicraft and production post-GDR. Earlier renditions also comprised fewer (and older) booths offering greater product variety. It was observed that the modern market offered comparatively more gastronomy, and most locals commented on the ubiquitousness of food and Imbiss (snack bars), alongside the general increase in prices. Where craft products were currently offered (like Erzgebirge products), they were frequently considered "alles mehr so industriell [more industrial]" than in the past [D9-2]. Specifically, artisan products like glass blowing were reported to have ceased, with the last stallholder passing away in recent years and no family member wishing to take the position on. Other artisan products in decline included hand-crafted brooms, hand-carved Erzgebirge, Swiss-carved church figures, and important/nationally significant crochet and lacework (consisting of five or six vendors previously). A pottery stallholder revealed that around 30 years prior there were 12 stalls selling pottery from areas including Lausitz and Thüringen, with this merchant now being the single operator selling such handicrafts.

Rationale for artisanal decline was pronounced multi-fold. Firstly, it was explained that the stallholders were generally of an older population, with few younger workers employed. That the central problem "lies in the younger generation" [D10] was often reported, whereby they shy away from difficult work in the cold, demand comparatively high prices to work despite being inexperienced, and/or prefer to work in heated huts of Glühwein stalls where they are supplied with free drinks and food. Secondly, the modern row positioning limits opportunity, with a difficult position in a comparatively obscure row returning only 20% of the intake compared with a highly prominent position. Additionally, it was explained that because of price increases for energy and other costs, craft traders cannot simply afford to run a stall, but, conversely, sausage traders can: "Wir Händler so mit unserem Handwerk hier, wir sagen immer, gegessen wird immer. Dafür ist immer Geld da. Aber für das hier, so, damit gehen wir kaputt. [We traders with our craft here, we always say, we always eat. There is always money for that. But for this, like this, we're going to break.]" [D7]. Finally, the COVID pandemic seemed to account for the final straw for the artisanal merchants: "Und durch die Pandemie sind sehr viele Stände weggefallen [And due to the pandemic, a lot of stands have been lost]" [D6], with many providers not returning post-pandemic for a multitude of reasons, including simply not being able to finance the business. Most interesting is that the social community of the stallholders ceased to exist post-COVID as previously, with "Auch die Geschäfte, die Händler, die jetzt noch da sind, die sind härter geworden [The shops and dealers that are still there, have also become tougher]". Vendors are now forced to look after themselves and become more commercially focused, which, as one stallholder reported "das ist das Hauptproblem geworden [that's become the main problem]" [D7].

Results of these issues are clearly observable. A stallholder explained that the Striezelmarkt could harbour around 260 individual stalls prior to 2008, and market organisers correspondingly received around 550 applications. For the 2023 rendition, of the 225 stall positions on offer, an application rate of approximately one-to-one ensured reasonable success for any potential future stallholder. It is obvious how limited supply of stallholders could create substantial shifts in product choice and availability. As an example, to sustain stall numbers as desired by management, it was reported that stall categories have been substantially diversified, with around five to seven around forty years ago, later increasing to 15 or 20, and comprising 44 different categories in 2023.

Aside from composition, the Striezelmarkt has undergone other changes. Comments were made that there were no dangerous situations in earlier times, so the market could be more tightly packed safely. It was observed that the recent inclusion of bollards, barriers, and police is now always a consideration "das schlummert ja am Hinterkopf [that's at the back of my mind]" [D9-2], which softens the mood and creates secondary thoughts for attendance.

Being a Christian-centred event, it is interesting to consider how locals consider the religious connections of the Christmas market over a temporal period. Comments were made that there was a significant decline in terms of visitors at religious services and events in an adjacent church over the last ten years, with the COVID pandemic accelerating such trends. It was even reported that current church attendance reflected GDR communistic state times—when church traditions were totally unsupported. A church figure made clear that church concerts were well attended by visitors, but people "sobald die irgendwas mit Gottesdienst oder beten oder mit Service, dann sind sie zurückhaltend [as soon as they have anything to do with worship or prayer or service, then they are reserved]" [D12], and also that the ambience and history is appreciated, but too much service and too much religion is considered a problem. How much this is directly connected to market visitation and associated transformation is debatable, but considering deep religious connections to the Striezelmarkt's opening ceremony, such trends are notable and worthy of attention.

When probed to predict future market iterations, a stallholder replied that there is real uncertainty resulting from the retirement of the older generations, with the example given of the sole remaining traditional lacework merchant being in her seventies. Another crafts stallholder highlighted the difficulty in attracting younger people to purvey traditional handicrafts because of the difficult working conditions. The perspective was made that stalls offering handicrafts will become significantly fewer in the future. If, as predicted, the continued trajectory of diminished handicrafts continues through future intergenerational change, one stallholder suggested that market organisers would need to incorporate an increased number of product categories and diversify the market even further to retain financial capability, further altering market composition.

Modernisation of the future market iterations was also viewed as a real issue. A stallholder projected increased modernisation throughout the market, with the result that as traditionalism becomes lessened, the recognisable spirit of the market "dass dann das gewisse Etwas weggeht [that the special something would then go away]" [D7]. It was anticipated that special seasonal products offered would be readily available (and more cheaply) for purchase in standard shops or even online (e.g., Amazon), resulting in the loss of the market's unique identity, the demise of respect for craftsmanship and quality of work, and a break in the depth of connection with respect to culture and community through loss of sentimental inherited products.

Finally, similar to a notion observed in Frankfurt, older Dresden local visitors reflected that the warmer winter temperatures as a result of climate change would make the market much less interesting, as cold temperatures are essential to the Christmas market experience; "wir gehen schon gerne, wenn es kalt ist [we like to go when it's cold]" [D9-2].

3.4. *Mainz*

The Mainz Historischer Weihnachtsmarkt is a relatively small, visually attractive market, comprising approximately 100 stalls within the Marktplatz and surrounds (Figure 4). Whilst being a comparatively recent addition in the historic calendar (since 1978), the current market rendition has links to past Christmas markets in Mainz centuries earlier. It is a popular stop for tourists travelling on Rhein River cruises, and also for locals for regionally distinctive gastronomy.

Figure 4. Christmas market on the Marktplatz, Mainz, 2023 (photograph DHRS).

The Mainz Christmas market was reported to have undergone several changes, again both short- and long-term. Over the short-term, a stallholder observed that there was a notable increase in visitor numbers over the past six years, and a corresponding increase in attendance of family groups. The comment was made that although people were generally more friendly in the past (and now wander around with a more morose temperament), if the stallholder were to engage them in a pleasant manner, they would be paradoxically happier than before, with a vitalised Christmas spirit. Explanations were offered in that visitors were more stressed now, with high prices resting heavily on their shoulders. Another comment was made that locals' behaviours have shifted since the pandemic, with a subconscious pursuit of distance between visitors now taking place, interrupting the carefree nature of a busy market experience.

Young locals highlighted problems concerning significant price increases, suggesting that a preferred option would be to look for cheaper market options in smaller towns or communities. Older locals also commented on this issue, and whilst one dismissed it as "people pay it because of the atmosphere" [M1], the majority of locals viewed this a major drawback of change. Older locals also commented on the perceived market footprint increase over time, and the increase in security which was considered a major concern.

Multiple locals reported that stall locations were perceived as highly consistent over recent years. Such certainty created an anticipation of food which was considered comforting and provided familiarity and structure which was satisfying. Long-term stallholders, however, observed that in the early market renditions, stalls were scattered around much more and not so formalised as a concentrated unit being centred on the square. One stallholder commented that the increase in footprint was considered acceptable, but it was clearly stipulated that further increases were undesirable and unsustainable.

Other changes in market composition were offered. It was reported that previous renditions offered themed events like performances specifically for children, but that these are now no longer presented. A long-term stallholder also commented that the older carousel used to be double story with horses to ride on, unlike the single-story ride currently presented. Similarly to Dresden, recent issues of the lack of broadcasted Christmas music were highlighted as being critical. Running sporadically, background Christmas songs of Mainz Weihnachtsmarkt were "einfach schön [just beautiful]" [M7] and lifted the mood and spirit among the people, and the cessation of music across 2023 was considered sadly missed.

Whilst stall numbers have increased temporally, it was revealed that stall variety had not—there was simply a larger number of stalls purveying similar wares/products. Glühwein stands, for example, had increased from four to eleven. A stallholder acknowledged that whilst the public generally like a larger market, the increase in Glühwein stands changes visitors' behaviour by allowing more alcohol to be consumed (through shorter queues), with the evening attracting more young people intending to become inebriated, with the market dynamic changed as a result. It was also observed that with the reduction in handicrafts, spatial expansion requires an increase in gastronomy stalls but with fewer people buying products at any given stall—reducing profits for stallholders and an overall 'loss' for everyone. More recently, the market was considered to have become more commercial in feel overall—more stressful, hectic, and profit-oriented.

Long-term stallholders remarked that the Christmas market had doubled in size, shifting from a totally compact market on the square to the more expansive, sprawling market village around twenty years ago. Also considering the long-term, it was repeatedly reported that the market was presently centred less on craftsmanship offerings (crafts, carvings) and focused more on consumption, with food stalls now being considered so prevalent that they were deemed overwhelming. The introduction of low quality, Chinese-manufactured Christmas presents at the expense of rural German craftmanship was frequently observed as being highly displeasing for a traditional Christmas market, with stalls like Käthe Wohlfahrt being denoted as a major culprit with its inception around 2011. The demise of craftsmanship stalls was considered directly affected due to the shift in handicraft business practices. With the provision of non-local Christmas ornaments at reduced prices, established stalls offering higher-quality local crafts became priced out, could not compete, and ultimately disappeared from the Christmas market. Examples were given whereby two well-established stalls selling Christmas tree baubles and many selling wooden knickknacks (kleine Nippes) had completely disappeared as a result. The power of Käthe Wohlfahrt was evidently pointed out by one stallholder, who explained that the company directly opposed instructions to set up on the lower outskirts of the market, and fought for a prominent position in which to trade—even so that a large traditional carousel had to be displaced: "Ja, wir haben immer zwei Karussells. Aber das war hier, wo die steht, war ein Karussellplatz [Yes, we always had two carousels. But that was here, where it [Käthe Wohlfahrt] stands, was a carousel place]" [M5-2].

Policy and governance were offered as central reasons for specific future projected renditions in Mainz. One stallholder suggested that management change to a more centralised department, such as 'Tourism Mainz' for example, would generate an entirely different market experience, with increased fees and commercialisation as a result. Issues of intergenerational recruitment were presented similarly to Dresden, with a stallholder offering that there will be continued succession problems with the craft stalls, with the opinion offered that if a given family is not large enough to continue, the stall will ultimately disappear.

Modernisation issues were also linked to anticipated future management processes and planning, with a shift from the traditional experience towards greater commercial development and expected concomitant tourist-centric exploitation. Stallholders gave the opinion that markets were expected to become more modern, with a foreseeable increase in potential conflicts between management and stallholders as organisers adjust, modify, and constantly innovate. Multiple opinions were given that if the market was allowed to increase its size further, its distinctive atmosphere and identity would dissipate, and as such (from a stallholder's perspective) the market was desirable to be maintained in its current form and style.

3.5. *Offenbach*

Being a relatively small market (32 stalls), Weihnachtsmarkt Offenbach is largely aimed at local visitors. Established in 1978, it consists of gastronomical-centred offerings, the stalls of which are highly unique, with some being highly decorated and others presented as food trucks.

Local visitors commented that programmes, facilities, and huts presented at the market have been largely consistent over the last ten years, but with a generally appreciated expansion in market size over the previous decade. However, considering the longer term, one local commented that the market has lost much of its character, with resultant diminished number of visitors, a halving of size since around 30 years ago, and a loss of the traditional Christmas smells—being "nichts mehr [nothing anymore]" [O3-1]. Other locals commented that the present rendition is "alles so kommerziell [all so commercial]" [O1-2], and although typical smells of Glühwein remain, the lack of Christmas music makes the market feel like nothing to do with Christmas anymore, and now deemed being "nur Geldmacherei [just a money-making scheme]" [O1-1]. It was commented that the lack of traditional Christmas bakery smells contributes to forming a depleted atmosphere, and that the addition of live carolling would be "ein Traum [a dream]" [O3-1] and would substantially contribute to the market befitting a 'wonderland'.

In line with other markets, long-term stall composition had reputedly changed over time, with the loss of handmade products such as lamb skins, handmade cups, and jewellery. Despite this, a long-term stallholder made clear that the smaller number of stalls presently tendered ensured a market which is not overloaded with multiples of each offering—a design which was suggested to be preferred by visitors.

Critically, the market space was reported to have diminished in quality since the pandemic, with a decline in general quality of the shopping district and increased crime in the region, with the result being that locals often tend to prefer to shop in nearby Hanau or Mainz. Specifically to the market, the visitor experience was considered to be somewhat lessened. Older locals reported that the atmosphere was completely lacking due to an absence of Christmas music as a result of increased GEMA royalty licence fees. Visual amenity was also affected by the installation of concrete blocks prohibiting driving attacks, also affecting atmosphere by engendering a background fear. Flowers decorated previous renditions of markets before a subsequent shift to other lighted decorations, with these now being considered under threat due to rising energy costs over the four-week operating period—an extended season to assist stallholders' profits in light of lost income due to pandemic closures. It was also observed that post-COVID, customer demographics have changed somewhat, now being more centred on the 20–30-year-old age bracket rather than the older generation—with the younger generations being more interested in the traditional experience.

When considering future projections, a stallholder in her forties expressed fears that alongside general closure of shops and shift to internet shopping, the reduction in people in the streets would result in a corresponding decline in visitor numbers, a decreasing market footprint, and a conceivable obsolescence of the market over a twenty-year time period. The opinion was offered that larger, more tourist-focussed markets such as Nuremberg and Frankfurt would continue to operate, but smaller markets such as Offenbach might be forced to discontinue.

4. Discussion

Despite unique findings for each of the five Christmas markets, themes of transition are observable across all sites. There was a notable shift away from traditional handicrafts towards gastronomy and mass-produced products due to (amongst other reasons) increased economic pressures, commercialisation, and problems of intergenerational recruitment (Table 2). Such shifts were observable in all markets, being similar across all town sizes and market histories. Indeed, the predicted loss of handicraft offerings would result in the demise of industry heritage representation throughout entire markets. This is known to be particularly important in Dresden, as active engagement with traditional products (e.g., lacework, glass-blowing, wood carvings) at the Striezelmarkt was deemed vital in communicating historic and social practices of the local area's industry to subsequent generations [32]. The importance of such handicrafts to the heritage landscape was not raised by respondents at the other four markets studied, but if similar sentiments are held there as well, then it would be critical to address any predicted demise.

Moreover, the concomitant proportional rise in alcohol stands has resulted in some markets (notably Mainz) feeling more like drinking venues than social functioning spaces, with traditional experiential formats being lost. Whilst explanations from the supply side have been identified earlier (high turnover, desirable products), questions arise regarding demand. Recent analysis of Christmas market layouts has revealed that consumption of gastronomic offerings has become a major focus of Christmas market visitation, and has even accelerated post-pandemic, as evidenced by the increase in such stalls across several German markets [45]. The question remains regarding what causes this increased demand. How much of the increased proportion of alcohol offerings is linked to intensified tourism or changing local visitor demographics? If it derives from increasing visitation by younger generations, then implications arise for older generations who typical hold onto 'traditional' heritage ideals. If it results from increased tourism (national or international), then ramifications for local visitors are obvious, evidenced by those who seek more 'traditional' options in smaller towns or communities.

Other transition themes emerge from amended government policy and law, such as authenticity issues of Birmingham's market caused by Brexit, with the end result being the provision of a stylised Christmas market, but not German or 'traditional' in essence. Expectations of modernity and 'improvement' in Mainz also suggest redrawn future market styles, with prospects seemingly focused more on commercialisation and tourist appeal rather than on imparted heritage values or cultural importance. Increased prices and lessened stall diversity has made markets feel more commercial, and stall layout variation has resulted in an altering of visual amenity and, by extension, the entire intangible atmosphere. Additionally, attitudes upon return post-pandemic have resulted in some markets having a lingering sense of fragmentation and diminished cohesion, alongside changed visitor behaviour to avoid peak periods of weekends and evenings—notably so for older age visitors. Importantly, whilst footprint expansion has often created larger markets deemed 'too much' than desired (Frankfurt, Mainz), status quo operations would naturally result in saturation through increased visitation.

Table 2. Data summary of the five studied Christmas markets—up arrows denote increase, down arrows denote decrease, horizonal denote similar continuation (for sensescape change: up arrows denote improved, down arrows denote diminished).

Issue		Birmingham	Frankfurt	Dresden	Mainz	Offenbach
Visitation	Visitors	↓ short term	↑	↑	↑ short term	↓ long term
	Profile				↑ younger and family groups	↑ younger
Stallholders	Ageing	Yes		Yes	Yes	
Stall numbers		↓ short term ↑ long term	↑ long term	↑ long term	↑ long term	↑ short term ↓ long term
Layout	Footprint	↓ short term ↑ long term	↑	↑	↑	↑ short term ↓ long term
	Laneway size		↑	↑		
	Layout change			Yes—soldier like to organic arrangement	Yes—more formalised and concentrated	
	Seating	↓				
Stall offerings	Variety	↓ management working on this	→	↓	→	↓ long term
	Foods	↑		↑	↑	
	Glühwein	↑ long term			↑	
	Crafts	↓ management working on this	↓	↓	↓	↓
Prices		↑ short term	↑	↑	↑	
Children's programmes		↑ long term		↓	↓	
Security presence		↑		↑	↑	↑
Commercialisation /Industrialisation			↑	↑	↑	↑
Modernisation		↑	↑			
Sensescape change	Soundscape		↓	↓	↓	↓
	Visualscape	↑	↓	↑ and ↓	Yes—not stated	↓
	Smellscape					↓
	Tactile (climate change)		↓	↓		
COVID-19 related		↑ visitor flow ↓ crowding ↓ stall numbers and offerings	↑ market footprint ↓ crowding	↓ artisanal products ↓ stallholder social community	Subconscious pursuit of visitor distancing	↑ younger generations ↓ older generations
Future projections	Projections	Continuation but not as an authentic German Christmas market	↑ sterile environment further layout and design modification	Uncertainty of artisanal offerings ↓ unique identity ↓ respect for craftsmanship and quality of work ↓ depth of connection with respect to cultural and community ↓ sentimental inherited products altered market composition	↑ centralised department ↓ traditional experience ↓ craft stalls ↑ commercial development ↑ tourist-centric exploitation ↑ fees ↓ distinctive atmosphere and identity	↓ visitor numbers ↓market footprint conceivable obsolescence of the market
	Expected causes	Effects of Brexit policy and immigration rules Ageing staff profile	Tighter regulations Local policy and management	Ageing stallholders Difficulty attracting younger workers Modernisation	Policy and governance Expectation of modernity Intergenerational recruitment	Shop closure Internet shopping

Most critical, however, is the decline in the intangible sensory and experiential landscape of the markets, notably pertaining to the visual, aural, tactile, and olfactory realms. Loss of traditional music (and indeed all music) is palpable, characteristic smells of baked goods/incense are conspicuously missed (notably so in Offenbach), visual amenity has changed with the introduction of container-style stalls and LEDs (Frankfurt), and tactile childhood encounters were largely absent (Dresden). These issues form critical discussion for future projections by multiple participants. That an organiser of the Frankfurt market suggested that their Christmas market will become 'much more boring' in the future speaks profoundly of stricter regulation, climate change, and modernisation. Conversely, that Birmingham's modern market revealed an improved visualscape raises further discussion on the status quo of predicted sensescapes, being not drawn into community expectations. Furthermore, as this study forms part of a larger project exploring multisensory heritage of Christmas markets, it would be doubtful if such elements would have come to light so fruitfully if our focus was solely on visual or physical/spatial components, rather than a holistic multisensory analysis.

The transitions highlighted above ultimately affect both authenticity and identity of a given market, with overall experiential feelings being more generic and less rooted in the traditional—a shift away from the 'magical atmosphere' to more functional, and possibly sterile, environments. If authenticity, traditional quality of work, and sensory richness continue to decline as predicted, the future experience of these cultural landscapes is at risk.

Importantly, the gradual change in offerings (to more alcohol and modern wares) perhaps diminishes any "Christmas-ness" of a Christmas market. This is additionally so considering the loss of Christmas music, a decline in religious connections, and lessened interactive Christmas-themed activities. Recognising changed values in the German population, such as the increase in importance of friends and, therefore, social engagement [46], we need to consider at what point will Weihnachtsmärkte be considered 'Christmas markets' in the future or simply just 'markets' that occur at Christmas time as they used to be in the pre-Bourgeois era. The whole essence of Christmas-ness at these markets appears to be diminishing—whilst there are some positive efforts to retain this existence, identified changes (whether perceived as positive or negative) are ultimately changing the fundamental anchor (Christmas-ness) of the markets. What does it mean if Christmas markets and the ephemeral cultural landscape that they represent are no longer about Christmas? Could this be conceived as a 'win' for living heritage or a 'loss' of traditional heritage? While numbers of practicing Christians in Europe are recognised to be decreasing [47], that does not necessarily equate to Christmas-style traditions being desired to be discarded. Indeed, there has been an almost doubling of church attendance on special holy days such as Christmas over the last twenty years in Germany, and a reported increase in religious importance across Germany [46]. Such consideration has significant implications for local heritage Christmas market landscapes, and indeed national and/or cross-national landscapes during the Advent period. This may be particularly so in areas of strong immigration incorporating multi-faith populations such as Offenbach, and although beyond the scope of this paper, presents fruitful avenues for future research.

The importance of a multisensory investigation and analysis demonstrated in this paper raises critical contemporary issues considering landscape heritage not previously addressed in the literature. Whilst recognising sample limitations, our study has merit and finds that Christmas markets represent multidimensional ephemeral cultural landscapes. Seen through the lens of 'traditional' spatially circumscribed tangible approaches to cultural landscapes, Christmas markets with their arrays of stalls and focal points (e.g., Christmas trees, nativity scenes, carousels) can be equated with temporally constrained miniature villages of a non-residential character. These miniature villages are erected within the confines of an urban open space which acts as a giant stage with the commonly historic buildings providing for a theatrical backdrop.

While the spatial positioning of the focal points tends to be static, providing a temporal consistency, the arrangement of the stalls in rows or as agglomerations with pedestrian circulation spaces echoes urban planning concepts. This is reinforced by the zoning of stalls, on the one hand ensuring a mix of products, while on the other relegating less 'Christmassy' offerings to the periphery. From an organiser's perspective, the ephemerality of a Christmas market not only allows adjustment with the zoning but also permits wholesale restructuring. That can be a dramatic redesign as in the case of the Dresden Strietzelmarkt, where a rigid linear, street-like arrangement of stalls in 2007 was changed into a conglomerate-type of arrangement in 2008, or this can be adjustments to the density of stalls in response to the COVID-19 pandemic [45]. The echoes of urban planning concepts resonate in the Frankfurt Christmas market, which has seen continual spatial expansion akin to urban sprawl. It extended from its traditional core on the Römerberg onto the Paulsplatz in 1973, the Neue Kräme and Liebfrauensberg in 1996, the Liebfrauenstrasse in 1997, the Zeil in 1999, the Mainkai in 2003, and finally the Roßmarkt in 2023 [29]. Even the concept of satellite towns is represented in Frankfurt with the establishment of the LGBTQIA+ 'Rosa Weihnachtsmarkt' (since 2012) [48].

Setting aside some tinkering with the spatial distribution of gastronomic offerings, the observed longitudinal structural changes discussed in this paper are unlikely to substantively alter the 'urban planning' aspects of the ephemeral cultural landscape that a Christmas market represents. As this paper has shown, however, Christmas markets represent another, distinct cultural landscape which finds expression in the intangible, multisensory realm. The multisensory realities of Christmas markets, which are recognisable and experientable with visual, audio, olfactory, and tactile senses, manifest themselves in the spatial domain as sightscapes, soundscapes, smellscapes, and touchscapes. Cumulatively, the sensescapes can find expression as a melange where individual senses compete for dominance, or as a palimpsest where stimuli for some senses dominate spatially, but where other sense stimuli persist in residual form.

The valuation of cultural landscapes, as that of all heritage assets, is subjective and shaped by the enculturation and the epistemology of the valuer [49]. The underlying tenet is that a valued asset is observable and that this observation can be replicated by others. When we are considering and evaluating ephemeral cultural landscapes, we assume that although these lack permanency, they have a modicum of chronological durability. While this holds true for the tangible construct of a Christmas market, the sensescapes are temporally more volatile. Air temperature, wind, time of day, and number of visitors will affect how soundscapes, smellscapes, and touchscapes, and even how sightscapes will be perceived. This volatility, coupled with the individual enculturation and the epistemology of the valuer, implies that no two visitors will have the same experience and perception of that cultural landscape in a given visit, and moreover, that it is unlikely that even the same visitors will experience the same sensescapes on repeat visits with the same market period, let alone between different years. Despite this, multiple visits would aggregate a sense of expectation of the intangible experience, especially if experienced over longer timeframes, of which any visit would entail anticipated sights, smells, and sounds of the market.

There is a third dimension that needs to be considered here, that of the Christmas markets past. While this does not apply to most tourists and visitors who experience a Christmas market for the first time (ever or at this location) in the here and now, it will affect the local residents. Their memoryscapes are subconscious lenses that will interpret and evaluate the reality of the present as 'traditional' and thus implicitly as 'correct', or alternatively as 'modern' and thus as implicitly alien and 'not quite right.'

Gradual change in the sensescapes, especially where the tangible component remains consistent, may go largely unnoticed or may be tolerated unless reflection is prompted (which the interviews for this research project may have achieved). Rapid or incisive change, on the other hand, especially when accompanied by structural change in the market layout or set up, may trigger feelings of nostalgia and, if significant structural change occurred affecting the tangible elements of the Christmas markets, also a feeling of solastalgia.

Christmas markets are an event to participate in, but they are also a cultural landscape to experience through traversing the sensescapes. While department stores and online shopping have long since supplanted Christmas markets as the major venues to purchase Christmas presents, they cannot replace the landscape and sensescape presented by Christmas markets.

From the perspective of the management of the cultural heritage assets and the cultural landscapes represented by the Christmas markets, it would be desirable that the tradition of Christmas markets as a whole be recognised by the German Commission for UNESCO [34] or, failing that, that selected Christmas markets be recognised by Ministers for Culture at the state level, such as Nuremberg (Bavaria), Dresden (Saxony), or Frankfurt (Hessia). Such future listings can be aided by systematic historic analyses of these markets as well as by sensescape studies and sentiment analyses. Elsewhere, we have reviewed the nature of material culture associated with such markets [31], as well as the extent to which this material culture is being collected by museums [50]. This paper has shown the contribution that sensescape studies and sentiment analyses can make to the understanding of ephemeral cultural event landscapes.

5. Conclusions

Christmas markets represent a cultural landscape en miniature, the structural elements being overlain by multisensorial elements, with each heritage landscape being unique. Themes of transition are observable across all markets, including a notable shift away from traditional handicrafts towards gastronomy due to increased economic pressures and problems of intergenerational recruitment. Other transition themes emerge from amended government policy and law, with commercialisation and tourist appeal being the central focus.

Most critical is the decline in the intangible sensory and experiential landscape of the German markets, ultimately affecting both authenticity and identity, with the experiential feelings shifting away from the 'magical atmosphere' to more functional, sterile environments. Alongside the notable decline in soundscapes and smellscapes, modernisation and commercialisation of German markets present future projections which lack unique identities and traditional experience, being devoid of a distinctive atmosphere, and in the case of small-town markets such as Offenbach, conceivably obsolescent. The decline in the sensory landscape and loss of identity presents German Christmas market futures which can credibly be construed as 'more boring' in detail and experience. As Christmas markets represent multidimensional ephemeral cultural landscapes characteristic of urban planning, sensescape changes may trigger feelings of nostalgia or the more negative feeling of solastalgia. However, Birmingham's Christmas market contrasts these findings, and although anticipated to be less authentic over the next ten years, management processes are actively engaged in processes to energise and revitalise Christmas market landscapes. Our findings highlight potential conflicts considering transition and development across these cultural landscapes in both Germany and the United Kingdom, and issues with respect to heritage and cultural integrity.

Supplementary Materials: The following supporting information can be downloaded at: https://www.mdpi.com/article/10.3390/land14112161/s1, Table S1: Participant demographics.

Author Contributions: Conceptualization, M.P. and D.H.R.S.; Methodology, M.P., D.H.R.S. and J.B.; Formal Analysis, M.P.; Investigation, M.P. and D.H.R.S.; Resources, M.P. and D.H.R.S.; Data Curation, M.P.; Writing—Original Draft Preparation, M.P. and D.H.R.S.; Writing—Review and Editing, M.P., D.H.R.S. and J.B.; Visualization, M.P.; Supervision, D.H.R.S. and J.B.; Project Administration, M.P. and D.H.R.S.; Funding Acquisition, M.P. All authors have read and agreed to the published version of the manuscript.

Funding: This research was funded by an Australian Government Research Training Program Scholarship (Charles Sturt University).

Institutional Review Board Statement: This research was approved by the Charles Sturt University Human Research Ethics Committee (Protocol number: H23548).

Data Availability Statement: Data not available/The data that has been used is confidential.

Acknowledgments: The primary author wishes to deeply thank the many participants of each Christmas market who offered their time, undertaking interviews to provide the dataset on which this paper is based.

Conflicts of Interest: The authors declare no conflicts of interest.

References

1. Papayannis, T.; Howard, P. Nature as heritage. *Int. J. Herit. Stud.* **2007**, *13*, 298–307. [CrossRef]
2. Taylor, K.; Lennon, J. Cultural landscapes: A bridge between culture and nature? *Int. J. Herit. Stud.* **2011**, *17*, 537–554. [CrossRef]
3. Windsor, G. The recognition of Aboriginal placenames in New South Wales. In *Aboriginal Placenames: Naming and Re-Naming the Australian Landscape*; Koch, H., Hercus, L., Eds.; ANU: Canberra, Australia, 2009; pp. 71–87.
4. Carter, P. Proper Names: Differences Between Aboriginal and Colonial Toponymy. In *Naming No Man's Land: Postcolonial Toponymies*; Springer: Cham, Switzerland, 2024; pp. 59–94.
5. Davis, L.; Thys-Şenocak, L. Heritage and scent: Research and exhibition of Istanbul's changing smellscapes. *Int. J. Herit. Stud.* **2017**, *23*, 723–741. [CrossRef]
6. Spennemann, D.H.R.; Duffy, D. The ephemeral cultural landscape of an Australian Federal election. *Land* **2025**, *14*, 1610. [CrossRef]
7. Mitin, I. Constructing urban cultural landscapes & living in the palimpsests: A case of Moscow city (Russia) distant residential areas. *Belgeo Rev. Belg. Géographie* **2018**, *4*. [CrossRef]
8. Wilson, D.R. Reading the palimpsest: Landscape studies and air-photography. *Landsc. Hist.* **1987**, *9*, 5–25. [CrossRef]
9. Spennemann, D.H.R. A disappearing cultural landscape: Managing the heritage of German-style land use and pug-and-pine architecture in Australia. *Land* **2025**, *14*, 1517. [CrossRef]
10. Aplin, G. World heritage cultural landscapes. *Int. J. Herit. Stud.* **2007**, *13*, 427–446. [CrossRef]
11. Cocks, M.; Vetter, S.; Wiersum, K.F. From universal to local: Perspectives on cultural landscape heritage in South Africa. *Int. J. Herit. Stud.* **2018**, *24*, 35–52. [CrossRef]
12. Spennemann, D.H.R. Conceptualizing a Methodology for Cultural Heritage Futures: Using Futurist Hindsight to make 'Known Unknowns' Knowable. *Heritage* **2023**, *6*, 548–566. [CrossRef]
13. Sinamai, A. Melodies of God: The significance of the soundscape in conserving the Great Zimbabwe landscape. *J. Community Archaeol. Herit.* **2018**, *5*, 17–29. [CrossRef]
14. Zhao, S. Aural landscape as an intangible heritage in Chinese classical gardens. *Stud. Hist. Gard. Des. Landsc.* **2017**, *37*, 242–249. [CrossRef]
15. Stibral, K.; Faktorová, V. The origins of the aesthetic appreciation and artistic depiction of the industrial landscape. *J. Herit. Tour.* **2021**, *16*, 151–163. [CrossRef]
16. Wierciński, H. Climbing heritage: The lines and aesthetics of rock climbing. *Anthropol. Today* **2021**, *37*, 10–13. [CrossRef]
17. He, Y.; Bowring, J. Five site visits over four centuries: Imagining pasts and futures for Yuanming Yuan Garden, Beijing, China. *Stud. Hist. Gard. Des. Landsc.* **2018**, *38*, 330–341. [CrossRef]

18. Sinha, A. Conservation of historic water systems in Champaner-Pavagadh, Gujarat, India. *Landsc. Res.* **2019**, *44*, 588–599. [CrossRef]
19. Alqadrie, A.M.; Martokusumo, W. The Other Sensory Approach on Exploring Urban Heritage Landscape. Case Study: Chinese Quarter of Semarang. *J. Infrastruct. Facil. Asset Manag.* **2021**, *3*, 147–162. [CrossRef]
20. Pérez-Gálvez, J.C.; Medina-Viruel, M.J.; Jara-Alba, C.; López-Guzmán, T. Segmentation of food market visitors in World Heritage Sites. Case study of the city of Córdoba (Spain). *Curr. Issues Tour.* **2021**, *24*, 1139–1153. [CrossRef]
21. Parker, M.; Spennemann, D.H.R.; Bond, J. Using destination reviews to explore tourists' sensory experiences at Christmas markets in Germany and Austria. *J. Herit. Tour.* **2023**, *19*, 172–200. [CrossRef]
22. Spennemann, D.H.R.; Parker, M. The changing face of German Christmas Markets: Historic, mercantile, social and experiential dimensions. *Heritage* **2021**, *4*, 1821–1835. [CrossRef]
23. Wölfle, F.; Schnorbus, L. *Weihnachtsmärkte-Charakterisierung der Besucher und Bedeutung Unterschiedlicher Faktoren Für Diese*; IUBH Discussion Papers-Tourismus & Hospitality; IUBH Internationale Hochschule: Erfurt, Germany, 2019.
24. Młynkowiak-Stawarz, A.; Bęben, R. Social, Cognitive and Traditional Motives in Christmas Market Tourism: A Case Study from Poland. *J. Environ. Manag. Tour.* **2024**, *15*, 629–639. [CrossRef]
25. Marcher, A.; Erschbamer, G.; Pechlaner, H. Escaping from the Event? Residents' Perception of Christmas Markets in Northern Italy. *Event Manag.* **2019**, *23*, 599–611. [CrossRef]
26. Brida, J.G.; Meleddu, M.; Tokarchuk, O. Use value of cultural events: The case of the Christmas markets. *Tour. Manag.* **2017**, *59*, 67–75. [CrossRef]
27. Broeckerhoff, A.; Galalae, C. Christmas markets–marketplace icon. *Consum. Mark. Cult.* **2022**, *25*, 79–90. [CrossRef]
28. Trandberg Jensen, M.; Fusté-Forné, F. "It's beginning to look a lot like Christmas": Exploring the sensescapes of food experiences in Tivoli. *Food Cult. Soc.* **2024**, *27*, 1394–1417. [CrossRef]
29. Spennemann, D.H.R. Commercial Christmas Cheer: The reestablishment and proliferation of Christmas Markets in post-World War II Germany. *SSRN Econ. Hist. Ejournal* **2025**. [CrossRef]
30. Spennemann, D.H.R.; Parker, M.; Bond, J. Marketing Christmas Markets: Emotive and commercial domains in promotional texts. *Int. J. Tour. Cities*, 2025; *in press*.
31. Spennemann, D.H.R. Intangible Heritage and its Associative Objects as exemplified by the Materiality of the Portable Material Culture of German Christmas Markets. *Heritage* **2024**, *7*, 3511–3553. [CrossRef]
32. Parker, M.; Spennemann, D.H.R.; Bond, J. A multisensory puzzle: An exploration into sensory and multisensory heritage perceptions across five European Christmas markets. 2025; *in press*.
33. UNESCO. *Basic Texts of the 2003 Convention for the Safeguarding of Intangible Cultural Heritage' for Its Protection and Promotion*; UNESCO: Paris, France, 2020.
34. Deutsche UNESCO Kommission. Immaterielles Kulturerbe in Deutschland. Available online: https://www.unesco.de/orte/immaterielles-kulturerbe/immaterielles-kulturerbe-in-deutschland (accessed on 23 October 2025).
35. Deutsche UNESCO Kommission. Immaterielles Kulturerbe in den Ländern. Available online: https://www.unesco.de/orte/immaterielles-kulturerbe/immaterielles-kulturerbe-in-deutschland/immaterielles-kulturerbe-in-den-laendern/ (accessed on 23 October 2025).
36. Ministerium für Bildung und Kultur Saarland. Immaterielles Kulturerbe. Available online: https://www.saarland.de/mbk/DE/portale/kulturportal/kulturpolitik/ike1/ikelandesverzeichnis.html (accessed on 23 October 2025).
37. Sächsisches Staatsministerium für Kultur und Tourismus. Immaterielles Kulturerbe. Available online: https://www.mkw.nrw/themen/kultur/kulturfelder/immaterielles-kulturerbe (accessed on 23 October 2025).
38. Ministerium für Kultur und Wissenschaft. Immaterielles Kulturerbe Inventar NRW. Available online: https://www.kulturland.sachsen.de/landesliste-des-immateriellen-kulturerbes-3999.html (accessed on 23 October 2025).
39. Bayerisches Staatsministerium der Finanzen und für Heimat. Bayerisches Landesverzeichnis des Immateriellen Kulturerbes. Available online: https://www.ike.bayern.de/verzeichnis/index.html (accessed on 23 October 2025).
40. Landesheimatbund Sachsen-Anhalt. Immaterielles Kulturerbe in Sachsen-Anhalt. Available online: https://landesheimatbund.de/ike/ (accessed on 23 October 2025).
41. Thüringer Ministerium für Bildung, W.u.K. Immaterielles Kulturerbe. Available online: https://bildung.thueringen.de/kultur/immaterielles-kulturerbe (accessed on 23 October 2025).
42. Yin, R.K. *Case Study Research: Design and Methods*, 5th ed.; Sage: Thousand Oaks, CA, USA, 2014; Volume 5.
43. Saldaña, J. *The Coding Manual for Qualitative Researchers*, 4th ed.; SAGE Publishing: Thousand Oaks, CA, USA, 2021.
44. Parker, M.; Spennemann, D.H.R.; Bond, J. The exploration of sensory and multisensory experiences of Hobart's Salamanca Market and their potential as intangible heritage. *J. Herit. Tour.* **2024**, *20*, 317–347. [CrossRef]
45. Spennemann, D.H.R.; Parker, M. "Endlich wieder Strietzeln": The restart of German Christmas Markets after the COVID-19 pandemic. *Int. J. Event Festiv. Manag.* **2025**, *16*, 509–529. [CrossRef]

46. Inglehart, R.; Haerpfer, C.; Moreno, A.; Welzel, C.; Kizilova, K.; Diez-Medrano, J.; Lagos, M.; Norris, P.; Ponarin, E. (Eds.) *World Values Survey: All Rounds—Country-Pooled Datafile Version*; JD Systems Institute: Madrid, India, 2022.
47. Molteni, F.; Biolcati, F. Religious Decline as a Population Dynamic: Generational Replacement and Religious Attendance in Europe. *Soc. Forces* **2023**, *101*, 2034–2058. [CrossRef]
48. Manfrahs, F. Immer was los: Innenstadt-Belebung durch den Aufbau zugkräftiger Veranstaltungen. In *Citymanagement: Innenstadt-Belebung Mit System—Starke Zentren Mit Erlebnisqualität Gestalten*; Manfrahs, F., Ed.; Springer Fachmedien Wiesbaden: Wiesbaden, Germany, 2020; pp. 185–204. [CrossRef]
49. Spennemann, D.H.R. The usefulness of the Johari Window for the Cultural Heritage Planning Process. *Heritage* **2023**, *6*, 724–741. [CrossRef]
50. Spennemann, D.H.R.; Parker, M. Collecting Christmas Commerce: Material Culture related to German Christmas Markets held in German Museums. *J. Muzeológia A Kultúrne Dedičstvo Museol. Cult. Herit.* 2025; *in press*.

land

Article

Linking Ecosystem Services, Cultural Identity, and Subjective Wellbeing in an Emergent Cultural Landscape of the Galápagos Islands

Jenny A. Quijozaca [1,†], Mateo Aguado [2,3,†] and José A. González [3,*]

[1] Asamblea Comunitaria San Cristóbal, San Cristóbal 200104, Ecuador; galapagos.jenny@gmail.com
[2] Research Group EDUCAMB, Faculty of Educational Sciences and Humanities, Universidad Internacional de La Rioja (UNIR), Av. de la Paz 137, 26006 Logroño, Spain; mateo.aguado@unir.net
[3] Social-Ecological Systems Laboratory, Department of Ecology, Universidad Autónoma de Madrid, Calle Darwin 2, 28018 Madrid, Spain
* Correspondence: jose.gonzalez@uam.es
† These authors contributed equally to this work.

Abstract: Punta Carola beach, located on San Cristóbal Island in the Galápagos Archipelago, exemplifies how island environments of recent human settlement can rapidly acquire cultural significance. Drawing on a survey of 201 residents, this study examines perceptions of the ecosystem services provided by Punta Carola, their relationship to subjective wellbeing, and attitudes toward alternative tourism and governance strategies. The findings reveal that the inhabitants of San Cristóbal recognize a wide range of tangible and intangible benefits that the natural environment of Punta Carola contributes to their quality of life, sense of belonging, and cultural identity, as well as recreational and aesthetic values. Life satisfaction was positively associated with perceptions of nature's contribution, the maintenance of local ecological knowledge, and household income, while formal education correlated negatively. Residents identified freshwater and artisanal fishing as the most critical ecosystem services, reflecting their centrality to local livelihoods and wellbeing. The results also underscore widespread criticism of large-scale tourism projects perceived as exclusionary or unsustainable and highlight the importance of participatory governance schemes that legitimize local values. Punta Carola thus emerges as a "cultural landscape of resistance", where external pressures catalyze identity construction and territorial rootedness. This case contributes to academic debates on socio-ecological resilience in fragile island contexts and offers actionable insights for inclusive planning in the Galápagos.

Keywords: cultural landscapes; ecosystem services; subjective wellbeing; local ecological knowledge; participatory governance; Galápagos; socio-ecological resilience

1. Introduction

The accelerated transformations currently affecting ecosystems worldwide have redefined the relationship between human societies and nature, generating new territories of meaning where ecological, social, and cultural processes converge [1,2]. Contemporary social–ecological systems not only reflect this dynamic co-evolution but also reveal how local communities build deep identity bonds with natural spaces over relatively short periods of time [3–5]. These processes challenge traditional conceptions of cultural heritage that privilege antiquity over the richness and depth of cultural and emotional meanings associated with a territory [6,7].

 https://doi.org/10.3390/land14112208

This conceptual evolution is supported by contemporary theoretical frameworks that recognize the multidimensionality of cultural landscapes [8]. In this paper, the term "cultural landscape" is used in a broad socio-ecological sense, referring to an integrated system of natural and cultural elements shaped by human interactions, meanings, and practices. Far from being conceived merely as physical environments, cultural landscapes represent lived spaces where uses, memories, knowledge, and relationships intertwine to give meaning to the territory [9]. Thus, the landscape is no longer a visual backdrop but becomes an active stage for everyday experiences, spatial organization, and community ties [7]. These everyday landscapes constitute a legitimate form of cultural heritage, even when they lack millenary or monumental histories [9].

In coastal and island contexts such as the Galápagos, where permanent human colonization is relatively recent [5], these symbolic bonds emerge rapidly and with unique depth. Recent research in coastal ecosystems [10] has shown that specific characteristics of marine environments—such as species diversity, beach type, and landscape attractiveness—significantly influence recreational preferences, cultural use patterns, and the wellbeing of local communities. In these territories, human practices such as artisanal fishing, tourism, or emerging activities like surfing interact with unique ecosystems, generating multiple goods and services that sustain not only subsistence and wellbeing but also the identity and cultural construction of local communities [1,11].

The case of Punta Carola beach, located on San Cristóbal Island in the Galápagos archipelago, constitutes a paradigmatic example of an emerging cultural landscape: a natural space that has acquired a dense set of social, affective, and political meanings for the local community in a relatively short historical period. Locally known as *"playa del amor"* (the beach of love), Punta Carola has historically served as a space for affective, social, and family encounters where generations have built deep emotional and community bonds. These shared experiences—from family gatherings to the forging of intimate relationships—have shaped a strong sense of territorial belonging within the community. Today, many of those who grew up closely linked to this place are leading its defense against external threats increasingly affecting Galápagos landscapes. Such threats include, among others, the impacts of tourism growth [12], the effects of climate change on biodiversity [13], and marine plastic pollution [14]. This symbolic attachment is intertwined with remarkable ecological richness, as Punta Carola harbors numerous marine and terrestrial species endemic to the archipelago [15], reinforcing its value as an emerging cultural landscape and as a contested social–ecological system.

Many conservation strategies have failed by disconnecting local communities from the stewardship of their territories, imposing "top-down" models that disregard their perceptions, knowledge, and territorial values [1]. Community survey methodologies have proven effective in this regard as tools to capture the sociocultural perception of the different services that ecosystems provide to people. These approaches reveal perceptions and territory–identity bonds that other quantitative or spatial methods often fail to detect [16,17]. Identifying and understanding how different social actors perceive ecosystem services has been shown to be fundamental for informed decision-making [18], particularly in areas where nature conservation is threatened by expanding economic activities [19]. Moreover, these methods are especially valuable for exploring the multidimensional nature of human wellbeing in cultural landscapes, including symbolic, spiritual, and affective aspects that transcend conventional economic assessments [20]. For this reason, it is urgent to promote methodologies that incorporate sociocultural perceptions into ecosystem service assessments, recognizing not only economic but also symbolic, identity, and affective values that shape human wellbeing in territories [10,21,22]. Landscape perception, mediated by direct experiences and specific cultural practices, is key to motivating pro-environmental attitudes and fostering responsible territorial appropriation by local inhabitants [23,24].

The application of these cultural convergence frameworks is particularly urgent in territories such as Punta Carola, where the accelerated construction of community cultural meanings occurs simultaneously with growing external extractive pressures. Research on cultural landscapes suggests that when local communities perceive threats to highly valued symbolic spaces, processes of territorial appropriation intensify [25], generating what could be termed "cultural landscapes of resistance." In such contexts, understanding community perceptions of ecosystem services is not merely an academic exercise but an essential political tool for legitimizing the territorial demands of local communities in the face of unsustainable development pressures.

In this study, we analyze the multidimensional role of Punta Carola as an emerging cultural landscape, drawing on the social perceptions of San Cristóbal's inhabitants. Specifically, we explore how these perceptions relate to subjective wellbeing, identity construction, and community strategies for protecting territorial and cultural values. Our specific objectives are: (SO1) to identify the sociocultural and subjective factors that best predict people's life satisfaction in the study area; (SO2) to assess how different sociodemographic and cultural variables influence residents' perceptions of the importance of various ecosystem services associated with the studied cultural landscape; and (SO3) to discuss the potential role of three alternative management and conservation strategies for Punta Carola within a cultural landscapes framework.

2. Materials and Methods

2.1. Study Area

San Cristóbal is one of the fifteen main islands that make up the Galápagos Archipelago, located in the Pacific Ocean, 960 km off the continental coast of Ecuador. The biodiversity of these volcanic islands has been favored by the convergence of three major ocean currents: the Humboldt Current from the south, the eastward South Equatorial Surface Current, and the subsurface Cromwell Current [26]. Less than two kilometers north of Puerto Baquerizo Moreno (the provincial capital), Punta Carola beach extends for 300 m and represents a coastal ecosystem of high ecological value, crucial for the conservation of numerous endemic and endangered species. Among the most notable are the Galápagos sea lion (*Zalophus wollebaeki*) and the marine iguana (*Amblyrhynchus cristatus*), both of which use the beach for reproduction [27,28].

San Cristóbal is the second most populated island in the archipelago, with 8155 inhabitants distributed across 2699 households [29]. Most of the population is concentrated in Puerto Baquerizo Moreno (89.4%), while only 10.6% reside in rural areas. More than three-quarters of San Cristóbal's residents are employed in the service sector, mainly through tourism. In contrast, the secondary sector (construction, manufacturing) accounts for around 8% of the population, while the primary sector (agriculture, livestock, fishing) comprises 6% [29]. According to data from the Galápagos National Park, the island received 78,406 tourists in 2024, more than half of whom visited Punta Carola. This represents a 147% increase compared to 2019, reflecting the structural transformation San Cristóbal has experienced in recent years, shifting from a traditional economy based on primary activities to a service-based model centered on tourism [30].

The continuous process of human colonization since 1866 has significantly altered the island's native vegetation, particularly in the southwest—where Puerto Baquerizo Moreno is located—and in the highlands [31]. According to estimates by Rivas-Torres et al. [32], 0.3% of the island's territory has already been converted to urban areas, mainly in the lowlands. For all these reasons, Punta Carola today constitutes not only a space of high biodiversity but also one of the most frequented sites on San Cristóbal, where tourist activities such as guided nature tours, surfing, photography, and snorkeling converge with the use and enjoyment of the space by the local community. This interaction, together with its proximity to the capital (just a ten min walk) and its location at the boundary between the Galápagos National Park (GNP) and various private areas (Figure 1), makes Punta Carola a key site for the sustainable management of the island. In recent years, these tensions have resulted in various social confrontations around the future of the site, where divergent perspectives on conservation, development, and community rights converge, highlighting the need to implement participatory governance approaches that integrate the ecological, cultural, and social dimensions of the territory.

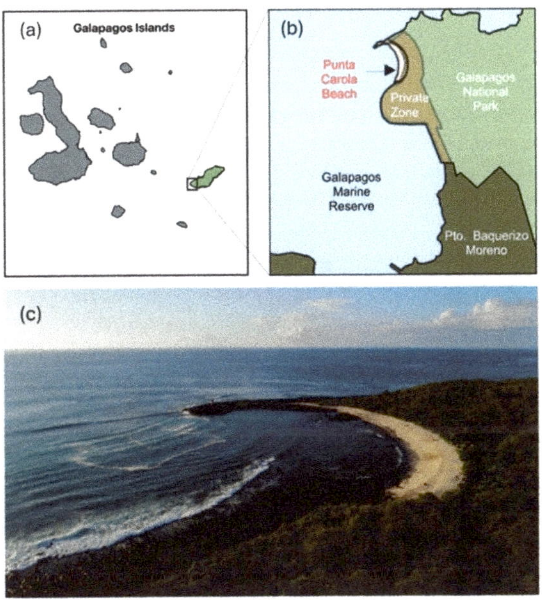

Figure 1. Study area: (**a**) the Galapagos Islands with San Cristóbal highlighted; (**b**) location of Punta Carola beach on the southwestern coast of the island; (**c**) photograph of Punta Carola beach.

2.2. Sampling Methods

We conducted a total of 201 surveys in two complementary phases: the first between 15 April and 15 May 2023, and the second between 28 June and 8 July 2024. The study population consisted of local inhabitants of San Cristóbal Island, with sampling restricted to individuals over 18 years old. Based on sociodemographic information for the island, we stratified the sampling by participants' gender and age, as well as their profile and occupation (including fishers, farmers, livestock breeders, merchants, nature guides, and public and private sector workers).

The survey (Appendix A) included questions on socioeconomic and sociodemographic aspects (#1–10), as well as subjective perception questions regarding ecosystem services (#11–14) and human wellbeing, which was addressed through subjective life satisfaction (#15). At the end of the survey, participants were invited to choose one of three possible management strategies previously designed by the research team (#16): (a) participatory management, (b) strict conservation, and (c) economic development. For the assessment of ecosystem services, respondents were provided with a closed list of 12 ecosystem services: four provisioning, four regulating, and four cultural services. Furthermore, the interviewer presented three visual panels listing, describing, and illustrating examples of the 12 ecosystem services (Table S1, Supplementary Materials).

2.3. Data Analysis

To analyze the factors influencing subjective life satisfaction (SO1), we performed a stepwise multiple regression to identify the factors that best explained life satisfaction in the study area. Seven independent variables were used to build the models: gender, age, length of residence on the island, educational level, monthly income, perception of the maintenance of local ecological knowledge (LEK), and perception of the extent to which nature contributes to wellbeing. The Akaike Information Criterion (AIC) was used to select the most parsimonious model. Cronbach's alpha was used to evaluate the internal consistency of the responses.

To analyze the effect of sociodemographic and cultural factors on the valuation of ecosystem services (SO2), we conducted a Redundancy Analysis (RDA), where perceptions of the importance of the 12 ecosystem services under study were used as dependent variables, and sociodemographic and cultural factors as explanatory variables. These explanatory factors included gender, age, length of residence on the island, educational level, perception of the maintenance of local ecological knowledge (LEK), monthly income, and whether respondents owned tourist accommodation facilities. Additionally, we evaluated the importance and vulnerability attributed to different ecosystem services by various actors using a scatter plot diagram, based on the frequency with which services were identified as important and vulnerable.

Finally, to explore the role of the three proposed management and conservation strategies in the study area (SO3), we conducted Kruskal–Wallis tests between the groups of participants who selected each of the strategies and both their valuations of ecosystem services and selected sociodemographic variables. All analyses were performed using XLSTAT 2014 software (Addinsoft, Paris, France).

3. Results

3.1. Factors Influencing Life Satisfaction

The multiple regression models showed that life satisfaction was explained by four key variables: (i) perception of nature's contribution to wellbeing, (ii) perception of the maintenance of LEK, (iii) monthly income, and (iv) education level. Among these, only education level had a negative influence on life satisfaction, such that respondents with higher levels of formal education tended to report lower satisfaction with life (Table 1). Other predictors used in the analyses—such as gender, age, and length of residence on the island—did not appear in any of the models predicting life satisfaction in the study area.

Table 1. Best multivariate regression model to estimate the effect of different factors on life satisfaction.

Explanatory Variables	Coefficient	t	p-Value
Intercept	0.102	0.416	0.678
Nature contribution to wellbeing	0.595	11.029	<0.0001
Perception of LEK maintenance	0.055	4.554	<0.0001
Monthly income	0.082	3.002	0.003
Education level	−0.041	−2.898	0.004
R^2		0.446	
Adjusted R^2		0.425	
AIC		−711.840	
F		22.161	
P		<0.0001	

3.2. Perception of Ecosystem Services

The ecosystem services perceived as most important (above-average values) were freshwater for consumption (selected by 56.2% of respondents), food from agriculture and livestock (41.8%), food from fishing, shellfish, and gathering (40.8%), and air purification (31.8%). In contrast, the services perceived as most vulnerable (above-average values) were freshwater for consumption (47.8%), food from fishing, shellfish, and gathering (35.3%), soil fertility and erosion control (29.9%), climate regulation (29.4%), and aesthetic enjoyment of landscapes, recreation, and tourism (25.9%) (Table S2, Supplementary Materials).

The analysis of social perceptions of ecosystem services in the study area also allowed us to identify those services recognized as critical by respondents, i.e., those simultaneously rated highly in terms of both their importance for wellbeing and their perceived vulnerability (understood as risk of degradation or loss). Two ecosystem services emerged as most critical: (i) freshwater for consumption and (ii) food from fishing, shellfish, and gathering (Figure 2).

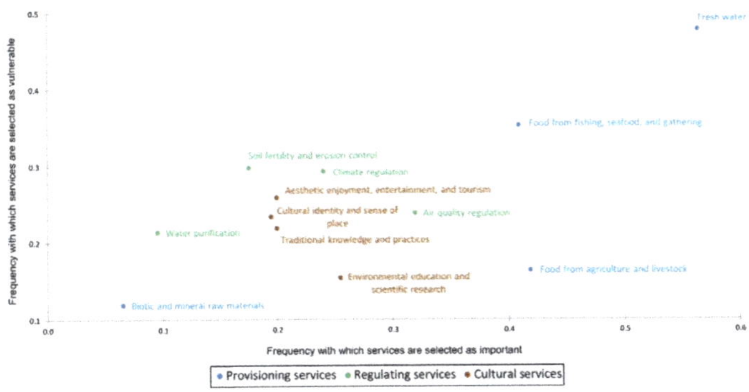

Figure 2. Scatter plot diagrams of the perceived importance and vulnerability of ecosystem services in the study area, both calculated as the frequency with which each service was selected as important/vulnerable by respondents.

The RDA revealed a significant association between respondents' sociodemographic characteristics and the relative importance they attributed to the studied ecosystem services ($p < 0.001$, 1000 permutations). The first three axes explained 76.5% of the total variance. Based on explained variance, total inertia, and eigenvalues, we focused on the first two axes as they captured the most relevant trends explaining differences in perceptions of ecosystem services (Figure 3).

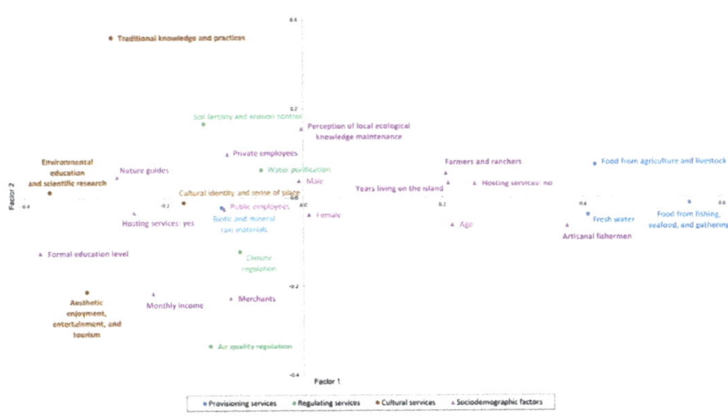

Figure 3. Biplot of the first two RDA axes, showing the relationships between perceived importance of ecosystem services (dependent variables) and sociodemographic variables (explanatory variables). Bold letters indicate variables with squared cosines higher than 0.40 and 0.10, respectively (Table S3, Supplementary Materials).

The first RDA axis (49.1% of total variance) revealed a contrast between three provisioning services (food from agriculture and livestock, food from fishing, shellfish, and gathering, and freshwater for human consumption) and two cultural services (environmental education and scientific research, and aesthetic enjoyment of landscapes, recreation, and tourism). We found that sociodemographic variables such as education level, ownership of accommodation services for tourists, length of residence on the island, and occupation influenced the importance respondents assigned to these services. Thus, individuals who had lived longer on the island, who did not own lodging services, and whose main occupation was in the primary sector (artisanal fishers, farmers, and livestock breeders) attributed higher importance to these provisioning services. In contrast, individuals with higher levels of formal education, who owned lodging services, and who worked as nature guides valued cultural services more.

The second RDA axis (16.0% of total variance) showed a contrast between the regulating service of air purification and the cultural service of traditional knowledge. Here, sociodemographic variables such as income, education level, occupation, and perception of LEK maintenance determined the importance attributed to these services.

3.3. Management and Conservation Strategies for Punta Carola

Among the three management strategies offered to participants in the survey, most opted for the participatory management model (55.2%), followed by strict conservation (39.3%) and economic development (5.5%). These groups differed significantly in some sociodemographic variables, such as education level (Kruskal–Wallis test, $\chi^2 = 12.1, p < 0.01$) and monthly income (Kruskal–Wallis test, $\chi^2 = 9.9, p < 0.01$) (Figure 4), as well as in their perception that the increase in tourism activities and hotel presence in Punta Carola could negatively affect human wellbeing (Kruskal–Wallis test, $\chi^2 = 23.3, p < 0.001$) (Figure 5).

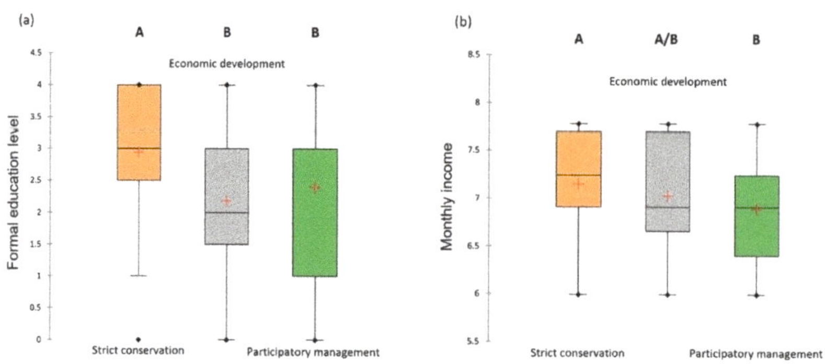

Figure 4. Differences across management strategies for respondents' education level (**a**) and monthly income (**b**). Box plots show the mean (red crosses), median (central thick lines), 25% and 75% quartiles (box width), and upper and lower limits (T). Different letters indicate significant differences among strategies (post hoc Dunn's tests, $p < 0.001$).

The valuations of the 12 ecosystem services varied considerably among respondents depending on the management strategy selected for Punta Carola (Figure S1, Supplementary Materials). Notably, significant differences were observed in the category of cultural services, whose valuation varied according to the selected management strategy (Kruskal–Wallis test, $\chi^2 = 6.6, p < 0.05$) (Figure 6).

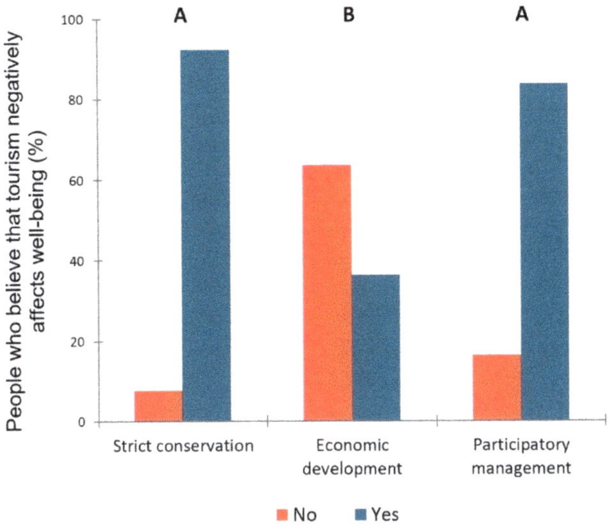

Figure 5. Differences across management strategies in the perception that increasing tourism activities and hotel presence in Punta Carola negatively affects wellbeing. Different letters indicate significant differences among strategies (post hoc Dunn's tests, $p < 0.001$).

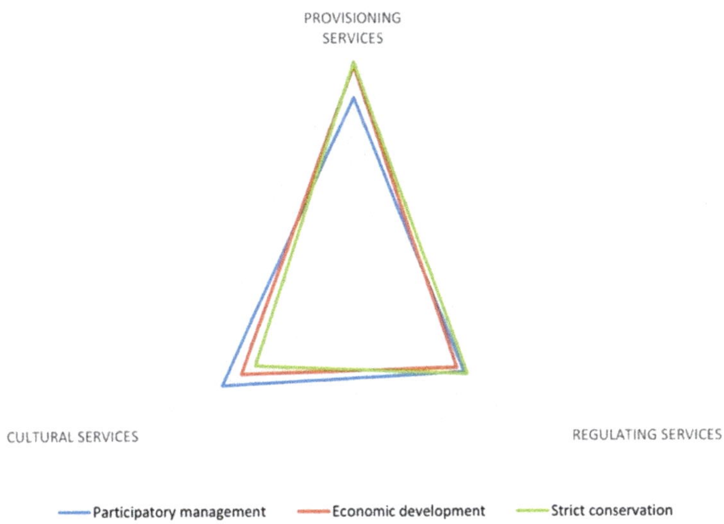

Figure 6. Radar diagram showing differences in the perception of provisioning, regulating and cultural services among the people selecting each management strategies.

4. Discussion

This study contributes to the academic debate on how cultural landscapes, even those with relatively recent human occupation, can acquire profound meanings over short periods of time [4,5]. Cases such as Punta Carola highlight that the cultural dimension of landscapes does not depend solely on the antiquity of settlement, but also on the intensity of shared experiences and the capacity of places to become references for identity, wellbeing, and community cohesion. In recent years, Punta Carola has been collectively appropriated as an intergenerational space for gathering, recreation, and environmental education, while simultaneously becoming a territory of resistance against threats derived from tourism and urbanization [12]. This dual process of accelerated identity construction coincides with what has been described in the literature on complex social–ecological landscapes, where external pressures intensify the symbolic appropriation of space [33]. In this context, the category of "cultural landscapes of resistance" acquires relevance, as it illustrates how external pressure can catalyze processes of territorial rootedness and defense that reinforce collective identity.

Regarding subjective wellbeing (SO1), our findings confirm the strong link between nature and life satisfaction [34–36]. The perception that nature contributes to wellbeing, along with the perception of maintaining LEK, were significant predictors of life satisfaction, in line with previous studies that emphasize the centrality of these connections in rural and traditional communities [21,37,38]. These results reinforce the idea that biodiversity conservation and the strengthening of local knowledge are essential not only for ecological sustainability but also for human wellbeing [39]. Moreover, our results suggest that educational and conservation policies could mutually benefit from incorporating LEK into curricula and recognizing its role in building collective wellbeing [40,41].

Another significant finding of this research was the negative association between formal education and subjective life satisfaction, indicating that respondents with higher levels of formal education tended to report lower personal life satisfaction than those with less education. This points to a possible "aspiration mismatch," whereby increased but unmet professional expectations may generate frustration [42,43]. Conversely, Zorondo-Rodríguez et al. [44] found a positive relationship between the perception of maintaining LEK and subjective wellbeing, which suggests that formal schooling, when not integrating community values and local knowledge, can erode traditional knowledge systems [45]. Indeed, research has shown that the intergenerational loss of LEK, accelerated by modernization, globalization, and urbanization, reduces not only social–ecological resilience but also wellbeing and social cohesion [46,47]. These patterns highlight the importance of preserving biocultural refugia that maintain LEK as an essential component for both socio-ecological resilience and human wellbeing [46,48]. Furthermore, they invite reflection on the need to design more contextualized educational policies that integrate traditional knowledge in dialog with scientific knowledge, fostering intercultural education that enhances wellbeing.

Although more nuanced, respondents' income showed a positive effect on life satisfaction. This result is consistent with previous studies in tribal, rural, and peasant communities in India [44], China [49,50], and Malaysia [51], which found a positive association between economic indicators such as income and subjective wellbeing [42]. However, other studies conducted in indigenous communities in Bolivia [52] and Ecuador [21] found that economic factors are not relevant in explaining subjective wellbeing. Faced with these disparities, some authors suggest that cultural contexts (such as the degree of integration into the market economy or the predominance of local versus foreign cultures) ultimately determine how people perceive the role of the economy in their wellbeing [21,49]. In the case of San Cristóbal Island, the positive income–wellbeing association may reflect the growing dependence of the local economy on tourism and services, which places monetary income in a more visible position within wellbeing perceptions.

Regarding ecosystem services (SO2), our results revealed that freshwater and artisanal fishing are perceived as the two most critical services [53,54], adding urgency to their management as they represent fundamental material pillars of community life on San Cristóbal Island. This finding underscores that conservation policies cannot be disconnected from the basic needs of the local population [55]. Integrating conservation strategies with the recognition of community rights and the guarantee of access to essential ecosystem services is indispensable for the sustainability of cultural landscapes [56].

In line with previous research, our results also showed that the sociocultural perception of ecosystem services varies significantly according to respondents' sociodemographic characteristics [18,19,21]. We found that provisioning services critical for human subsistence in rural settings, such as food and water, were more highly valued by actors linked to the primary sector (farmers, ranchers, and fishers), who directly depend on them [57,58]. Conversely, and consistent with earlier studies, we found that cultural services traditionally associated with urban demand, such as environmental education, scientific research, aesthetic enjoyment of landscapes, recreation, and tourism [18], were more highly valued by actors working as nature guides, those linked to tourism, and respondents with higher levels of formal education [59,60].

This diversity of valuations coincides with previous findings showing that cultural services associated with scientific knowledge and education are mainly valued by researchers, managers, tourists, and urban residents, whereas basic provisioning services are predominantly valued by local fishers, farmers, and ranchers whose livelihoods depend on their management through traditional ecological knowledge [17–19]. If not adequately managed, these diverse perceptions may lead to tensions over land use, potentially jeopardizing local conservation goals [22] and resulting in policies that disproportionately affect specific stakeholder groups [61]. Incorporating analyses of the social dimension of ecosystem services into land management can promote a more equitable provision of nature's contributions to people while reducing conflicts among stakeholders [62].

In terms of governance and management strategies (SO3), the majority preference for a participatory model in Punta Carola (55.2%) reflects a clear community orientation toward inclusive processes that articulate biodiversity, wellbeing, and cultural rights [63,64]. This finding aligns with Munévar-Quintero et al. [65], who stress that citizen participation is a fundamental right and an essential criterion for environmental governance, indispensable for legitimizing protected areas [66,67]. Along these lines, the Strategic Institutional Plan of the Galápagos National Park Directorate has for several years recognized the potential of environmental education and social participation as strategic pillars to achieve archipelago sustainability [68].

The differences observed regarding education, income, and tourism reinforce the idea that the perception and choice of territorial management strategies are strongly conditioned by social factors [69,70]. Likewise, our results revealed that the valuation of ecosystem services (mainly cultural) also changes depending on preferred management strategies. These findings are consistent with Schmidt et al. [16], who showed that sociocultural valuations of ecosystem services directly influence land use preferences and, consequently, the choice of territorial management models. Altogether, these results suggest that integrating the cultural and heritage values that structure local identity constitutes a key driver of nature conservation [71]. In Punta Carola, those who supported participatory management were also those who placed higher value on cultural ecosystem services, confirming the importance of collective memory and daily practices in ecosystem conservation and sustainability.

The finding that uncontrolled tourism negatively affects human wellbeing resonates with broader debates on the need to balance local niches of sustainable development with external pressures. This balance has been identified as essential to sustaining both biodiversity and local identities [72]. In this sense, López-Santiago et al. [1] criticize that traditional conservation policies have privileged biophysical control while marginalizing local perceptions, creating a disconnection between communities and ecosystem stewardship. Overcoming this deficit requires recognizing that cultural landscapes are living systems where biodiversity, wellbeing, and identity are interwoven. In Punta Carola, conflicts generated by uncontrolled tourism show that touristification can trivialize historical memory and deprive residents of their sense of belonging, as documented in other Latin American cultural landscapes [73].

Similarly, the sociodemographic differences observed among those who support strict conservation, those who prefer more developmental approaches, and those who advocate participatory management reveal that there is no single vision for the future of Punta Carola, but rather a field of tensions where different expectations and life trajectories converge. This finding coincides with recent proposals on cultural convergence frameworks in cultural landscape conservation, which emphasize the need to integrate spiritual, aesthetic, and community dimensions with technical expertise to avoid both the romanticization of nature and the commodification of territory [74].

The case of Punta Carola also illustrates a central aspect of the global debate: the tension between conservation and economic development in fragile territories. The growing pressure of unsustainable tourism, exemplified by projects such as the "Eco-Hotel San Cristóbal", threatens both biodiversity and the cultural value of the site. This situation demonstrates that legal and planning frameworks are insufficient if not accompanied by effective mechanisms of citizen participation [75]. At the same time, it shows how external threats can reinforce community rootedness and foster processes of identity construction, turning cultural landscapes into arenas of territorial resistance.

5. Conclusions

Punta Carola beach illustrates how contemporary island landscapes can act as laboratories for understanding the co-evolution between nature and society. Its dual condition as a site of exceptional biodiversity and a place of deep symbolic rootedness positions Punta Carola as an emerging "cultural landscape of resistance", where external pressures accelerate processes of identity construction and reinforce community cohesion. This case highlights the importance of recognizing that cultural landscapes are not static, but dynamic systems in which ecological, cultural, and social dimensions are intertwined. The effective conservation of such fragile and meaningful territories requires governance frameworks that move beyond traditional top–down approaches. Our results emphasize the need for participatory governance schemes that acknowledge and legitimize local perceptions and values, explicitly integrating the links between ecosystem services, subjective wellbeing, and cultural identity. While these findings provide a solid basis for inclusive planning, the dynamic nature of this landscape warrants a continuous research agenda: we suggest prioritizing long-term analyses of how perceptions evolve under changing tourism pressures, comparative studies with other island contexts, and, crucially, qualitative investigations into the design and implementation of effective participatory governance mechanisms. Only through such integrative and inclusive approaches will it be possible to harmonize biodiversity conservation with the quality of life of local communities and to secure a sustainable future for Punta Carola and other vulnerable island landscapes.

Supplementary Materials: The following supporting information can be downloaded at: https://www.mdpi.com/article/10.3390/land14112208/s1, Table S1: Panels used to describe and explain the ecosystem services; Table S2: Ecosystem services according to their frequency of importance and vulnerability; Table S3: Resulting Factor Scores from RDA; Figure S1: Radar chart for the 12 ecosystem services for the three management strategies studied.

Author Contributions: Conceptualization: M.A. and J.A.Q.; methodology: M.A. and J.A.Q.; formal analysis: M.A. and J.A.G.; investigation: M.A., J.A.Q. and J.A.G.; writing—original draft preparation: M.A. and J.A.Q.; writing—review and editing: M.A. and J.A.G.; supervision: J.A.G., J.A.Q. and M.A. contributed equally to this work. All authors have read and agreed to the published version of the manuscript.

Funding: This research received no external funding.

Data Availability Statement: The datasets generated during the current study are not publicly available due to confidentiality agreements with interviewees, but they are available from the corresponding author upon reasonable request.

Acknowledgments: We are deeply grateful to the Asamblea Comunitaria San Cristóbal, as well as to the fisher, farming, and merchant families, whose knowledge, experiences, and commitment have been essential for the development of this work. We also acknowledge the broader community of San Cristóbal, whose participation and voices continue to inspire processes of conservation and territorial defense. We gratefully acknowledge the National Institute of Statistics and Censuses (INEC) of Ecuador and the Ministry of Agriculture and Livestock—Galápagos District, for providing access to the databases that supported the analysis presented in this study. During the preparation of this manuscript, the authors used ChatGPT (OpenAI, GPT-5, San Francisco, CA, USA, 2025) for assistance with English language editing and style refinement. The authors have reviewed and edited all the content and take full responsibility for the final version of this publication.

Conflicts of Interest: The authors declare no conflicts of interest.

Appendix A

English summary of the questionnaire used in face-to-face surveys. The original questionnaire was administered in Spanish and included more questions. Only the questions pertinent to this research paper are presented here.

1. Do you live in Galápagos? Yes/No
2. How many years have you been living on the island?
3. Where were you born?
4. How old are you?
5. What is your sex? Female/Male
6. Which is the highest education level that you have reached
   - Illiterate
   - Basic studies/Primary studies
   - Secondary studies
   - Technologies/Undergraduate studies
   - Higher education
7. Currently, what is your professional status or your main source of income?
   - Artisanal fishing
   - Agriculture/Livestock raising
   - Public employee
   - Private employee.
   - Business and market
   - Nature guide
8. In which of these intervals does your monthly income fall?
   - Less than $400
   - Between $400–$800
   - Between $800–$1200
   - Between $1200–$1600
   - Between $2000–$2400
   - More than $2400
9. Do you or your family provide any accommodation facilities (hotel, hostel, etc.) for tourists? Yes/No

10. Do you believe that the increase in tourism activities and the presence of a hotel complex in the Punta Carola area could negatively affect your well-being and that of your family? Yes/No
11. On a scale from 0 to 5, how much do you think nature contributes to your quality of life?
12. Which of the following 12 benefits provided by nature would you say are the 3 most important for your well-being and quality of life?
    - Food from agriculture and livestock
    - Food from fishing, seafood, and gathering
    - Fresh water
    - Biotic and mineral raw materials
    - Climate regulation
    - Air quality regulation
    - Water purification
    - Soil fertility and erosion control
    - Environmental education and scientific research
    - Aesthetic enjoyment, recreation, and tourism
    - Traditional knowledge and practices
    - Cultural identity and sense of place
13. And which of these would you say are the 3 that are in more danger of disappearing or degrading in the following years (the most vulnerable ones)?
14. Do you think ancestral stories and knowledge about nature from your parents and grandparents are still present in your life?
    - Yes, completely
    - Yes, although some has been lost
    - Yes, although much has been lost
    - No
15. On a scale from 0 to 5, how would you rate your satisfaction with your life?
16. Which of the following management strategies would you prefer for Punta Carola?
    - The local community and institutions work together for the sustainability of Punta Carola. They develop a comprehensive participatory management plan for the area that respects the ecosystem's dynamics and biodiversity. Sustainable local tourism is promoted, and the community is educated about the importance of conserving nature.
    - Urban development is allowed in Punta Carola. Economic gains from construction and tourism increase, but ecological degradation and damage to biodiversity also rise.
    - Punta Carola is declared part of the National System of Protected Areas of Ecuador. Construction is prohibited, access is limited, and tourism is controlled. Regulation by the State preserves biodiversity and the ecosystem's ecological integrity.

References

1. López-Santiago, C.A.; Aguado, M.; González Nóvoa, J.A.; Bidegain, I. Evaluación sociocultural del paisaje, una necesidad para la planificación y gestión sostenible de los sistemas socioecológicos. In *Claves para la Gestión Humana de los Sistemas Naturales*; Cerda, C.J., Palacios-Agundez, J.M., Eds.; Ocho Libros Editores: Santiago, Chile, 2019; pp. 105–137. Available online: https://repositorio.uam.es/handle/10486/720046 (accessed on 20 September 2025).
2. Berkes, F.; Folke, C.; Colding, J. (Eds.) *Linking Social and Ecological Systems: Management Practices and Social Mechanisms for Building Resilience*; Cambridge University Press: Cambridge, UK, 2000.

3. Brown, G.; Raymond, C. The relationship between place attachment and landscape values: Toward mapping place attachment. *Appl. Geogr.* **2007**, *27*, 89–111. [CrossRef]
4. Bedoya, J.M.L. El Ecuador y los Paisajes Culturales en la Gestión del Territorio. *Cad. Naui* **2023**, *12*, 148–174.
5. González, J.A.; Montes, C.; Rodríguez, J.; Tapia, W. Rethinking the Galapagos Islands as a complex social-ecological system: Implications for conservation and management. *Ecol. Soc.* **2008**, *13*, 1–26. [CrossRef]
6. Taylor, K. Landscape and meaning: Context for a global discourse on cultural landscapes values. In *Managing Cultural Landscapes*; Routledge: Abingdon, UK, 2012; pp. 21–44.
7. Mughal, M.A.Z. Landscape, space, and time: Navigating the cultural landscape through socio-spatial and socio-temporal organization in rural Pakistan. *GeoJournal* **2023**, *88*, 6175–6187. [CrossRef]
8. Burlingame, K. Landscape geographies: Interdisciplinary landscape research and a new framework to apply landscape as method. *Landsc. Res.* **2025**, *50*, 754–765. [CrossRef]
9. Melnick, R.Z. Protecting rural cultural landscapes: Finding value in the countryside. *Landsc. J.* **1983**, *2*, 85–96. [CrossRef]
10. Reke, A.; Ruskule, A.; Lees, L.; Kotta, J.; Veidemane, K.; Kõivupuu, A.; Herkül, K.; Aps, R.; Jänes, H.; Barboza, F.R. Linking coastal cultural ecosystem services to human well-being and leisure preferences: Insights from the Baltic Sea region. *Ecosyst. People* **2025**, *21*, 2530103. [CrossRef]
11. Díaz, S.; Fargione, J.; Chapin, F.S., III; Tilman, D. Biodiversity loss threatens human well-being. *PLoS Biol.* **2006**, *4*, e277. [CrossRef] [PubMed]
12. Burbano, D.V.; Meredith, T.C. Effects of tourism growth in a UNESCO World Heritage Site: Resource-based livelihood diversification in the Galapagos Islands, Ecuador. *J. Sustain. Tour.* **2021**, *29*, 1270–1289. [CrossRef]
13. Dueñas, A.; Jiménez-Uzcátegui, G.; Bosker, T. The effects of climate change on wildlife biodiversity of the Galapagos Islands. *Clim. Change Ecol.* **2021**, *2*, 100026. [CrossRef]
14. Jones, J.S.; Porter, A.; Muñoz-Pérez, J.P.; Alarcón-Ruales, D.; Galloway, T.S.; Godley, B.J.; Santillo, D.; Vagg, J.; Lewis, C. Plastic contamination of a Galapagos Island (Ecuador) and the relative risks to native marine species. *Sci. Total Environ.* **2021**, *789*, 147704. [CrossRef] [PubMed]
15. Bensted-Smith, R. (Ed.) *A Biodiversity Vision for the Galapagos Islands*; Charles Darwin Foundation and World Wildlife Fund: Puerto Ayora, Ecuador, 2002. Available online: https://www.researchgate.net/profile/Robert-Bensted-Smith/publication/358445842_A_biodiversity_vision_for_the_Galapagos_Islands_Charles_Darwin_Foundation_and_World_Wildlife_Fund_Ed_R_Bensted-Smith/links/6202eda0c2d279745e7629ec/A-biodiversity-vision-for-the-Galapagos-Islands-Charles-Darwin-Foundation-and-World-Wildlife-Fund-Ed-R-Bensted-Smith.pdf (accessed on 18 September 2025).
16. Schmidt, K.; Walz, A.; Martín-López, B.; Sachse, L. Testing socio-cultural valuation methods of ecosystem services to explain land use preferences. *Ecosyst. Serv.* **2017**, *26*, 270–288. [CrossRef]
17. López-Santiago, C.A.; Oteros-Rozas, E.; Martín-López, B.; Plieninger, T.; González Martín, E.; González, J.A. Using visual stimuli to explore the social perceptions of ecosystem services in cultural landscapes: The case of transhumance in Mediterranean Spain. *Ecol. Soc.* **2014**, *19*, 27. [CrossRef]
18. Martín-López, B.; Iniesta-Arandia, I.; García-Llorente, M.; Palomo, I.; Casado-Arzuaga, I.; Amo, D.G.D.; Gómez-Baggethun, E.; Oteros-Rozas, E.; Palacios-Agundez, I.; Willaarts, B.; et al. Uncovering ecosystem service bundles through social preferences. *PLoS ONE* **2012**, *7*, e38970. [CrossRef]
19. Bidegain, I.; Cerda, C.; Catalán, E.; Tironi, A.; López-Santiago, C. Social preferences for ecosystem services in a biodiversity hotspot in South America. *PloS ONE* **2019**, *14*, e0215715. [CrossRef]
20. Kelemen, E.; García-Llorente, M.; Pataki, G.; Martín-López, B.; Gómez-Baggethun, E. Non-monetary techniques for the valuation of ecosystem services. In *OpenNESS Ecosystem Services Reference Book*; Potschin, M., Jax, K., Eds.; EC FP7 Grant Agreement no. 308428; 2016; pp. 1–5. Available online: https://connectingnature.oppla.eu/sites/default/files/uploads/sp-non-monetary-valuation.pdf (accessed on 18 September 2025).
21. Aguado, M.; González, J.A.; Bellott, K.S.; López-Santiago, C.; Montes, C. Exploring subjective well-being and ecosystem services perception along a rural–urban gradient in the high Andes of Ecuador. *Ecosyst. Serv.* **2018**, *34*, 1–10. [CrossRef]
22. Plieninger, T.; Dijks, S.; Oteros-Rozas, E.; Bieling, C. Assessing, mapping, and quantifying cultural ecosystem services at community level. *Land Use Policy* **2013**, *33*, 118–129. [CrossRef]
23. Menatti, L.; Da Rocha, A.C. Landscape and health: Connecting psychology, aesthetics, and philosophy through the concept of affordance. *Front. Psychol.* **2016**, *7*, 571. [CrossRef] [PubMed]
24. García-Llorente, M.; Martín-López, B.; Iniesta-Arandia, I.; López-Santiago, C.A.; Aguilera, P.A.; Montes, C. The role of multifunctionality in social preferences toward semi-arid rural landscapes: An ecosystem service approach. *Environ. Sci. Policy* **2012**, *19–20*, 136–146. [CrossRef]
25. Hanaček, K.; Tran, D.; Landau, A.; Sanz, T.; Thiri, M.A.; Navas, G.; Del Bene, D.; Liu, J.; Walter, M.; Lopez, A.; et al. "We are protectors, not protestors": Global impacts of extractivism on human–nature bonds. *Sustain. Sci.* **2024**, *19*, 1789–1808. [CrossRef]

26. Palacios, D.M. Factors influencing the island-mass effect of the Galápagos Archipelago. *Geophys. Res. Lett.* **2002**, *29*, 2134. [CrossRef]
27. Denkinger, J.; Gordillo, L.; Montero-Serra, I.; Murillo, J.C.; Guevara, N.; Hirschfeld, M.; Fietz, K.; Rubianes, F.; Dan, M. Urban life of Galapagos sea lions (*Zalophus wollebaeki*) on San Cristobal Island, Ecuador: Colony trends and threats. *J. Sea Res.* **2015**, *105*, 10–14. [CrossRef]
28. Wikelski, M.; Nelson, K. Conservation of Galápagos marine iguanas (*Amblyrhynchus cristatus*). *Iguana* **2004**, *11*, 190–197.
29. Instituto Nacional de Estadística y Censos (INEC). *Perfil Territorial. Provincia Galápagos, Cantón San Cristóbal: Resultados del Censo de Población y Vivienda 2022 [Tablero Interactivo]*; INEC: Quito, Ecuador, 2022. Available online: https://cubos.inec.gob.ec/AppCensoEcuador/ (accessed on 20 September 2025).
30. Lucero, R.; Mera, M. Impacto del Turismo en el Desarrollo Económico Sostenible de la Isla San Cristóbal, Galápagos. *Maest. Soc.* **2024**, *21*, 2074–2084.
31. Gordillo, J.G.; Kendrick, A.W. Isla San Cristóbal. *Not. Galápagos* **1989**, *48*, 40–48.
32. Rivas-Torres, G.F.; Benítez, F.L.; Rueda, D.; Sevilla, C.; Mena, C.F. A methodology for mapping native and invasive vegetation coverage in archipelagos: An example from the Galápagos Islands. *Prog. Phys. Geogr.* **2018**, *42*, 83–111. [CrossRef]
33. Mons, S.; Rodríguez, N.L. The complex socio-ecological landscape in Latin America: Transdisciplinary knowledge production to address diversity. *Eur. Rev. Lat. Am. Caribb. Stud.* **2023**, *115*, 1–23. [CrossRef]
34. Nisbet, E.K.; Zelenski, J.M.; Murphy, S.A. Happiness is in our nature: Exploring nature relatedness as a contributor to subjective well-being. *J. Happiness Stud.* **2011**, *12*, 303–322. [CrossRef]
35. Nisbet, E.K.; Zelenski, J.M. Nature relatedness and subjective well-being. In *Encyclopedia of Quality of Life and Well-Being Research*; Springer International Publishing: Cham, Switzerland, 2023; pp. 1–9.
36. MacKerron, G.; Mourato, S. Happiness is greater in natural environments. *Glob. Environ. Change* **2013**, *23*, 992–1000. [CrossRef]
37. Delgado, L.E.; Rojo Negrete, I.A.; Torres-Gómez, M.; Alfonso, A.; Zorondo-Rodríguez, F. Social-ecological Systems and Human Well-Being. In *Social-ecological Systems of Latin America: Complexities and Challenges*; Delgado, L., Marín, V., Eds.; Springer: Cham, Switzerland, 2019. [CrossRef]
38. Das, M.; Das, A.; Seikh, S.; Pandey, R. Nexus between indigenous ecological knowledge and ecosystem services: A socio-ecological analysis for sustainable ecosystem management. *Environ. Sci. Pollut. Res.* **2022**, *29*, 61561–61578. [CrossRef] [PubMed]
39. Majumdar, K.; Chaterjee, D. Perception of subjective well-being of the Lodha Tribe in West Bengal. *Contemp. Voice Dalit* **2025**, *17*, 227–241. [CrossRef]
40. Ardoin, N.M.; Bowers, A.W.; Gaillard, E. Environmental education outcomes for conservation: A systematic review. *Biol. Conserv.* **2020**, *241*, 108224. [CrossRef]
41. Kimmerer, R.W. Weaving traditional ecological knowledge into biological education: A call to action. *BioScience* **2002**, *52*, 432–438. [CrossRef]
42. Reyes-García, V.; Babigumira, R.; Pyhälä, A.; Wunder, S.; Zorondo-Rodríguez, F.; Angelsen, A. Subjective wellbeing and income: Empirical patterns in the rural developing world. *J. Happiness Stud.* **2016**, *17*, 773–791. [CrossRef]
43. Bednarik, Z. *The Relationship Between Education and Happiness: Findings from the North Central and Northeast Regions*; North Central Regional Center for Rural Development: West Lafayette, IN, USA, 2024. [CrossRef]
44. Zorondo-Rodríguez, F.; Grau-Satorras, M.; Kalla, J.; Demps, K.; Gómez-Baggethun, E.; García, C.; Reyes-García, V. Contribution of natural and economic capital to subjective well-being: Empirical evidence from a small-scale society in Kodagu (Karnataka), India. *Soc. Indic. Res.* **2015**, *127*, 919–937. [CrossRef]
45. McCarter, J.; Gavin, M.C. Perceptions of the value of traditional ecological knowledge to formal school curricula: Opportunities and challenges from Malekula Island, Vanuatu. *J. Ethnobiol. Ethnomedicine* **2011**, *7*, 38. [CrossRef]
46. Gómez-Baggethun, E.; Mingorría, S.; Reyes-García, V.; Calvet, L.; Montes, C. Traditional ecological knowledge trends in the transition to a market economy: Empirical study in the Doñana natural areas. *Conserv. Biol.* **2010**, *24*, 721–729. [CrossRef]
47. Gómez-Baggethun, E.; Corbera, E.; Reyes-García, V. Traditional ecological knowledge and global environmental change: Research findings and policy implications. *Ecol. Soc.* **2013**, *18*, 72. [CrossRef]
48. Barthel, S.; Crumley, C.L.; Svedin, U. Biocultural refugia: Combating the erosion of diversity in landscapes of food production. *Ecol. Soc.* **2013**, *18*, 71. [CrossRef]
49. Wu, R.; Tang, H.; Lu, Y. Exploring subjective well-being and ecosystem services perception in the agro-pastoral ecotone of northern China. *J. Environ. Manag.* **2022**, *318*, 115591. [CrossRef] [PubMed]
50. Zhang, W. Social capital, income and subjective well-being: Evidence in rural China. *Heliyon* **2022**, *8*, e08705. [CrossRef] [PubMed]
51. Howell, C.J.; Howell, R.T.; Schwabe, K.A. Does wealth enhance life satisfaction for people who are materially deprived? Exploring the association among the Orang Asli of Peninsular Malaysia. *Soc. Indic. Res.* **2006**, *76*, 499–524. [CrossRef]
52. Masferrer-Dodas, E.; Rico-Garcia, L.; Huanca, T.; TAPS Bolivian Study Team; Reyes-García, V. Consumption of market goods and wellbeing in small-scale societies: An empirical test among the Tsimane' in the Bolivian Amazon. *Ecol. Econ.* **2012**, *84*, 213–220. [CrossRef]
53. Zhang, G.; Zhang, X.; Xie, L.; Zhang, Q.; Liu, D.; Wu, H.; Li, S. Perceived Importance and Bundles of Ecosystem Services in the Yangtze River Middle Reaches Megalopolis, China. *Front. Environ. Sci.* **2021**, *9*, 739876. [CrossRef]

54. Pellowe, K.E.; Leslie, H.M. Ecosystem service lens reveals diverse community values of small-scale fisheries. *Ambio* **2021**, *50*, 586–600. [CrossRef]
55. Chan, K.M.A.; Shaw, M.R.; Cameron, D.R.; Underwood, E.C.; Daily, G.C. Conservation planning for ecosystem services. *PLoS Biol.* **2006**, *4*, e379. [CrossRef]
56. Hartel, T.; Fischer, J.; Câmpeanu, C.; Milcu, A.I.; Hanspach, J.; Fazey, I. The importance of ecosystem services for rural inhabitants in a changing cultural landscape in Romania. *Ecol. Soc.* **2014**, *19*, 42. [CrossRef]
57. Temesgen, H.; Wu, W. Farmers' value assessment of sociocultural and ecological ecosystem services in agricultural landscapes. *Sustainability* **2018**, *10*, 703. [CrossRef]
58. Thiemann, M.; Riebl, R.; Haensel, M.; Schmitt, T.M.; Steinbauer, M.J.; Landwehr, T.; Fricke, U.; Redlich, S.; Koellner, T. Perceptions of ecosystem services: Comparing socio-cultural and environmental influences. *PLoS ONE* **2022**, *17*, e0276432. [CrossRef]
59. Lima, F.P.; Bastos, R.P. Perceiving the invisible: Formal education affects the perception of ecosystem services provided by native areas. *Ecosyst. Serv.* **2019**, *40*, 101029. [CrossRef]
60. Casado-Arzuaga, I.; Madariaga, I.; Onaindia, M. Perception, demand and user contribution to ecosystem services in the Bilbao Metropolitan Greenbelt. *J. Environ. Manag.* **2013**, *129*, 33–43. [CrossRef]
61. Jorda-Capdevila, D.; Rodríguez-Labajos, B. An ecosystem service approach to understand conflicts on river flows: Local views on the Ter River (Catalonia). *Sustain. Sci.* **2015**, *10*, 463–477. [CrossRef]
62. Castro, A.J.; Vaughn, C.C.; Julian, J.P.; García-Llorente, M. Social demand for ecosystem services and implications for watershed management. *J. Am. Water Resour. Assoc.* **2016**, *52*, 209–221. [CrossRef]
63. Cruz, C.E.; Zizumbo, V.L.; Chaisatit, N. La gobernanza ambiental: El estudio del capital social en las Áreas Naturales Protegidas. *Territorios* **2019**, *40*, 29–51. [CrossRef]
64. Martín, G. Ante las metas 2030 del convenio de biodiversidad. ¿Es la gobernanza de las áreas protegidas una alternativa a la gobernanza política tradicional? *Rev. Univ. Geogr.* **2021**, *30*, 111–144.
65. Munévar-Quintero, C.; Valencia-Hernández, J.G.; Hernández-Gómez, N.; Aguirre-Fajardo, A.M.; Ramírez-Ríos, M. Gobernanza participativa en la conformación del Sistema Nacional de Áreas Protegidas de la ecorregión Eje Cafetero, Colombia. *Jurídicas* **2023**, *20*, 139–157. [CrossRef]
66. Palomo, I.; Martín-López, B.; López-Santiago, C.; Montes, C. Participatory scenario planning for protected areas management under the ecosystem services framework: The Doñana social-ecological system in southwestern Spain. *Ecol. Soc.* **2011**, *16*, 23. [CrossRef]
67. Palomo, I.; Montes, C.; Martin-Lopez, B.; González, J.A.; Garcia-Llorente, M.; Alcorlo, P.; Mora, M.R.G. Incorporating the social–ecological approach in protected areas in the Anthropocene. *BioScience* **2014**, *64*, 181–191. [CrossRef]
68. Dirección del Parque Nacional Galápagos (DPNG). *Plan Institucional DPNG 2021–2025*; Dirección del Parque Nacional Galápagos: Puerto Ayora, Ecuador, 2021. Available online: https://www.galapagos.gob.ec/wp-content/uploads/2024/01/Plan_institucional_dpng_2021-2025-final.pdf (accessed on 18 September 2025).
69. Cortés-Capano, G.; Hanley, N.; Sheremet, O.; Hausmann, A.; Toivonen, T.; Garibotto-Carton, G.; Soutullo, A.; Di Minin, E. Assessing landowners' preferences to inform voluntary private land conservation: The role of non-monetary incentives. *Land Use Policy* **2021**, *109*, 105626. [CrossRef]
70. Lee, C.L.; Wang, C.H.; Lee, C.H.; Sriarkarin, S. Evaluating the public's preferences toward sustainable planning under climate and land use change in forest parks. *Sustainability* **2019**, *11*, 3149. [CrossRef]
71. Walter, R.K.; Hamilton, R.J. A cultural landscape approach to community-based conservation in Solomon Islands. *Ecol. Soc.* **2014**, *19*, 41. [CrossRef]
72. Plieninger, T.; van der Horst, D.; Schleyer, C.; Bieling, C. Sustaining ecosystem services in cultural landscapes. *Ecol. Soc.* **2014**, *19*, 59. [CrossRef]
73. Ibarlucea, L. Patrimonio cultural y turismo en una ciudad patrimonio mundial: Encuentros (y desencuentros) en Colonia del Sacramento, Uruguay. In *Gestión del Patrimonio: Paisajes Culturales y Participación Ciudadana*; Galceran, C., Giordano, F., Eds.; Comisión del Patrimonio Cultural de la Nación, Intendencia de Río Negro, Universidad CLAEH: Montevideo, Uruguay, 2018; pp. 221–238.
74. Djabarouti, J.; Ren, Y. Cultural convergence in heritage landscape conservation: A comparative study of Chinese and English traditions. *Arts Commun.* **2024**, *2*, 1923. [CrossRef]
75. Riba-Hernández, L. Participación ciudadana para la conservación de los paisajes culturales de la UNESCO en América Latina: Crítica descolonial para el tránsito entre la teoría y la práctica. In *Gestión del Patrimonio: Paisajes Culturales y Participación Ciudadana*; Galceran, C., Giordano, F., Eds.; Comisión del Patrimonio Cultural de la Nación, Intendencia de Río Negro, Universidad CLAEH: Montevideo, Uruguay, 2018; pp. 97–114.

land

MDPI

Article

Spatial Structure of Settlements in Mainland China in the Early 20th Century

Raorao Su [1] and Zhen Zhao [2,*]

[1] School of Geography and Remote Sensing, Guangzhou University, Guangzhou 510000, China; raosu@gzhu.edu.cn
[2] School of History, Renmin University of China, Beijing 100872, China
* Correspondence: zhaozhen@ruc.edu.cn

Abstract

Settlements and settlement systems are key arenas of human–environment interaction, and reconstructing their spatial patterns is essential for understanding historical socio-environmental dynamics. Using the *Complete Map of the Great Qing Empire* (1905), this study employs digital extraction and spatial-statistical analysis to examine the nationwide settlement system of late Qing China. The results reveal that: (1) The system features dispersed high-level settlements and highly clustered low-level ones; provincial and prefectural cities follow administrative divisions, while counties, towns, and villages display strong spatial self-organization. (2) Mid-to high-level systems exhibit hierarchical fractures, whereas low-level settlements conform to Zipf's law, highlighting the regularity and universality of grassroots networks. (3) Road accessibility, slope, and elevation significantly influence settlement hierarchy, whereas river proximity plays a limited role—indicating greater dependence on transportation and terrain adaptability. Overall, the study elucidates the spatial structure and formative mechanisms of the Qing settlement system and provides empirical insights into the evolution of surface patterns and regional resilience since the modern era.

Keywords: Qing dynasty China; spatial structure; historical GIS (HGIS); Zipf's law; hierarchical distribution; environmental factor

1. Introduction

In geography, a settlement refers to a community of people living in a place, and a settlement system is the hierarchical arrangement of such communities by size, population, and services within a region. Meanwhile, settlements and settlement systems have long been the subject of multidisciplinary attention. As the primary spatial carriers of human activities on the Earth's surface, their patterns and evolutions often embody profound social, cultural, economic, and political significances, while also serving as important indicators of environmental changes [1–3]. Long-term studies of settlements can help us understand surface patterns and the underlying interactions between human-social systems, holding significant value in revealing surface processes, social resilience, human-land relationships, spatial economic structures, and cultural patterns [4].

Existing studies have explored the relationships between settlements and socio-environmental dynamics across various regions and periods. Mortensen (1970s) examined early settlements and population dynamics in Egypt, integrating burial and anthropological data to reveal the profound influences of environmental and social factors on settlement selection and population structure [5]. Zhang et al. (2005), drawing on archeological and environmental data from the Yangtze River Delta, demonstrated the decisive role of climate and hydrological changes in the development of Neolithic cultures, cultural interruptions, and the rise and fall of civilizations [6]. Shennan (1993), combining archeological and chronological data, unveiled major transformations in Central European settlements and social structures from 3500 to 1500 BC, including population fluctuations, shifts in gender relations, and the linkages between metallurgical developments and the emergence of chiefdoms [7]. Easterlin (1976), using the northern United States as a case study, revealed the phased expansion and stagnation of farm populations under land resource constraints, thereby validating the applicability of Malthusian population theory in historical contexts [8]. Li et al. (2022), taking the mountainous regions of western Hunan as an example, illustrated the spatial characteristics of 18th-century settlement expansion and its socio-cultural roots, driven by policy shifts, population dynamics, and agricultural innovations [9].

From the perspective of current research, most scholars have focused on the spatiotemporal evolution of settlements and populations, or leveraged them to investigate multidimensional themes—such as social, environmental, economic, and cultural dimensions—in historical periods. However, there remains insufficient attention to the internal structures and interconnectivities of settlement systems themselves [10–12]. In fact, as early as the last century, Skinner (1964) demonstrated the integrative role of market networks in traditional agricultural societies through an analysis of China's rural marketing system, emphasizing that markets function not only as economic mechanisms but also as critical linkages in village-community social structures and broader societal integration [13]. This perspective offers important methodological insights for examining the spatial structures of historical settlement systems. The diversity of Chinese history, coupled with its abundant documentary records, provides unparalleled conditions for such research. In recent years, with the rapid advancement of interdisciplinary research, studies of settlement evolution have expanded beyond traditional pattern analysis to encompass broader themes such as cultural and environmental adaptation [14]. The temporal dimension remains a fundamental aspect of geographical inquiry [15] and is equally crucial in settlement geography and development studies [16]. Recently, scholars have increasingly emphasized the importance of uncovering or reconstructing settlement patterns using new data sources and methods [17–19]. Historical maps, in particular, have emerged as valuable data sources for revealing spatial structures prior to the remote-sensing era (before the 1960s). With the growing integration of digital humanities and artificial intelligence, their potential is being further recognized and explored [20–22].

Building on this foundation, the present study employs the preserved *Complete Map of the Great Qing Empire* (*Da Qing Digguo Quantu*, 1905) from the late Qing Dynasty to digitally reconstruct the settlement system across mainland China at the turn of the 20th century. Specifically, this study addresses three core questions: (1) What were the spatial distribution characteristics of the settlement system in late Qing China? (2) How were settlement hierarchies shaped by environmental and topographic factors? (3) How can historical maps, such as the *Complete Map of the Great Qing Empire*, be digitally reconstructed to derive reliable historical settlement data and assess their accuracy?

By addressing these questions, this study not only reconstructs the hierarchical structure of historical settlements on a national scale but also provides a methodological framework for integrating historical cartographic data into quantitative land-system analyses.

This contributes to the broader field of land science by linking spatial reconstruction, environmental adaptation, and socio-political organization within a long-term human–land relationship perspective.

2. Materials and Methods

2.1. Map Digitization and Data Extraction

The Complete Map of the Great Qing Empire was published by the Commercial Press in Shanghai in the 31st year of the Guangxu reign (1905), comprising 25 maps in total, including one national map and 24 provincial and regional maps. As one of the significant achievements in late Qing cartography, it is regarded both as a representative of the transition from traditional Chinese cartography to modern cartography and as a typical case in the compilation of provincial maps following the Opium Wars [23,24]. Its content is relatively detailed, employing up to 15 distinct symbols for settlements, while also marking elements such as treaty ports, railways, and telegraph lines, thereby reflecting the latest changes in the social and economic landscape of the time.

One of the objectives of this study is to digitally extract settlement information from the map. Referencing established processing methods from prior research, the central meridian was set at 116.4283° [22], with latitudes based on the original map notations. Under the ArcGIS 10.8 environment, manual georeferencing and rectification were performed for all map sheets. Historical maps were rectified using second-and third-order polynomial geometric transformations. For each province, 7–31 ground control points (GCPs) were selected from each map. The provincial RMS registration errors are all less than 0.0139, with a national composite RMS of 0.0619. Most RMS values are below 0.01, indicating satisfactory geometric consistency sufficient for province-level analyses of land use and topography. Detailed transformation parameters and accuracy statistics are listed in Appendix A Table A1.

Subsequently, extraction was conducted according to the legend standards, categorized as provincial capitals—prefectural cities—directly administered departments—directly administered subprefectures—subprefectures—counties (including indigenous subprefectures and indigenous departments), treaty ports, postal stations, general settlements, and larger settlements (Figure 1) (Appendix A). Calculations indicate that the average positioning error for provincial capitals is approximately 14 km, with notably larger deviations for certain frontier cities such as Urumqi and Lhasa; detailed results are provided in Appendix B.

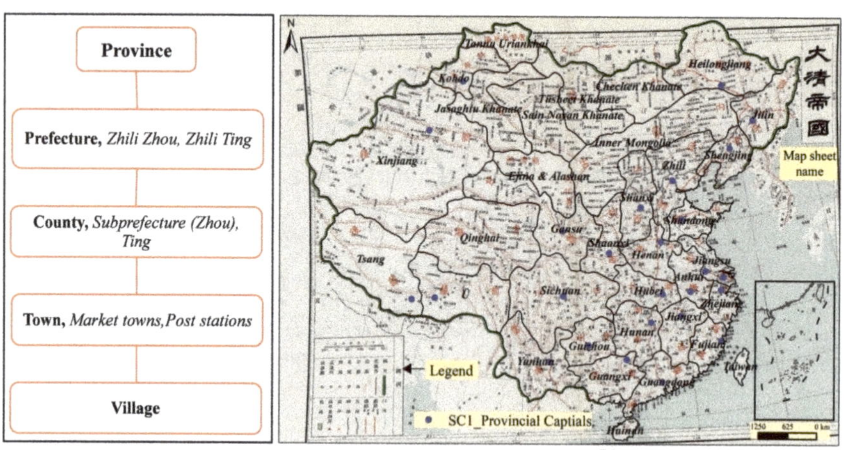

(a) (b)

Figure 1. Research Path (**a**) and Digitization Effect (National Map Sheet) (**b**) (The basemap shows the visualization of the nationwide map sheets after georeferencing of the atlas).

2.2. Other Data Sources

Global DEM data were sourced from GDEMM2024: 30 Arcsec Global Digital Elevation Merged Model 2024, a suite for Earth relief [25]. Road data were sourced from GROADS v1 (Global Roads Open Access Data Set) (Center for International Earth Science Information Network—CIESIN—Columbia University & Information Technology Outreach Services—ITOS—University of Georgia, 2013). River network data were sourced from the HydroSHEDS database (https://www.hydrosheds.org/, accessed on 15 September 2025).

2.3. Research Methods

2.3.1. Data Preparation and Settlement Extraction

Settlement locations were digitized from the *Complete Map of the Great Qing Empire* according to the legend classification (provincial capitals, SC1; prefectural seats, SC2; counties, SC3; towns, SC4; and villages, SC5). Each settlement was assigned to one of five hierarchical levels (SC1–SC5), forming the basis of subsequent spatial and statistical analysis.

2.3.2. Environmental Variable Assignment

Distance to Rivers and Roads: Using the Near Analysis tool in ArcGIS, the Euclidean distance from each settlement to the nearest river and road was calculated.

Topographic Factors: Elevation (GDEMM2024) and slope (derived from DEM using ArcGIS Slope function) were extracted for each settlement point via the Extract Multi Values to Points tool.

The resulting settlement-level database thus contained location, hierarchy, and four environmental variables: distance to river, distance to road, elevation, and slope.

2.3.3. Spatial and Statistical Analysis

Spatial Structure: Kernel density estimation (KDE), Average Nearest Neighbor (ANN), and Local Moran's I were applied to identify settlement clustering and spatial autocorrelation at different hierarchical levels.

Rank–Size Analysis: Settlement counts by province were fitted to Zipf's law to assess hierarchical organization.

Statistical Tests: One-way ANOVA and Tukey's HSD were used to compare environmental factors across settlement hierarchies.

Regression Analysis: Ordered logistic regression was applied to evaluate the relative effects of river distance, road distance, elevation, and slope on settlement hierarchy, with variance inflation factor (VIF) diagnostics to test multicollinearity.

Together, these methods enable an integrated understanding of the spatial distribution, hierarchical organization, and environmental determinants of settlements in late Qing China.

2.3.4. Assessment of Map-Scale-Dependent Biases

To quantitatively assess the potential bias caused by variable map scales across submaps of the *Complete Map of the Great Qing Empire*, this study conducted an additional regression analysis linking the number of village-level settlements (SC5) with map scale and environmental–socioeconomic factors. Specifically, Map Scale, Estimated Settlements 1930s (Number of settlements at the village and town level) [26], Household_1912 [27], Cropland_1908 [28], Mean Elevation, and Mean Slope were selected as explanatory variables, while ln(SC5) served as the dependent variable. All variables were log-transformed to reduce magnitude differences and approximate linearity. Both scatterplots and regression analyses were used to visualize and quantify the effect of map scale.

To test robustness, two models were constructed: (1) all provinces; and (2) excluding Fujian, Guizhou, and Guangxi, where unusually large numbers of mapped villages were observed. Multiple regression models were then applied to evaluate the significance and stability of scale effects, with variance inflation factor (VIF) diagnostics performed to avoid multicollinearity.

3. Results

3.1. Settlement Distribution Pattern

Based on kernel density estimation (KDE) and local spatial autocorrelation analysis (Local Moran's I), settlements of different hierarchical levels in mainland China in the early 20th century exhibited significant spatial distribution differences (Figure 2a–e).

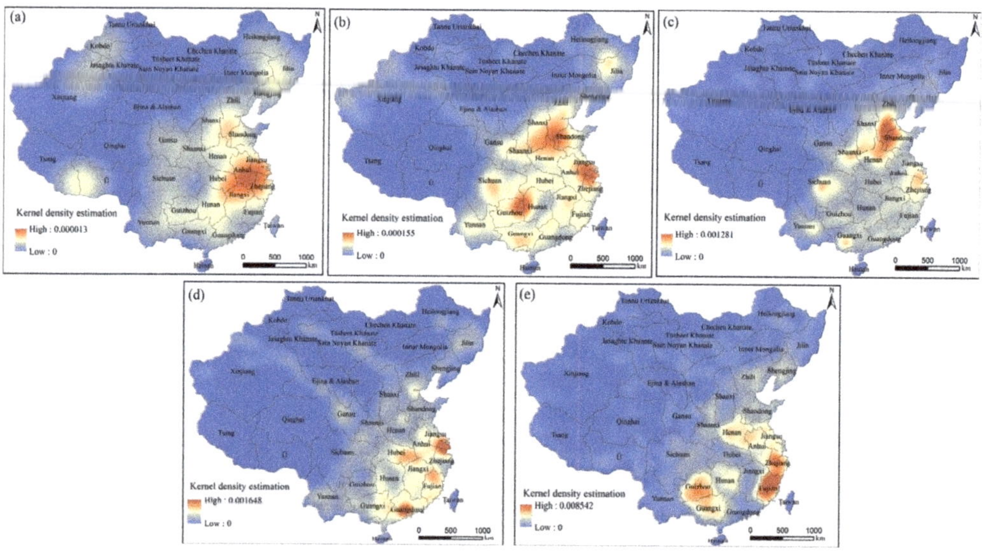

Figure 2. KDE Distributions of Different Types of Settlements. (**a**) represents SC1—level settlements; (**b**) represents SC2; (**c**) represents SC3; (**d**) represents SC4; and (**e**) represents SC5.

SC1 (provincial level, Figure 2a) is concentrated in the Yangtze River Delta, while forming a secondary dense belt in the North China Plain. Henan—Shandong—Zhejiang—Anhui constitute the core agglomeration area. The Mongolian Plateau exhibits a typical low-low agglomeration area. As China's economic core region since the Song Dynasty, the Yangtze River Delta aggregates the highest hierarchical settlement clusters (Figure 3a).

SC2 (prefectural level, Figure 2b) is primarily distributed in the middle and lower reaches of the Yellow River, the Yangtze River Delta, the Sichuan Basin, and the Yunnan-Guizhou Plateau. The provinces and regions exhibiting high-high agglomeration areas form an arc-shaped belt oriented from northeast to southwest, extending from Shaanxi to Yunnan, thereby presenting a typical central corridor pattern (Figure 3b). This distribution reflects the pivotal role of prefectural seats within the regional administrative network.

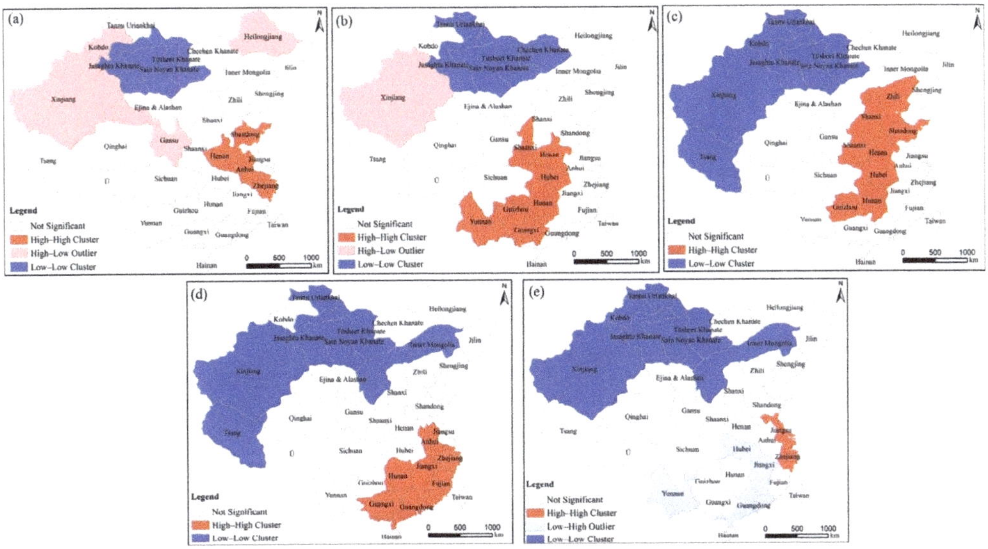

Figure 3. Local Moran's I Cluster Analysis Results for Different Types of Settlements. (**a**) represents SC1−level settlements; (**b**) represents SC2; (**c**) represents SC3; (**d**) represents SC4; and (**e**) represents SC5.

SC3 (county-level, Figure 2c) is primarily concentrated in the North China Plain, Sichuan Basin, and Yangtze River Delta. According to the Local Moran results (Figure 3c), its agglomeration centers exhibit high overlap with the secondary settlement provinces and regions, forming a northeast-southwest orientation from Zhili to Guizhou. This pattern underscores the hierarchical interconnectivity between prefectures and counties. SC4 (towns and postal stations, Figure 2d) more prominently reflects economic functions, with distributions centered in the Yangtze River Delta, Pearl River Delta, middle reaches of the Yangtze River basin, and Fujian coastal areas. In China's southeastern coastal provinces and regions, it displays typical high-high agglomeration, whereas the northwestern regions feature sparse distributions, forming low-low clusters (Figure 3d).

SC5 (village-level, Figure 2e) is primarily concentrated in eastern China, particularly along the southeastern coast, southwestern regions, and the provinces and regions in the middle and lower reaches of the Yellow River, demonstrating the characteristic that provinces with larger map scales have more detailed records. According to the Local Moran analysis (Figure 3e), it still exhibits a high-high zone centered on the Yangtze River Delta, while the northwestern regions form a low-low zone, reflecting spatial disparities in economic development.

3.2. Settlement Clustering Characteristics

To elucidate the spatial agglomeration characteristics of settlements in mainland China at the end of the Qing Dynasty (as reflected in the mapped areas), this study employs the Average Nearest Neighbor Index (NNR) and Moran's I spatial autocorrelation to analyze settlements of different settlement hierarchies, respectively.

The results indicate that the NNR values for settlements of different hierarchical levels differ significantly (Table 1). For SC1, the NNR is 1.2054, with a Z value of 2.0037 and a significance level of $p < 0.05$, suggesting that provincial capitals exhibit a slightly dispersed distribution nationwide, primarily due to the limited number of provincial capitals and their adherence to administrative division regulations. In contrast, SC2 ($NNR = 0.7838$),

SC3 (NNR = 0.7474), SC4 (NNR = 0.5513), and SC5 (NNR = 0.6124) all demonstrate significant clustered distributions, with the NNR values decreasing and the clustering intensity strengthening as the hierarchical level declines. Notably, SC4 and SC5 exhibit Z values of −34.5313 and −71.1171, respectively, underscoring their highly concentrated distributions and reflecting the strong dependence of grassroots settlements on natural environments and transportation networks. Although the NNR value of SC5 (villages) is slightly higher than that of SC4 (towns), this pattern reflects the broader spatial spacing of agricultural villages rather than a data artifact. As noted by Zhao, historical Chinese settlements exhibited a hierarchical differentiation in which large villages gradually evolved into towns ("ju cun wei zhen," 巨村为镇, meaning "large villages becoming towns") [29]. Villages were constrained by arable land availability and terrain conditions, whereas towns were more concentrated along transportation corridors and trade nodes, leading to a relatively tighter clustering pattern.

Table 1. NNR Analysis Results.

Indicator	SC1 (Provincial)	SC2 (Prefectural)	SC3 (County)	SC4 (Town)	SC5 (Village)
NN Ratio	1.2054	0.7838	0.7474	0.5513	0.6124
NN Z Score	2.0037	−7.1027	−18.3609	−34.5313	−71.1171
p	0.0451	0	0	0	0

Notes: $p = 0$ denotes $p < 0.001$.

The Moran's I index further validates this trend (Table 2). The Moran's I values are 0.3818 for SC1, 0.3686 for SC2, 0.5548 for SC3, 0.5900 for SC4, and 0.3637 for SC5, all significant at the 1% level. This indicates that all five hierarchical levels of settlements exhibit overall spatial positive autocorrelation, manifesting as a clustered pattern characterized by "like neighbors like." Among them, SC3 and SC4 exhibit the highest Moran's I values, demonstrating the strongest spatial clustering and suggesting that, within the late Qing settlement system, the spatial organization at the county- and town-levels was the most compact.

Table 2. Spatial autocorrelation index Analysis Results.

Indicator	SC1 (Provincial)	SC2 (Prefectural)	SC3 (County)	SC4 (Town)	SC5 (Village)
Index	0.3818	0.3686	0.5548	0.5900	0.3637
Z Score	3.7797	3.6335	5.4251	5.7584	3.6576
p	0.0002	0.0003	0	0	0.0003

Notes: $p = 0$ denotes $p < 0.001$.

In summary, this study draws the following conclusions: the settlement system at the end of the Qing Dynasty exhibits a significant spatial distribution characteristic of "high hierarchical levels being relatively dispersed and low hierarchical levels being highly clustered." The distributions of provincial and prefectural cities are predominantly influenced by administrative division policies, whereas counties, towns, and villages demonstrate evident spatial self-organization features, which are closely associated with resource accessibility, transportation convenience, and population carrying capacity in the context of an agricultural society.

3.3. Hierarchical Structure and Rank-Size Analysis

In terms of quantity indicators, the slopes for the number of settlements in SC2–SC4 range from $q \approx 2.52$–3.67, while R^2 values are generally low (<0.22) (Figure 4). The steep slopes indicate a pronounced hierarchical gradient, reflecting strong primacy and concentration of higher-level settlements in a few dominant provinces. However, the low R^2

values reveal weak internal continuity and poor hierarchical coherence, suggesting that the middle- and upper-level settlement systems were fragmented rather than smoothly ordered. In other words, although the hierarchy was steep, the internal structure lacked stability and completeness across provinces.

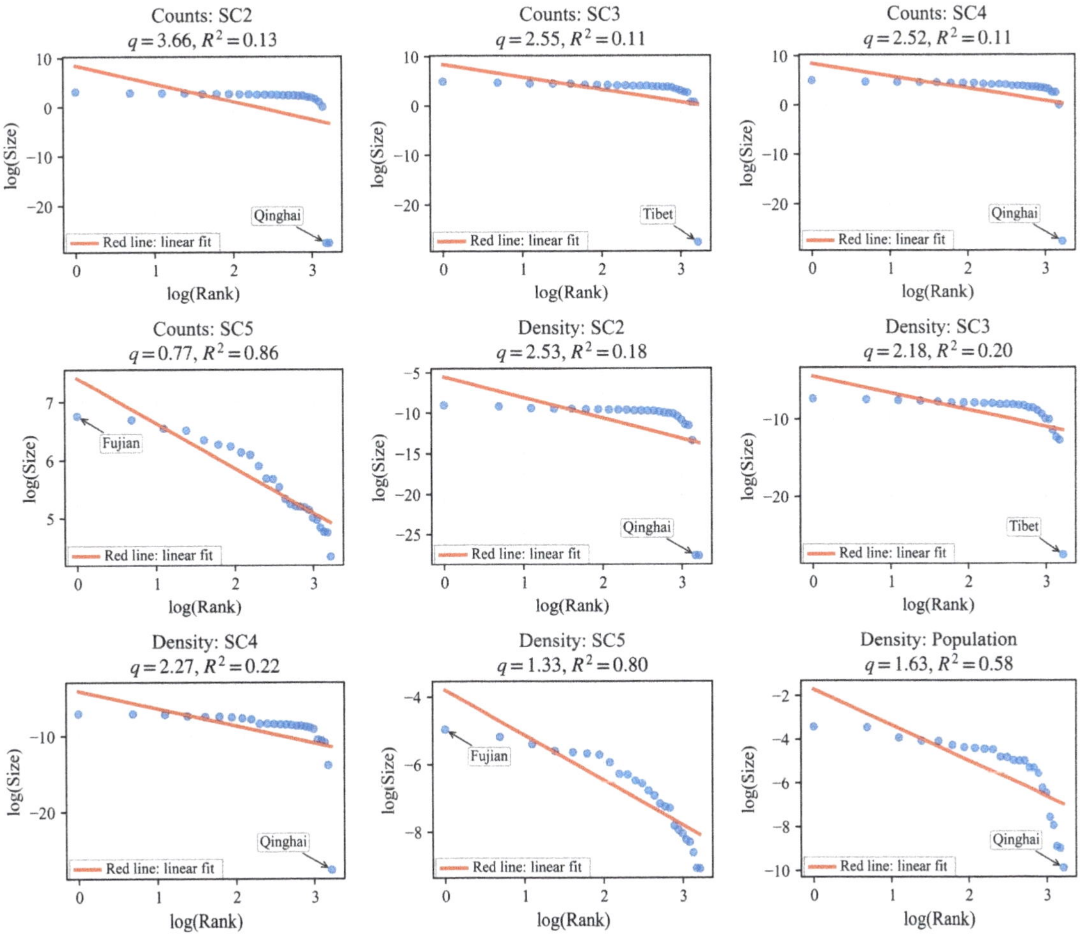

Figure 4. Rank-Size Results for Settlements Based by Province. Outer Mongolia was excluded from the Rank-Size Analysis. For provinces with zero counts (e.g., Qinghai and Tibet at certain levels), a small constant (10^{-12}) replaced zeros to allow logarithmic transformation without impacting rank order or model stability.

For SC5, the slope is $q = 0.77$ with $R^2 = 0.86$, showing a good power-law fit (Figure 4). This indicates a flatter rank-size curve and a more even distribution of numerous small settlements, rather than a strict adherence to Zipf's law ($q \approx 1$). The pattern suggests that grassroots settlements formed a more continuous and spatially balanced system, highlighting the overall regularity and completeness of the settlement network in densely populated regions.

In terms of density indicators, the q values for SC2–SC4 are approximately 2.27–2.53, with low R^2 values (0.18–0.22) (Figure 4), indicating substantial spatial density variations among mid- to high-level settlements and strong regional unevenness. In contrast, SC5 ($q = 1.33$, $R^2 = 0.80$) and population density ($q = 1.63$, $R^2 = 0.58$) exhibit better power-law fits (Figure 4). At the national scale, population distribution adheres to a stronger power-law regularity.

In summary, the Rank-Size analysis reveals that the high-level (SC2–SC3) and SC4 settlement systems are imperfect, characterized by severe hierarchical fractures and regional unevenness. In contrast, the grassroots settlement (SC5) system is relatively complete, with its distribution more closely conforming to Zipf's law. Provincial population distributions also largely adhere to Zipf's law, with the corresponding Rank-Size slope for settlement populations being $q = 1.14$ and $R^2 = 0.60$, approaching the classic Zipf's law ($q \approx 1$). This indicates a tendency toward regularization in the spatial densities of grassroots settlements and population distributions. It further underscores the supportive role of population dynamics in the hierarchization of the settlement system. Overall, these findings elucidate the "duality" of the late Qing Chinese settlement system: imbalance in the upper hierarchies, juxtaposed against a vast and highly regular grassroots structure.

3.4. *Influencing Factors of Settlement Hierarchical Distribution*

3.4.1. Relationship Between Settlement Hierarchy and Environmental Factors

From the statistical results and box plot comparisons of settlement levels and environmental factors, settlements of different hierarchical levels exhibit certain regular differences in terms of distance to rivers, distance to roads, slope, and elevation (Figure 5).

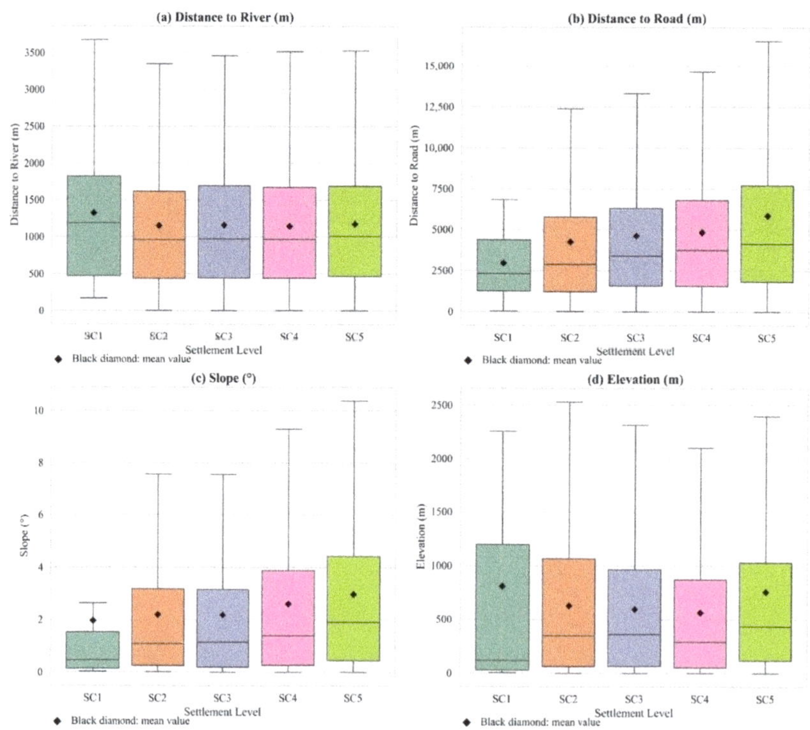

Figure 5. Relationships between Various Settlement Types and Environmental Factors: (**a**) River Distance, (**b**) Road Distance, (**c**) Slope, and (**d**) Elevation.

1. River Distance The average distances of settlements at various levels to rivers are roughly between 1100 and 1300 m, with no significant differences. The box plots show that the distributions for the five levels are relatively similar, with medians maintained at around 1 km. This indicates that whether settlements are provincial cities, prefectural

cities, or grassroots settlements, they all tend to be built along rivers, making rivers a common factor in site selection for settlements of different levels in the early 20th century.

2. Road Distance There are significant differences in road accessibility among different hierarchical levels of settlements. SC1 averages about 3 km from roads, with prefectural and county-level settlements gradually increasing, and SC5 having the highest average distance to roads, reaching 5–6 km. The box plots show that as the level decreases, the mean and median of the distribution overall rise, indicating that higher-level settlements are closer to major transportation routes. This aligns with the functions of provincial and prefectural cities as transportation hubs and administrative centers in the late Qing Dynasty.

3. Slope The lower the settlement level, the greater the slope at its site. The medians for SC1 and SC2 are both below 2°, indicating that high-level settlements are mostly distributed in plains or basins with gentle terrain; meanwhile, for SC5, the median is close to 2°, the 75th percentile is near 4.5°, and some even exceed 20°, reflecting that grassroots settlements are more distributed in hilly and mountainous areas.

4. Elevation Differences in elevation among different levels of settlements are relatively obvious. SC1 has an average elevation of about 800 m, with a wide distribution range; meanwhile, SC3 and SC4 have medians generally below 400 m; SC5, although with an average elevation of about 760 m, has a large distribution span, with some villages located in the Tibetan Plateau region above 6000 m. Overall, high-level settlements are more concentrated in mid-to-low elevation plains and basins, while low-level settlements are more dispersed, including more mountainous samples (Figure 6).

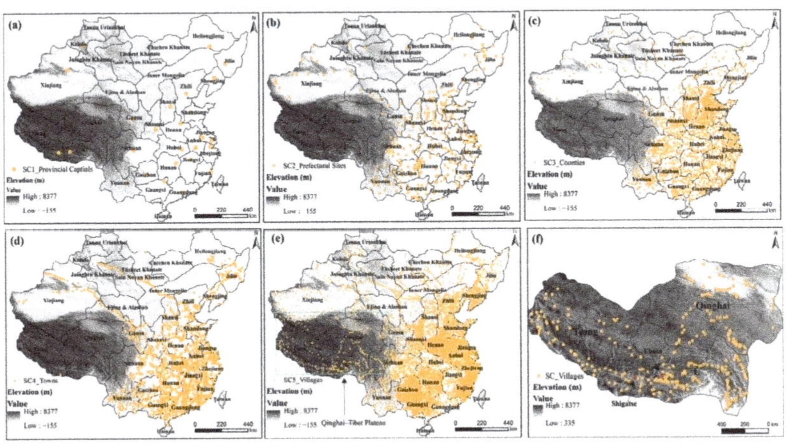

Figure 6. Spatial Overlay Visualization of Settlements at Different Levels and Elevation. (**a**) represents SC1−level settlements; (**b**) represents SC2; (**c**) represents SC3; (**d**) represents SC4; and (**e**) represents SC5; (**f**) represents the settlement distribution on the Qinghai–Tibet Plateau.

Overall, high-level settlements (provincial SC1, prefectural SC2) exhibit greater dependence on areas with convenient transportation and gentle terrain for site selection, whereas mid- to low-level settlements (towns SC4, villages SC5) demonstrate stronger terrain adaptability, with broader distribution ranges and particularly higher proportions in high-elevation and mountainous regions (Figure 6). This underscores the significant coupling relationship between the settlement hierarchy and the natural environment in late Qing China: high-level settlements leverage rivers, roads, and plains to perform administrative and economic core functions, while low-level settlements more prominently reflect the survival and adaptation of human habitations amid complex natural terrains.

3.4.2. Differences and Driving Factors Between Settlement Hierarchy and Environmental Factors

The results of one-way ANOVA indicate that settlements of different hierarchical levels exhibit significant differences in road distance ($F = 18.85$, $p < 0.001$), slope ($F = 23.87$, $p < 0.001$), and elevation ($F = 22.29$, $p < 0.001$), whereas the differences in river distance are not significant ($F = 0.58$, $p = 0.676$) (Table 3).

Table 3. One-way ANOVA and Tukey HSD results for environmental factors across settlement hierarchies.

Variable	F	*p*-Value	Significant Difference (Tukey HSD)
NEAR_DIST (River)	0.58	0.676	n.s.
NEAR_DIST (Road)	18.85	<0.001	SC5 vs. SC2, SC3, SC4
Slope	23.87	<0.001	SC5 vs. SC2–SC4
Elevation	22.29	<0.001	SC5 vs. SC3, SC4

Further Tukey's HSD post hoc multiple comparisons reveal (Table 3): (1) In terms of road distance, SC5 differs significantly from SC2–SC4, reflecting that low-level settlements are farther from major transportation trunk lines; (2) In terms of slope, SC5 exhibits significant differences from all other levels, indicating that villages are more frequently distributed in hilly areas with steeper slopes; (3) In terms of elevation, the distribution of SC5 is significantly higher than that of SC3 and SC4, suggesting stronger adaptability of villages to high-elevation terrains; (4) River distance shows no significant differences among levels, indicating that river-dependent siting is a universal feature of settlements, without manifesting hierarchical distinctions.

To further elucidate the influence of environmental factors on settlement hierarchy, this study constructed an ordered logistic regression model, the results of which validate this trend: road distance ($coef = 3.12 \times 10^{-5}$, $p < 0.001$), slope ($coef = 0.044$, $p < 0.001$), and elevation ($coef = 0.0001$, $p < 0.001$) all exert significant positive effects on settlement hierarchy, while river distance remains insignificant (Appendix C). This indicates that the hierarchical structure of the settlement system in late Qing China is primarily constrained by transportation accessibility and terrain conditions, with high-level settlements concentrated in regions featuring convenient transportation and gentle topography, whereas low-level settlements are more extensively distributed in mountainous and plateau areas.

3.5. *Evaluation of Map Scale Effects on Settlement*

The varying scales of sub-maps in the *Complete Map of the Great Qing Empire* may introduce spatial bias in settlement extraction, particularly for low-level settlements (SC5), which are more sensitive to mapping detail. Due to the SC5 in the Qinghai-Tibet Plateau and Mongolian Plateau being mostly drawn along roads, they were not included in the analysis.

To evaluate this effect, regression and visualization analyses were performed using ln(SC5) as the dependent variable and six explanatory factors: ln(Scale), ln(Estimated Settlements_1930s), ln(Household_1912), ln(Cropland_1908), ln(Mean Elevation), and ln(Mean Slope).

Scatterplots revealed that Fujian, Guizhou, and Guangxi were distinct outliers, with abnormally high SC5 values relative to their scales (Figure 7). Across all provinces, ln(Scale) and ln(SC5) show a significant negative relationship ($R^2 = 0.36$, $p < 0.01$), indicating that smaller-scale maps tend to contain fewer recorded villages. Household also shows a positive correlation ($R^2 = 0.20$, $p < 0.05$), while other factors were not significant.

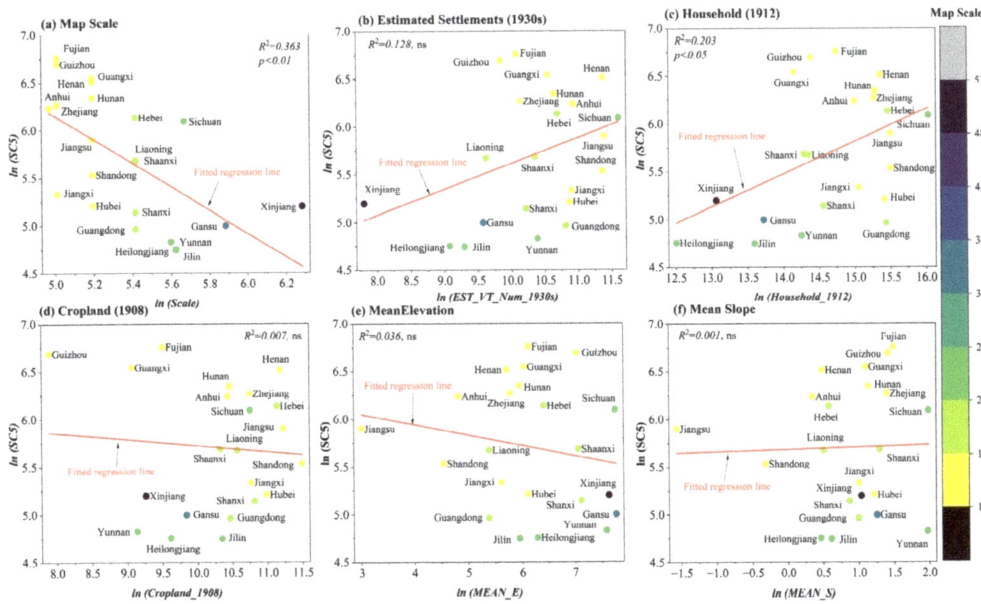

Figure 7. Scale-bias visualization across all provinces. Scatterplots showing the relationships between village–level settlements (ln SC5) and six explanatory variables: (**a**) Map Scale, (**b**) Estimated Settlements (1930s), (**c**) Household 1912, (**d**) Cropland 1908, (**e**) Mean Elevation, and (**f**) Mean Slope.

When the three outlier provinces were excluded, the correlations strengthened markedly (Figure 8): Scale–SC5 (R^2 = 0.27, $p < 0.05$), Estimated Settlements 1930s–SC5 (R^2 = 0.27, $p < 0.05$), Household–SC5 (R^2 = 0.41, $p < 0.01$), and Cropland–SC5 (R^2 = 0.30, $p < 0.05$). This confirms that extreme scale deviations concentrated in a few provinces may distort national patterns.

Multiple regression analysis (Table 4) using ln(Scale), ln(Household_1912), ln(Cropland_1908), ln(Mean Elevation), and ln(Range Slope) further validates this pattern. For the full sample (N = 22), the model shows R^2 = 0.52 (Adj. R^2 = 0.37, F = 3.49, p = 0.025), with a negative but statistically weak coefficient for Scale (–0.81, p = 0.16). After excluding the three provinces (Table 5) (N = 19), R^2 = 0.43 (Adj. R^2 = 0.21, p = 0.15), and the Scale coefficient diminishes (–0.34, p = 0.59). Thus, map scale effects are primarily localized rather than systemic, and removing the outliers does not substantially alter the national model ($\Delta R^2 < 0.1$).

Table 4. Regression results for ln (SC5)—full sample.

	Coefficients	Std. Error	t-Statistic	p-Value	Lower 95%	Upper 95%
Intercept	7.898302	4.18124	1.888986	0.077158	−0.96553	16.76214
ln(Scale)	−0.80862	0.554454	−1.45841	0.164079	−1.98401	0.366769
ln(Household_1912)	0.439647	0.233485	1.882975	0.078015	−0.05532	0.934614
ln(Cropland_1908)	−0.37685	0.208302	−1.80917	0.089251	−0.81843	0.064726
ln(MEAN_E)	0.091347	0.22162	0.412179	0.685678	−0.37847	0.561161
ln(RANGE_S)	−0.3108	0.597981	−0.51975	0.610352	−1.57847	0.95686

Multiple R = 0.7223, R^2 = 0.5217, Adj. R^2 = 0.3723, F = 3.491, p = 0.0253, N = 22.

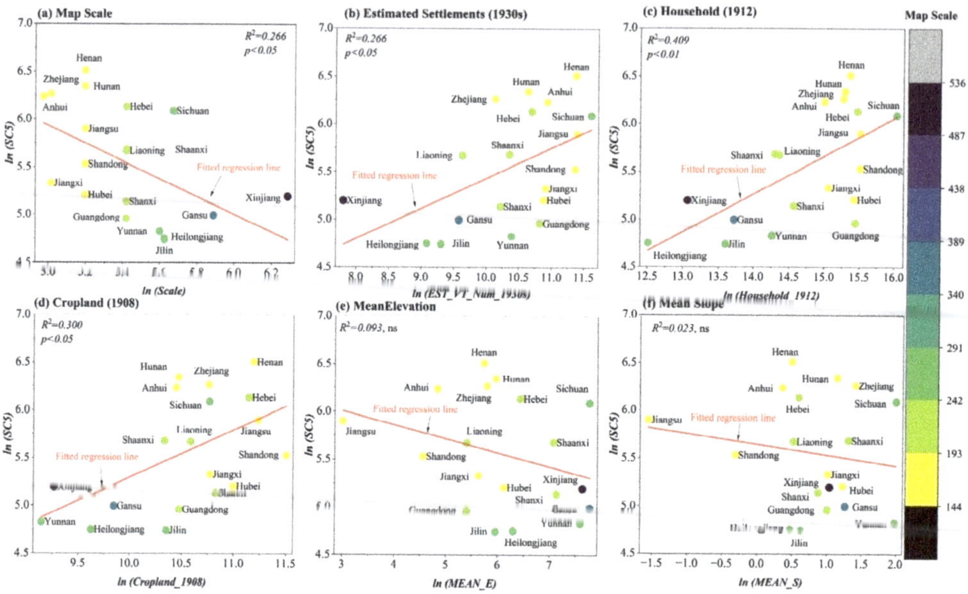

Figure 8. Scale−bias visualization excluding Fujian, Guizhou, and Guangxi. Same as Figure 7 but with three provinces (Fujian, Guizhou, Guangxi) removed. Scatterplots showing the relationships between village−level settlements (ln_SC5) and six explanatory variables: (**a**) Map Scale, (**b**) Estimated Settlements (1930s), (**c**) Household 1912, (**d**) Cropland 1908, (**e**) Mean Elevation, and (**f**) Mean Slope.

Table 5. Regression results for ln (SC5)—excluding Fujian, Guizhou, and Guangxi.

	Coefficients	Std. Error	t-Statistic	*p*-Value	Lower 95%	Upper 95%
Intercept	2.044514	5.556055	0.367979	0.718811	−9.95861	14.04764
ln(Scale)	−0.34021	0.624014	−0.54519	0.594854	−1.6883	1.007895
ln(Household_1912)	0.2867	0.28223	1.015837	0.328241	−0.32302	0.896421
ln(Cropland_1908)	0.088281	0.412501	0.214015	0.833856	−0.80287	0.979435
ln(MEAN_E)	0.02329	0.246014	0.09467	0.926021	−0.50819	0.55477
ln(RANGE_S)	0.019006	0.718003	0.02647	0.979284	−1.53215	1.570157

Multiple $R = 0.6558$, $R^2 = 0.4301$, Adj. $R^2 = 0.2109$, $F = 1.962$, $p = 0.1519$, $N = 19$.

Overall, these results suggest that map scale correlates with spatial variation in SC5 but acts more as a derivative indicator linked to topographic complexity and socioeconomic conditions rather than an independent driver. Consequently, while scale bias exists, it does not significantly distort the national-level conclusions on settlement hierarchies. The detailed regression coefficients and statistics are provided in Tables 4 and 5 and Appendix D. For clarity, Tables 4 and 5 summarize the results for the full and reduced samples, while additional models examining the determinants of map scale and household distribution are reported in the Appendix D.

4. Discussion

This section discusses how the empirical results address the study's main questions regarding the spatial distribution, hierarchical structure, and environmental determinants of settlements in late Qing China.

4.1. Causes of Formation of Settlement Distribution Patterns

This study indicates that by the end of the Qing Dynasty (late 19th to early 20th century), the Chinese settlement system had evolved to a critical period of social transformation marked by increasingly close connections between China and the world [30]. The spatial pattern of the national settlement system exhibits the following characteristics: high-level settlements are distributed relatively dispersed, whereas mid-to low-level settlements demonstrate high clustering. This disparity primarily stems from three aspects:

First, the establishment of high-level settlements such as provinces and prefectures adheres to the distribution pattern of national administrative divisions, placing greater emphasis on spatial equilibrium in political jurisdiction rather than optimal selections based on economic and natural conditions. This aligns closely with the insights from studies on Chinese historical political geography. Existing research indicates that "*mountains and rivers convenient*" (shanchuan xingsheng, 山川形胜) and "*interlocking teeth*" (quanyan jiaocuo, 犬牙交错) constituted the two major principles of administrative divisions during historical periods in China. The former delineates administrative divisions based on natural geographical conditions, yet geomorphic areas such as basins are prone to fostering local decentralization that opposes central authority [31]. Consequently, "artificial administrative divisions" (such as the multiple provinces of the Yuan Dynasty) that span multiple natural geomorphic areas and are dominated by central power emerged as necessary choices [31]. Therefore, the settlement system in China at the turn of the 20th century can also be interpreted from the perspective of spatial power.

Second, low-level settlements are more dependent on natural environments and transportation conditions, particularly the constraining effects of road accessibility, terrain slope, and elevation, which give rise to evident self-organizing clustering phenomena in social behaviors and economic activities. This aligns with G. William Skinner's interpretation of the traditional Chinese settlement system, integrating Christaller's Central Place Theory [13,32]. Furthermore, transportation networks represent a crucial aspect of China's national spatial organizational capacity during historical periods (including the transmission of information and the transfer of materials [33]), while the importance of rivers has relatively declined. This also explains why river distance did not exhibit significance in the statistical tests of this study.

4.2. Explanation of Settlement Hierarchical Differences

Through the systematic digitization of the *Complete Map of the Great Qing Empire*, this study reveals, at the national scale, the coexistence of hierarchical fractures and underlying regularities within the settlement system. On one hand, Rank-Size analysis indicates that the mid- to high-level settlement system is imperfect, lacking an ideal power-law distribution and exhibiting a pronounced "fractured" pattern, which is largely attributable to the spatial operations of political power. On the other hand, low-level settlements and population distributions more closely conform to Zipf's law, reflecting the universality and orderliness of the underlying network. These findings suggest that during the social transformation process in modern China, the administratively dominated upper hierarchy displayed imbalance and vulnerability, whereas the grassroots settlement network played a more enduring role in sustaining social operations and population carrying capacity.

Christaller proposed the Central Place Theory in the early twentieth century (1933) [32], which emphasizes that settlements of different hierarchical levels form a spatially ordered system based on service capacity and market range. The case of early-twentieth-century China illustrates a comparable pattern: higher-order settlements such as provincial and prefectural centers functioned as major central places, offering diversified services over broader areas; meanwhile, lower-order settlements, such as villages, served as minor central places with limited functions, smaller service radii, and denser distributions. The location and spacing of these centers were essentially determined by their threshold and range.

In late Qing China, higher-level settlements (SC1–SC3) were concentrated at administrative and transportation nodes, performing regional coordination and market functions analogous to Christaller's higher-order centers. In contrast, lower-level settlements (SC4–SC5) exhibited a more uniform and spatially continuous pattern, providing localized services within agrarian landscapes.

However, the observed hierarchy also departs from the idealized hexagonal regularity envisioned by Christaller. The steep rank-size slopes (high q values) and fragmented mid-level structures reveal the influence of administrative boundaries, topographic constraints, and agrarian production systems, which shaped a state-regulated rather than purely market-driven central-place network. This indicates that the Qing settlement hierarchy embodied both functional centrality and territorial governance, adapting central-place logic to a national social-economic and political context.

Furthermore, the various cartographic elements, including the scale, also embody hierarchy. Text, image, and the construction of regional knowledge are essential components of Humanistic Geography. As Buttimer argues, humanistic geography concerns not only how people interpret their everyday "lifeworlds," but also how academic and cartographic representations project particular "mindscapes" [34]. In this sense, the variations in map scale observed in this study also reflect deeper socio-economic and geographical factors, particularly their dependence on provincial area and administrative delineation. The provincial maps of the *Complete Map of the Great Qing Empire* thus function as visual confirmations of the political spatialization of power, where the delineation of provinces and counties represents a human-constructed geography deeply embedded in administrative and cultural logics.

4.3. Limitations and Prospects

It should be noted that this study still has several limitations. First, the *Complete Map of the Great Qing Empire* features varying scales across its sub-maps, with some regions (such as the Tibetan Plateau and northwestern frontiers) exhibiting missing place names or uncertain boundaries, potentially leading to uneven settlement data. Second, the research primarily relies on natural environmental and transportation factors, without incorporating the socio-economic, social, and institutional factors prevalent at the time. Future studies could introduce richer multi-source historical geographic data (such as population censuses and railway transportation records) for comprehensive analysis, thereby further deepening the interpretation of the causal mechanisms underlying settlement patterns. Additionally, systematic application of spatial georeferencing and error correction methods in future research would enhance data precision and the reliability of results.

Although the map-scale heterogeneity among sub-maps inevitably introduces minor regional bias—particularly in southwestern provinces—our regression analysis (Section 3.5) shows that such bias has limited statistical significance after excluding extreme cases. Therefore, while scale differences should be acknowledged as a data limitation, they do not systematically alter the main spatial or hierarchical conclusions of this study.

5. Conclusions

Overall, these findings directly address the research questions outlined in the Introduction and extend both theoretical and empirical understandings of settlement hierarchy and human–land interaction in historical China.

Utilizing digitized data from the *Complete Map of the Great Qing Empire* (1905), this study systematically reconstructs and analyzes the spatial structure and formative mechanisms of the settlement system in mainland China at the end of the Qing Dynasty. The main conclusions are as follows:

1. The late Qing settlement system displays a typical dual structure—relatively dispersed high-level settlements and highly clustered low-level ones—reflecting the combined effects of administrative division and environmental–transportation conditions.

2. Rank–Size analysis reveals pronounced hierarchical steepness and regional unevenness among mid- to high-level settlements, whereas the grassroots system exhibits a more even and continuous pattern approximating Zipf's law.

3. Statistical analyses identify transportation accessibility, slope, and elevation as major determinants of settlement hierarchy, while river proximity plays a minor role. These results suggest that the organization of settlements in late Qing China was primarily shaped by transportation connectivity and terrain adaptability.

In summary, the late Qing settlement system embodied a clear duality: the administratively dominated upper hierarchy was fragmented and vulnerable, whereas the self-organizing grassroots network was extensive and resilient.

By bridging spatial–statistical analysis with historical cartography, this study not only reconstructs the macro-scale structure of China's pre-modern settlements but also extends central-place and humanistic perspectives to an imperial agrarian context. It thus contributes to the broader field of land science by demonstrating how digitized historical maps can serve as quantitative evidence for exploring long-term human–land systems and the spatial evolution of socio-environmental organization.

Author Contributions: Conceptualization, R.S. and Z.Z.; methodology, R.S. and Z.Z.; software, R.S.; validation, R.S. and Z.Z.; formal analysis, R.S.; investigation, R.S. and Z.Z.; resources, R.S. and Z.Z.; data curation, R.S.; writing—original draft preparation, R.S.; writing—review and editing, Z.Z.; visualization, R.S.; supervision, Z.Z.; project administration, R.S. and Z.Z.; funding acquisition, R.S. and Z.Z. All authors have read and agreed to the published version of the manuscript.

Funding: This research has been funded by the major project program of the National Social Science Fund of China, grant number 20&ZD230. The work was also funded by the Postdoctoral Research Project of Guangzhou Municipality (625136-07).

Data Availability Statement: The associated data set in the study is available upon request.

Acknowledgments: The authors gratefully acknowledge the constructive feedback from the anonymous reviewers and the helpful guidance from the Academic Editor, which significantly contributed to improving this manuscript.

Conflicts of Interest: The authors declare no conflicts of interest.

Appendix A

This study is based on the basic information of each map sheet in the atlas. Different levels of settlements were digitized and extracted in ArcGIS 10.8, and the areas of provincial-level administrative units were calculated. At the same time, the scales of the different map sheets, as well as the control points for each map sheet during registration along with RMS error and Transformation Order, were recorded.

Table A1. RMSE and Georeferencing Accuracy Summary.

Province	Transformation Order	Ground Control Points	Total RMS Error
Zhili	2	21	0.00751812
Shengjing	2	9	0.00377645
Jilin	2	7	0.002
Heilongjiang	3	10	3.26×10^{-12}
Shandong	2	9	0.00297079
Shanxi	2	10	0.00623676
Henan	3	11	0.00277099
Jiangsu	3	13	0.00246838
Anhui	3	14	0.00000001
Jiangxi	3	17	0.00271372
Fujian	3	20	0.00277796
Zhejiang	3	24	0.00244408
Hubei	2	26	0.00747131
Hunan	2	31	0.00577058
Shaanxi	2	31	0.00688398
Gansu	2	23	0.0106866
Xinjiang	3	16	0.011003
Sichuan	3	18	0.00594711
Guangdong	3	13	0.00304336
Guangxi	3	13	0.00328411
Yunnan	3	15	0.00769711
Guizhou	3	11	0.00188859
Inner Mongolia and Outer Mongolia	3	10	2.26×10^{-12}
Qinghai and Tibet	3	17	0.0138929
County	3	21	0.0618996

Table A2. Basic Information of Each Map Sheet.

Province	SC5	SC4	SC3	SC2	SC1	Scale (×10,000)	Area/km^2
Jiangsu	365	90	50	12	2	180	102,494.4435
Anhui	510	87	46	10	1	144	149,050.1794
Zhejiang	526	50	56	7	1	150	93,384.8759
Jiangxi	207	107	64	13	1	150	183,326.3604
Hunan	569	90	57	13	1	180	216,480.976
Hubei	183	119	58	11	1	180	187,872.6896
Fujian	862	103	46	8	1	150	122,891.2128
Guangdong	143	163	62	15	1	225	210,036.9283
Guangxi	695	99	74	14	1	180	210,664.7611
Yunnan	125	75	70	15	1	270	353,762.7998
Guizhou	806	40	49	18	1	150	177,312.9463
Sichuan	442	66	115	24	1	290	574,116.7976
Zhili	461	64	143	19	1	225	326,837.6076
Henan	673	42	94	11	1	180	182,741.147
Shandong	253	30	88	11	1	180	135,146.5951
Shanxi	171	50	93	18	1	225	240,530.1796
Shaanxi	294	34	74	10	1	225	192,225.9077
Gansu	148	66	41	14	0	360	465,410.8025
Xinjiang	181	40	17	15	1	536	1,587,280.443
Liaoning	291	25	22	3	1	225	158,322.2775
Jilin	115	50	14	13	1	277	285,886.4389
Heilongjiang	116	13	2	5	1	277	469,045.0727
Qinghai	76	0	2	0	0	500	674,112.408
Tibet	189	1	0	0	2	500	1,042,683.343
Inner Mongolia	179	13	29	1	0	720	671,899.3172
Outer Mongolia	317	45	52	4	1	720	2,009,539.735

Appendix B. Offset Error Assessment Based on Provincial Administrative Points (SC1)

Based on a comparison between the *Complete Map of the Great Qing Empire* (1905) and modern geographic coordinates (2020) (National Catalogue Service for Geographical Information, (https://www.webmap.cn/commres.do?method=result100W, accessed on 15 September 2025), this study extracted 25 provincial capital sample points for error verification. The results (Table A2) indicate:

(1) Overall Error Level

The average spatial deviation is 13.82 km, with a standard deviation of approximately 9.16 km. The median error is 12.96 km, suggesting that most provincial capitals have deviations between 10 and 20 km. The minimum error occurs in Guilin (approximately 1.44 km), while the maximum error is observed in Urumqi (approximately 42.90 km).

(2) Error Distribution Characteristics

The Shapiro–Wilk test for normality yields W = 0.898, $p = 0.017 < 0.05$, indicating that the error distribution significantly deviates from normality. The one-sample t-test results in $t = 7.54$, $p < 0.001$, suggesting that the average error is significantly greater than 0, i.e., the map exhibits systematic positional bias.

Table A3. Offset Error Assessment Based on Provincial Administrative Points (SC1).

ID_Name_1905	X_1905	Y_1905	X_2020	Y_2020	ΔX	ΔY	Error_Deg	Error_km
Anqing	117.0224423	30.53577301	117.1099265	30.53468488	−0.087484212	0.001088137	0.087490979	8.396228753
Beijing	116.428851	39.89554265	116.7184397	39.9037508	−0.289588727	−0.008208151	0.28970503	24.78201378
Changsha	112.7852863	28.25089544	112.9778777	28.1176484	−0.192591341	0.133247045	0.234192655	23.99364324
Chengdu	104.0162228	30.67543945	104.0739775	30.65382477	−0.057754699	0.021614672	0.061666841	6.031477
Fuzhou	119.4141842	26.02749088	119.2902992	26.10348401	0.123884951	−0.075993129	0.1453356	14.98494339
Guangzhou	113.4252021	23.12833714	113.261191	23.13494289	0.164011114	−0.006605752	0.164144088	16.81440642
Guilin	110.1880771	25.24365196	110.175127	25.238142	0.012950135	0.005509957	0.014073579	1.44046739
Guiyang	106.556203	26.53360341	106.701798	26.603715	−0.14559501	−0.070111588	0.161596849	16.45480218
Hangzhou	120.0852292	30.28499581	120.148136	30.26970368	−0.06290674	0.015292135	0.064738762	6.285609116
Jilin	126.8578992	43.75287698	126.5434374	43.83580482	0.314461856	−0.082927842	0.325212678	26.93378201
Jinan	116.948567	36.60572913	117.0152951	36.66950549	−0.066728028	−0.063776362	0.092304139	9.257492083
Kaifeng	114.466899	34.92684577	114.3017044	34.79879922	0.165194595	0.128046551	0.209009984	20.7355411
Kunming	102.8314418	25.10092784	102.7087542	25.04927972	0.122687594	0.051648116	0.13311564	13.63607103
Lanzhou	103.8521813	36.13957498	103.8239565	36.06137196	0.028224822	0.078203018	0.083140559	9.042035328
Lhasa	91.01862424	29.68747549	91.115889	29.651787	−0.097264764	0.035688492	0.103605515	10.21293029
Nanchang	115.8130994	28.56999339	115.8081269	28.6411754	0.004972427	−0.07118201	0.071355473	7.904027636
Nanjing	118.7805656	32.06052635	118.7579787	32.06376979	0.0225869	−0.003243443	0.022818588	2.162977744
Qiqihar	123.8271074	47.19768643	123.9115462	47.35251839	−0.084438809	−0.154831959	0.176359995	18.36106655
Shenyang	123.4987109	41.85937562	123.4252886	41.8337045	0.073422303	0.025671125	0.077780725	6.73139648
Shigatse	88.72011203	29.25381927	88.87854907	29.2704495	−0.158437037	−0.016630221	0.159307435	15.50875132
Suzhou	120.5211469	31.29703707	120.5811607	31.30120784	−0.060013776	−0.004170769	0.060158529	5.732285544
Taiyuan	112.5033247	37.91656901	112.572592	37.81352301	−0.069267227	0.103046	0.124162904	12.96014929
Ürümqi	87.52520206	43.41263797	87.62580371	43.79177926	−0.100601652	−0.379141294	0.392261155	42.90025457
Wuhan	114.1657855	30.55402568	114.3361849	30.54883684	−0.170399406	0.005188844	0.170478391	16.35966227
Xi'an	108.8632998	34.26565175	108.9495464	34.26709703	−0.086246662	−0.00145287	0.086258771	7.94453985

Box plots and histograms further illustrate that the overall errors are concentrated between 10 and 20 km, but certain cities (such as Urumqi, Lhasa, Beijing, and Jilin) represent outlier deviation points.

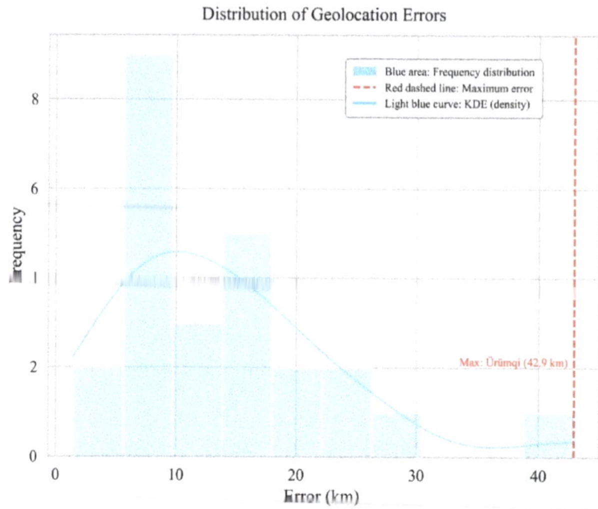

Figure A1. Histogram of Geographic Error Distribution for Provincial Settlements (SC1).

Figure A2. Box Plot of Geographic Error Distribution for Provincial Settlements (SC1).

(3) Geographic Differences

Inland provincial capitals (such as Nanjing, Hangzhou, and Chengdu) generally exhibit lower errors (<10 km), reflecting higher cartographic accuracy in the Yangtze River Basin and North China Plain regions.

Frontier and western provincial capitals (such as Urumqi, Lhasa, Kunming, and Guiyang) show relatively larger errors, indicating that mapping conditions in frontier areas during the late Qing Dynasty were limited, leading to reduced positional accuracy.

Appendix C. Detailed Results of the Ordered Logistic Regression Model for Settlement Hierarchy

Table A4. Ordered Logistic Regression Results for Environmental Determinants ofSettlement Hierarchy.

Variable	coef	std err	z	$p > \|z\|$	[0.025	0.975]
NEAR_DIST (River)	-1.066×10^{-5}	2.31×10^{-5}	−0.461	0.645	-5.6×10^{-5}	3.47×10^{-5}
NEAR_DIST (Road)	3.118×10^{-5}	4.38×10^{-6}	7.118	0.000	2.26×10^{-5}	3.98×10^{-5}
SlopeGDEMM	0.0441	0.008	5.624	0.000	0.029	0.060
GDEMM2024 (Elevation)	0.0001	2.91×10^{-5}	3.641	0.000	4.9×10^{-5}	0.000
/cut1	−5.8197	0.199	−29.193	0.000	−6.210	−5.429
/cut2	0.9139	0.075	12.119	0.000	0.766	1.062
/cut3	0.6172	0.029	21.487	0.000	0.561	0.673
/cut4	−0.2173	0.025	−8.857	0.000	−0.265	−0.169

The ordered logistic regression model assesses the impact of environmental factors on settlement hierarchy (SC1–SC5 as the ordinal dependent variable), assuming proportional odds. Predictors include river distance (NEAR_DIST), road distance (NEAR_DIST), slope (SlopeGDEMM), and elevation (GDEMM2024), with positive coefficients indicating higher hierarchy for greater values (e.g., longer road distances). Cut points (/cut1 to /cut4) denote thresholds between levels. All variables showed low multicollinearity (VIF < 5). Model fitted using statsmodels in Python 3.13.5.

It should be noted that the settlement hierarchy (SC1–SC5) was coded in ascending order (i.e., larger numerical values represent lower hierarchical levels). Therefore, the positive coefficient of "NEAR_DIST (Road)" in the ordered logistic regression does not indicate that higher-level settlements are farther from roads. Instead, it means that as the distance to roads increases, the probability of a settlement belonging to a lower-level category (e.g., SC4–SC5) increases. This finding is consistent with the ANOVA and boxplot results (Figure 6b), which show that higher-level settlements are closer to roads, while lower-level ones tend to be farther away.

Appendix D

To examine the effects of socio-economic and topographic factors on the variation in cartographic scales among provinces, multiple linear regression analyses were conducted with the logarithm of map scale (ln_Scale) as the dependent variable. The explanatory variables include population (ln_Household_1912), cropland area in 1908 (ln_Cropland_1908), mean elevation (ln_MEAN_E), slope range (ln_RANGE_S), and provincial area (ln_Area).

Results indicate that the model is statistically significant ($R^2 = 0.865$, adjusted $R^2 = 0.823$, $F = 20.58$, $p < 0.001$), suggesting that these factors collectively explain most of the inter-provincial variation in map scale. Among the predictors, ln_Area has the strongest and most significant positive effect ($\beta = 0.432$, $p < 0.001$), implying that provinces with larger spatial extents tend to have maps drawn at larger scales. Socio-economic and terrain factors show the expected directions but are not statistically significant at the national level, reflecting their relatively weaker influence compared with administrative spatial extent.

Table A5. Regression results for ln (Scale).

	Coefficients	Std. Error	t-Statistic	p-Value	Lower 95%	Upper 95%
Intercept	0.46456	1.167902	0.397773	0.696054	−2.01128	2.940401
ln(Household_1912)	−0.07894	0.057072	−1.38309	0.185634	−0.19992	0.042051
ln(Cropland_1908)	0.086875	0.04991	1.740608	0.100945	−0.01893	0.19268
ln(MEAN_E)	0.039111	0.056498	0.692257	0.498701	−0.08066	0.158881
ln(RANGE_S)	−0.12928	0.158732	−0.81448	0.427325	−0.46578	0.207212
ln(Area)	0.431831	0.075592	5.712676	3.21×10^{-5}	0.271584	0.592079

Multiple $R = 0.9303$, $R^2 = 0.88655$, Adj. $R^2 = 0.8234$, $F = 20.5844$, $p = 1.87544 \times 10^{-6}$, $N = 22$.

Table A6. Regression results for ln (Area).

	Coefficients	Std. Error	t-Statistic	p-Value	Lower 95%	Upper 95%
Intercept	12.71729	2.127919	5.976396	0.000015	8.227771	17.20681
ln(Household_1912)	−0.38826	0.157047	−2.47227	0.024286	−0.7196	−0.05692
ln(Cropland_1908)	0.230662	0.150048	1.53726	0.142633	−0.08591	0.547235
ln(MEAN_E)	0.221663	0.173117	1.28042	0.217593	−0.14358	0.586908
ln(RANGE_S)	0.501207	0.494569	1.013421	0.325067	−0.54224	1.544657

Multiple $R = 0.7863$, $R^2 = 0.66183$, Adj. $R^2 = 0.5284$, $F = 6.8830$, $p = 0.0017$, $N = 22$.

Table A7. Regression results for ln (Household).

	Coefficients	Std. Error	t-Statistic	p-Value	Lower 95%	Upper 95%
Intercept	13.85549	1.481152	9.354537	1.52×10^{-8}	10.75541	16.95558
ln(MEAN_E)	−0.74828	0.245076	−3.05327	0.006541	−1.26123	−0.23533
ln(RANGE_S)	1.686002	0.746555	2.258377	0.035877	0.123445	3.24856

Multiple $R = 0.5765$, $R^2 = 0.3323$, *Adj.* $R^2 = 0.2620$, $F = 4.7285$, $p = 0.0215$, $N = 22$.

Overall, the regression analyses demonstrate that map-scale variation across provinces primarily reflects structural differences in spatial extent (Area), while socio-economic and topographic factors exert secondary effects. This supports the interpretation that cartographic scale heterogeneity in the *Complete Map of the Great Qing Empire* arises from both administrative spatial delineation and human cartographic practices, rather than from purely natural or socio-economic causes.

References

1. Fisher, J.S.; Mitchelson, R.L. Forces of change in the American settlement pattern. *Geogr. Rev.* **1981**, *71*, 298–310. [CrossRef]
2. McLeman, R.A. Settlement abandonment in the context of global environmental change. *Glob. Environ. Chang.* **2011**, *21*, S108–S120. [CrossRef]
3. Carrión, J.S.; Fuentes, N.; González-Sampériz, P.; Sánchez Quirante, L.; Finlayson, J.C.; Fernández, S.; Andrade, A. Holocene environmental change in a montane region of southern Europe with a long history of human settlement. *Quat. Sci. Rev.* **2007**, *26*, 1455–1475. [CrossRef]
4. Kowalewski, S.A. Regional settlement pattern studies. *J. Archaeol. Res.* **2008**, *16*, 225–285. [CrossRef]
5. Mortensen, B. Change in the settlement pattern and population in the beginning of the historical period. *Ägypten Levante* **1991**, *2*, 11–37.
6. Zhang, Q.; Zhu, C.; Liu, C.-L.; Jiang, T. Environmental change and its impacts on human settlement in the Yangtze Delta, PR China. *Catena* **2005**, *60*, 267–277. [CrossRef]
7. Shennan, S.J. Settlement and social change in central Europe, 3500–1500 BC. *J. World Prehist.* **1993**, *7*, 121–161. [CrossRef]
8. Easterlin, R.A. Population change and farm settlement in the northern United States. *J. Econ. Hist.* **1976**, *36*, 45–75. [CrossRef]
9. Li, Y.; Ye, Y.; Fang, X.; Zheng, X.; Zhao, Z. Settlement expansion influenced by socio-cultural changes in western Hunan mountainous areas of China during the eighteenth century. *J. Hist. Geogr.* **2022**, *78*, 22–34. [CrossRef]
10. Pumain, D. Settlement systems in the evolution. *Geogr. Ann. Ser. B Hum. Geogr.* **2000**, *82*, 73–87. [CrossRef]

11. Artursson, M.; Bech, J.-H.; Earle, T.; Fuköh, D.; Kolb, M.J.; Kristiansen, K.; Ling, J.; Mühlenbock, C.; Prescott, C.; Sevara, C. Settlement structure and organisation. In *Organizing Bronze Age Societies: The Mediterranean, Central Europe, and Scandanavia Compared*; Earle, T., Kristiansen, K., Eds.; Cambridge University Press: Cambridge, UK, 2010; pp. 87–121.
12. Bura, S.; Guérin-Pace, F.; Mathian, H.; Pumain, D.; Sanders, L. Multiagent systems and the dynamics of a settlement system. *Geogr. Anal.* **1996**, *28*, 161–178. [CrossRef]
13. Skinner, G.W. Marketing and social structure in rural China, Part I. *J. Asian Stud.* **1964**, *24*, 3–43. [CrossRef]
14. Liu, P.; Zeng, C.; Liu, R. Environmental adaptation of traditional Chinese settlement patterns and its landscape gene mapping. *Habitat Int.* **2023**, *135*, 102808. [CrossRef]
15. Williams, M. "The apple of my eye": Carl Sauer and historical geography. *J. Hist. Geogr.* **1983**, *9*, 1–28. [CrossRef]
16. Wang, J.; Zhang, Y. Analysis on the evolution of rural settlement pattern and its influencing factors in China from 1995 to 2015. *Land* **2021**, *10*, 1137. [CrossRef]
17. Leyk, S.; Uhl, J.H.; Connor, D.S.; Braswell, A.E.; Mietkiewicz, N.; Balch, J.K.; Gutmann, M. Two centuries of settlement and urban development in the United States. *Sci. Adv.* **2020**, *6*, eaba2937. [CrossRef]
18. Jochem, W.C.; Leasure, D.R.; Pannell, O.; Chamberlain, H.R.; Jones, P.; Tatem, A.J. Classifying settlement types from multi-scale spatial patterns of building footprints. *Environ. Plan. B Urban Anal. City Sci.* **2021**, *48*, 1161–1179. [CrossRef]
19. Ahn, Y.; Leyk, S.; Uhl, J.H.; McShane, C.M. An integrated multi-source dataset for measuring settlement evolution in the United States from 1810 to 2020. *Sci. Data* **2024**, *11*, 275. [CrossRef]
20. Chiang, Y.-Y.; Leyk, S.; Knoblock, C.A. A survey of digital map processing techniques. *ACM Comput. Surv.* **2014**, *47*, 1–44. [CrossRef]
21. Liu, Q.; Gong, Z.; Li, B.; Yin, J. MHSI-Net: An interpretable multi-modal deep learning network for street identification in historical maps. *Int. J. Geogr. Inf. Sci.* **2025**, 1–32. [CrossRef]
22. Han, Z.Q.; Yang, X.; Liu, M.; He, G.P. Spatial variation of mapping accuracy of the area south of the Great Wall depicted on overview maps of imperial territories in the Kangxi reign. *J. Tsinghua Univ. Philos. Soc. Sci.* **2021**, *36*, 25–33+205–206. [CrossRef]
23. Liao, K.; Yu, C. *A History of Cartography in Modern China*; Shandong Education Press: Jinan, China, 2008; pp. 103–124.
24. Editorial Committee. *The History of Chinese Surveying and Mapping*; Surveying and Mapping Press: Beijing, China, 1995; pp. 144–152.
25. Abrykosov, O.; Ince, E.S.; Förste, C. GDEMM2024: 30 Arcsec Global Digital Elevation Merged Model 2024, a Suite for Earth Relief. GFZ Data Services. 2024. Available online: https://gfzpublic.gfz.de/pubman/faces/ViewItemFullPage.jsp?itemId=item_5027077_1 (accessed on 15 September 2025).
26. Statistics Division, Ministry of the Interior. Statistical Charts: Provincial Distribution of Towns and Villages by Population Density. *Q. J. Intern. Stat.* **1937**, *2*, 1. Available online: https://www.cnbksy.cn/search/detail/facce5d60f98812222a2ea59abdfe01d/7/6912898400439e278c3c3762 (accessed on 15 September 2025).
27. Liang, F. *Zhongguo Lidai Hukou, Tiandi, Tianfu Tongji (Statistical Records of Households, Land, and Taxation in Chinese History)*; Zhonghua Book Company: Beijing, China, 2008; pp. 368–373.
28. Wang, Y.-C. *Land Taxation in Imperial China, 1750–1911*; Harvard University Press: Cambridge, MA, USA, 1973; pp. 24–25.
29. Zhao, S. Villager and Townspeople: Location and Identity in Zezhou, Shanxi in the Ming–Qing Periods. *Qing Hist. J.* **2009**, *75*, 1–18.
30. Karl, R.E. Creating Asia: China in the world at the beginning of the twentieth century. *Am. Hist. Rev.* **1998**, *103*, 1096–1118. [CrossRef]
31. Zhou, Z. *History of China's Local Administrative System*; Shanghai People's Press: Shanghai, China, 2014; pp. 226–249.
32. Christaller, W. *Central Places in Southern Germany*; Baskin, C.W., Translator; Prentice-Hall: Englewood Cliffs, NJ, USA, 1966; pp. 1–202.
33. Editorial Committee. *History of Chinese Highway Transportation*; People's Communications Press: Beijing, China, 1994; pp. 1–11.
34. Buttimer, A. Humanistic Geography. In *International Encyclopedia of the Social & Behavioral Sciences*; Smelser, N.J., Baltes, P.B., Eds.; Pergamon: Oxford, UK, 2001; pp. 7062–7067. [CrossRef]

land

MDPI

Article

Compositional and Structural Evaluation of Historic Public Parks: A Design Analysis

Albert Fekete [1,*], Aikaterini Gkoltsiou [2], Tatjana Prusac [1], Agata Porębska [3], Ece Beyza Özdemir [4], Asli Taşkın [5], Levente Béla Haleczky [1] and Dániel Komes [1]

1 Institute of Landscape Architecture Urban Planning and Garden Art, Hungarian University of Agriculture and Life Science, 1118 Budapest, Hungary; tatjanaprusac2000@gmail.com (T.P.); haleczky.levente@gmail.com (L.B.H.); komes.daniel@phd.uni-mate.hu (D.K.)
2 Laboratory of Floriculture and Landscape Architecture, School of Plant Sciences, Agricultural University of Athens, 118 55 Athens, Greece; katerinagkoltsiou@aua.gr
3 Institute of Environmental Engineering, Department of Landscape Architecture, Warsaw University of Life Sciences, 02-787 Warsaw, Poland; agataporebska22@gmail.com
4 Department of Urban Design and Landscape Architecture, Bilkent University, Ankara 06800, Türkiye; e.ozdemir@ug.bilkent.edu.tr
5 Department of Landscape Architecture, Faculty of Architecture, İstanbul Technical University, İstanbul 34367, Türkiye; taskina21@itu.edu.tr
* Correspondence: fekete.albert@uni-mate.hu; Tel.: +36-205482979

Abstract: The article outlined a method to examine the compositional development of some well-known 19th century European public parks with rich historic character. It focused on investigations forming the foundation for a research and design approach ensuring diversity and similarity of the sites, contexts and design problems. Backgrounds and principles that underpin the relation between conservation and development in historical settings of parks are addressed. The survey and comparative analysis of old maps, designs, descriptions and documentation of the current conditions of the investigated parks primarily through remote sensing methods provided an overview of the development and state of preservation of key compositional aspects: spatial structure, design, tree canopy coverage, road and water system, visual connections, dendrological species and forms during the last two centuries. This design analysis approach served as a research method supporting new landscape architecture challenges in conservation and restoration in a changing ecological environment and with new requirements from contemporary planning and design.

Keywords: landscape architecture; public parks; precedent analysis; history and design

1. Introduction

"The issue of 'public parks' refers to a special type of parks that emerged in Europe from the 18th century on and was at that time associated with the industrial revolution and people moving to the cities" [1]. Fleming defines 'public park' as a park made for and usually maintained at the expense of an urban population [2]. Three aspects form the core of what a 'public park' is or should be. First, access for the general public; second, the goals of recreation and the improvement of social conditions; third, the contribution to public health in some form [1]. In the next decades several historic urban parks in Central and Eastern Europe will have to be redesigned due to changes in ecological environment, social needs, functional developments and inadequate maintenance. How can new challenges for contemporary landscape architecture be implemented in future plans? To answer this question, a research project was initiated within the framework of Restoration of Historic

Landscapes and Sites course by the Institute of Landscape Architecture, Urban Planning and Garden Art in Budapest.

In this article we present some student works started at the beginning of Spring semester 2025, focusing on basic investigations that form the foundation for a research and design approach in which the diversity and similarity of the sites, contexts and design problems can be secured. The material will serve to develop a research methodology for redesign of 19th century public parks in the Region based on European historical analogies.

2. Materials and Methods

The problem is framed around four European case studies—all cases are 19th century urban parks—with differences and similarities in cultural backgrounds, size, topography and design approaches, selected from a list of 30 city parks. Part of the research problem is defining the basic historical characteristics of these historic urban parks, following the evolution of the park's structure, spatial development and vegetation's stock. The aim was to develop a comparative research and investigation methodology that is suitable as a basis for sustainable maintenance or even for the redesign of the parks in the near future.

2.1. Research Questions

The research questions in this article focus on three specific theoretical issues:

- What was the role and importance of city parks in the 19th century and what is it nowadays?
- Can we identify the social, urban and ecological changes that drove modifications in the main compositional elements of city parks (spatial structure, visual connections, functions, tree canopy coverage proportions and plant compositions)?
- How can we relate the development of historical parks to the new challenges of landscape architecture such as climate change, loss of biodiversity, water management and the creation of healthy environments?

2.2. Research Approach

Landscape design can be researched in different ways, considering both design and research as ways of acquiring new knowledge, even though the methods differ [3–6]. Deming and Swaffield for instance address to this topic in two of their research strategies: "experimental strategies" and "projective design" [7]. In experimental strategies, they focus on investigation of landscape perception and preference by using different techniques. In projective design, they deal with design as research.In our paper we follow the "Research on design" approach. This is the type of research that takes place after the design process has been finished, including the historical analysis of plans, typological studies, evidence-based research and methodological investigations.

This research approach is based on the principles of the case study approach [8–10]. The research method adapted for this project implies the dynamics of urban landscape, including comparative analysis between old and new publications and descriptions (books, articles, plant lists, etc.), graphical representations (maps, designs, drawings and gravures, pictures, photo documentations, etc.) and remote sensing analysis based on Google Earth representations. Research materials comprise earlier research on these parks from landscape architecture and other disciplines.

3. Results

In the first step of the research, we conducted an indirect investigation based on comparison of old and current maps, designs, descriptions, articles and the documentation of the actual conditions of 30 European parks through the involvement of 24 landscape architecture students from different countries, enrolled in the international Master of Landscape Architecture and Garden Art Program at MATE University, Hungary. In this indirect investigation, we included well-known urban parks from 15 different European countries, collecting and analyzing similar basic data regarding each location: the name and size of the park, the urban context, the design period/data and the designer, the structure, main functions, features, plant use and compositional principles of the parks.

More detailed research has been conducted at the second level in the case of four selected urban parks: Birkenhead Park from Liverpool (UK), Battersea Park from London (UK), Városliget Park from Budapest (HU) and Kalemegdan Park from Belgrade (SRB)

From this basic survey, we selected four parks for a deeper investigation due to their basic characteristics. We considered both qualitative and quantitative selection criteria, which were the following:

I. There was enough available archival materials and current data in order to evaluate their historical character, development and current conditions;
II. They were representative of Western European and Eastern Europe parks in equivalent proportion;
III. They were considered representative as historical urban public parks both in their origin country and in the European context;
IV. They were established in the first part or middle of the 19th century (the typical time period when public parks started to be opened);
V. They have an area over 50 ha;
VI. It was possible to have on-site research access during the research period (Figure 1).

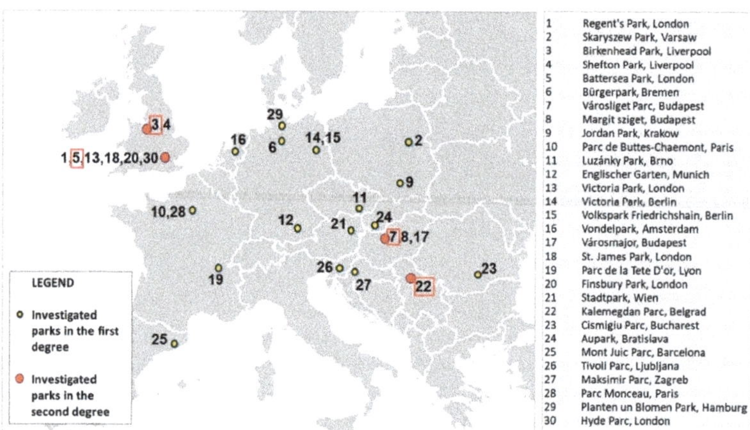

Figure 1. List and locations of the 30 city parks investigated in the general survey and the four case studies selected for more detailed analysis: Birkenhead Park, Liverpool (UK); Battersea Park, London (UK); Városliget Park, Budapest (HU), Kalemegdan, Belgrade (SRB). Source: Prepared by the Authors.

4. Discussion

4.1. Design Analysis of the Selected Parks

Social needs, economic growth and climate change undoubtedly affect the function, form and use of public parks. Historical functions of public parks have largely remained, but new functions have also emerged and, consequently, new types of uses and users. Functions as part of the landscape as a natural system become more explicit and extended [11–14]. The issue of public health plays a major role in the functioning of public parks [15]. Not only the parks maintain their traditional role of 'lungs' of the city, but this becomes even more important due to the rising air pollution and urban heat island effect [16–22]. Water surfaces, fountains and flowing water have a mitigating effect on the urban climate, a principle already practiced in the design of historical gardens [23]. Social aspects, accessibility, integration, community involvement and democratic use are more and more highlighted in the functional re-evaluation of historical public parks [24–28].

During the research at the selected four locations, our investigation and analysis highlighted functions, main design elements and artistic composition, space structure and canopy coverage, plant composition, visual connections and water and road systems at each location.

As the space limitations of this article do not allow us to present a detailed history of all four selected parks, we will focus on the comparative analyses carried out according to the criteria described above.

The basic parameters and a general bird's-eye view of each park are presented in Figures 2 and 3.

	BIRKENHEAD	BATTERSEA	VÁROSLIGET	KALEMEGDAN
Location	Liverpool	London	Budapest	Belgrade
Designer	Sir Joseph Paxton	Sir James Pennethorne/ John Gibson	Christian Heinrich Neb-bien	Emilijan Josimović
Establishment	1847	1858	1809	1867
Area	90 ha	83 ha	122 ha	57 ha

Figure 2. The basic parameters of the selected parks. Source: prepared by the Authors using [29–37].

Figure 3. The bird's-eye view of the selected parks. Source: [38–41]. (**a**) Birkenhead Park; (**b**) Battersea Park; (**c**) Városliget Park; (**d**) Kalemegdan Park.

4.2. Functional Analysis

The following four designated parks are analyzed both in terms of their individual functions within each park and as part of a larger complex, contributing to broader goals and serving various social and other purposes.

a. Birkenhead Park, Liverpool, UK

Designed by Joseph Paxton and opened in 1847, Birkenhead Park set a significant precedent in urban landscape architecture. Originally, the park was created to encourage passive recreation, such as walking, sitting, and picnicking, reflecting the 19th-century ideals of public health, social order, and moral improvement (Figure 4). Remarkably, Birkenhead Park has remained largely true to its original design, with only minor changes to its layout and function. Its picturesque landscape continues to charm visitors, supported by the careful preservation of historic buildings and architectural features. By the late 19th century, the park's role expanded to include active recreational spaces such as cricket grounds, football pitches, and an archery lawn, evolving into a multifunctional public landscape (Figure 5). This flexibility was especially evident during World War II, when parts of the park were repurposed as allotments to support food production [29–31].

Figure 4. The 1858 zonation of Birkenhead Park. The park was dominated mainly by areas designated for general recreation, but a few smaller areas with cultural purposes can also be found on the map. Source: prepared by the Authors.

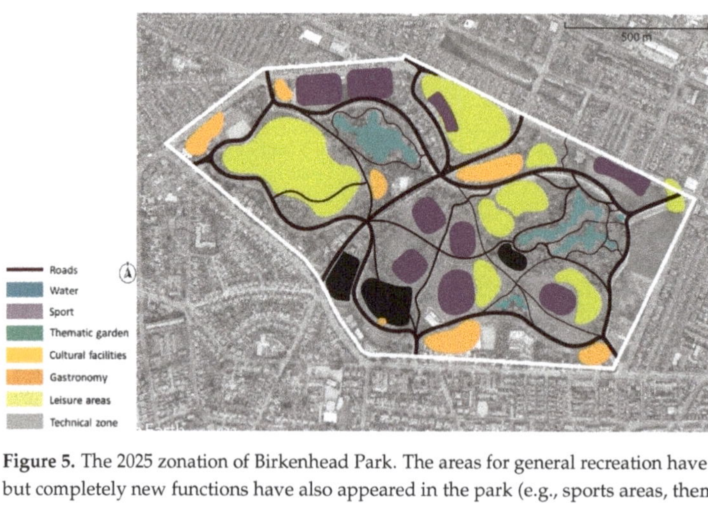

Figure 5. The 2025 zonation of Birkenhead Park. The areas for general recreation have been retained, but completely new functions have also appeared in the park (e.g., sports areas, thematic gardens). Source: prepared by the Authors.

b. **Battersea Park, London, UK**

Established in 1858 along the banks of the River Thames, Battersea Park reflected a Victorian vision of civic improvement, public health and social reform. The site, formerly a marshy common associated with stagnant water, taverns and gambling, was transformed into a green and healthy environment as part of a wider public park movement aimed at improving urban living conditions [32]. Its global function was to provide rest, fresh air and moral uplift to the working classes, offering relief from the overcrowded and polluted city environment [29]. The original design by landscape architect James Pennethorne followed the Victorian park typology and featured winding paths, carriage rides, ornamental gardens, and an artificial lake. In its early years, Battersea Park was primarily a space for passive leisure and ornamental recreation, such as boating, strolling and attending concerts at the central bandstand (Figure 6). Over time, the park adapted to changing urban needs. After its transfer to the Metropolitan Board of Works, new functions were introduced, including tennis courts, bowling greens and athletics facilities, reflecting growing interest in organized sport [33]. During both World Wars, the park served military and civil purposes. A major functional shift occurred during the 1951 Festival of Britain, when the Pleasure Gardens brought in fairgrounds, exhibitions and cafés, expanding the park's role in leisure and entertainment. Today, Battersea Park is a multifunctional public space with significant ecological value and a wide range of activities including sports, art, tourism and cultural events (Figure 7).

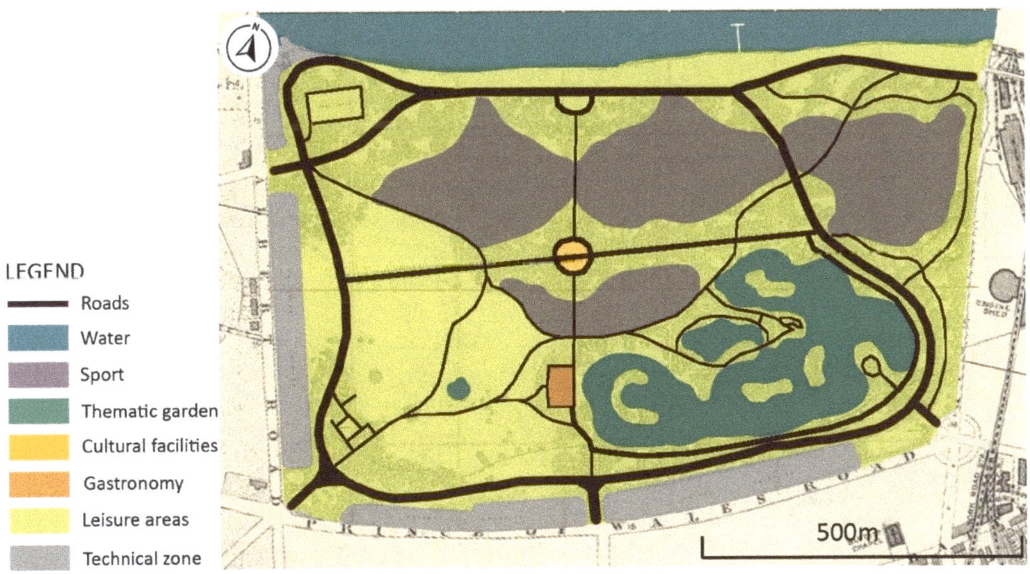

Figure 6. The 1858 zonation of Battersea Park. The park is dominated by areas for general recreation. There is also a significant amount of water surface. Source: prepared by the Authors.

Figure 7. The 2025 zonation of Battersea Park. The water surfaces in the park have largely retained their original form for more than 150 years. However, the use of the park has become more nuanced, with catering and gastronomic functions and thematic gardens gaining ground alongside general recreational areas. Source: prepared by the Authors.

c. **Városliget Park, Budapest, HUN**

The designer of Városliget Park was Heinrich Nebbien, a renowned landscape architect who won the commission in a competition in 1813. The park's construction works began in 1817 and were personally supervised by the designer [35]. Városliget is a very popular park thanks to the variety of activities and attractions it offers. It serves many functions related to the social life of city residents and tourists, including walking, leisure, education, gastronomy, and active recreation. The park was originally designed following the principles of landscape style parks, focusing mainly on walking and leisure [34] (Figure 8). Over time it developed into a multifunctional space. Moreover, the City Park was supposed to provide Budapest with a large amount of green space that helps the city adapt to and even mitigate climate change. The main function of the park was to purify the air, provide a valuable area for water infiltration and retention, which is rare in the city center, and reduce the urban heat island effect. The recreational functions of the park have changed over time, reflecting the development of public spaces in Europe as well as shifting trends and the needs of residents. In the second half of the nineteenth century, the park offered a much wider range of recreational opportunities (Figure 9) compared to its original design. New infrastructure and architectural features were built, along with temporary structures for the Millennium Exhibition. As a result, the park began to provide new attractions, including an ice rink in winter and a boating pond in summer, a circus, restaurants and cafés, a zoo and a museum. In later years, these elements were further developed, and new ones were created, such as permanent buildings replacing the temporary exhibition structures, new museums and a botanical garden [35].

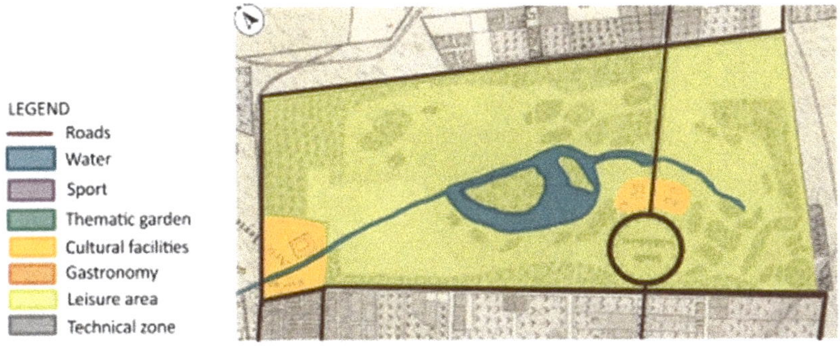

Figure 8. The 1830 zonation of Városliget park. Leisure was the focal point for park users shortly after the opening of the City Park of Budapest. Some cultural institutions had already appeared in this early phase. Source: prepared by the Authors.

Figure 9. The 2025 zonation of Városliget park. Today, the park is characterized by greatly increased cultural functions, thematic garden use and catering establishments. Accordingly, the park's paths have multiplied, covering the entire area. Source: prepared by the Authors.

d. Kalemegdan Park, Belgrade, SRB

With the gradual loss of Kalemegdan's military relevance in the second half of the 19th century, Emilijan Josimović, Belgrade's first urban planner, introduced a visionary urban concept in his 1867 regulatory plan, proposing the transformation of the former military base to a publicly accessible green area. Within this framework, Kalemegdan was conceived as a vital component of a "green belt" surrounding the historical city core, designed to promote public health and urban esthetics [36]. Kalemegdan's transformation began with the initial development of the Upper Kalemegdan in the late 19th century. Over time, the park expanded to include Lower Kalemegdan, the Inner Fortress, and Lower Town, gradually shaping Kalemegdan we know today. This evolution reflected not only spatial growth but also a shift in function. Park was initially developed as a landscape style park, primarily intended for walking, relaxation and passive leisure, in line with the design trends of the time (Figure 10). However, over the years, its role has evolved to meet the increasing demands of modern urban life, and the park gradually expanded its functions to include spaces for sports, tourism, cultural events and ecological benefits. The ecological role of Kalemegdan is particularly important, as it serves as the primary green space for both the city center and the surrounding residential areas. Kalemegdan's elevated position and open exposure to prevailing winds help reduce air pollution, while its rich vegetation supports temperature regulation and ecological balance [37]. It functions as both a historic symbol and a modern facility, home to the Military Museum, Belgrade Zoo, observatory, numerous monuments, and cultural institutions (Figure 11).

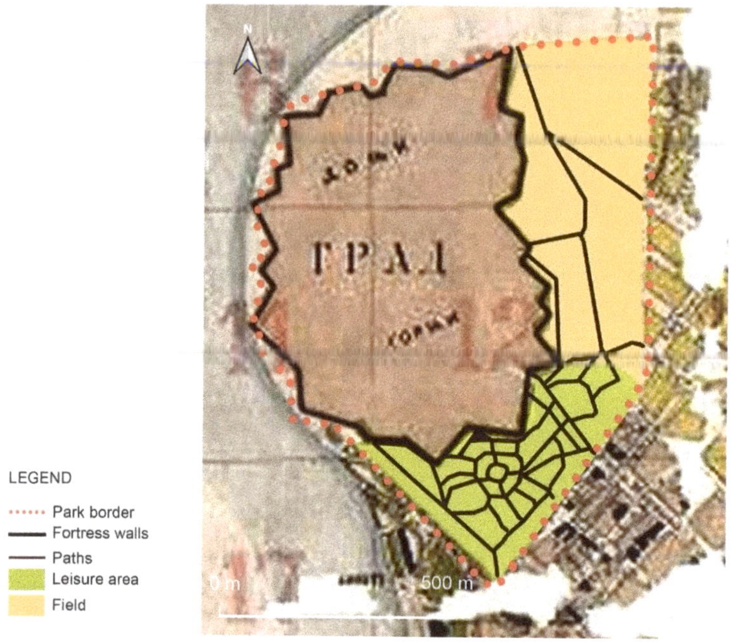

Figure 10. The 1867 zonation of Kalemegdan park. A relatively dense pathway system and exclusively general leisure areas characterize the park structure. Source: prepared by the Authors.

Figure 11. The 2025 zonation of Kalemegdan Park. Nowadays a considerably increased park area, an extremely dense traffic network, sports areas, thematic gardens and various facilities make Kalemegdan more diverse from both a functional and structural point of view. Source: prepared by the Authors.

4.3. Compositional Analysis

a. **Birkenhead Park, Liverpool, UK**

Birkenhead Park's layout, architectural language and landscape design reflect the English Picturesque design. This landscape technique aimed to evoke natural beauty while carefully guiding user movement and visual experience [42]. The park exemplifies this tradition through its deliberate modification of topography, views and planting, all crafted to promote an immersive visual and spatial experience. The park composition is characterized by a flowing, undulating terrain with strategically placed artificial mounds and valleys, particularly around the lakes, which were excavated to create sculptural landforms [30]. Key design elements of the park's layout include the following:

I. Curvilinear carriage drives and footpaths, avoiding axial geometry.
II. Short-range views and carefully framed vistas.
III. Higher-intensity planting around water features to enhance visual drama.
IV. Tree screening along the edges of the park, reinforcing a "green oasis" effect and concealing nearby urban fabric.

The intention of this design was to simulate the esthetic of the English countryside while providing the urban public a space for passive recreation. Paxton's vision extended beyond planting and path systems to a collection of architectural features placed as visual anchors and focal points. The most iconic structures include The Grand Entrance, Swiss Bridge and Boat House, Park Lodges, etc. The pastoral ideal was expressed through extensive lawns, meandering lakes, layered vegetation and varied tree species. Birkenhead Park is framed by a residential belt of villas and terraces, many of which were designed to harmonize with the park's appearance. Like many historic parks, Birkenhead Park has had to adapt to the pressures of globalization, rising urban populations, and evolving public expectations. No longer serving solely as a place for passive recreation, it has transformed into a multifunctional space.

b. **Battersea Park, London, UK**

Similar to Birkenhead Park, Battersea Park reflects the English picturesque style through its deliberate interplay of scenic planting, curved paths, and focal points. Irregularly shaped water bodies, especially the boating lake, provide scenic contrast to the open lawns and heavily planted tree clusters. The use of exotic plantations across different zones of the park demonstrates Victorian horticultural ambitions, while functionally separated areas for sport, leisure, and strolling reflect the period's approach to organizing recreation in a structured yet naturalistic setting. The park's architectural elements are distributed across the landscape as visual points, many of which have survived in altered or restored forms. At its entrances, original gate lodges once anchored each access point: Sun Gate Lodge, West Lodge, Rosary Lodge and Chelsea Lodge, although the last one was destroyed by fire. These structures reflect the formal thresholds of Victorian parks. One of the park's earliest structures, the rifle yard, located near the northwest gate, dates back to the park's initial construction and currently houses the park's offices. In the park's center lies the Victorian bandstand, situated at a key junction of paths. This placement was intentional: in the nineteenth century, the bandstand served as both a programmatic centerpiece and a social gathering point, reflecting the Victorian ideal of cultural enrichment through public leisure. A notable example of Victorian engineering within the park is the Grade II listed Pump House, built in 1861 to supply water to the lake and to the adjacent cascades, a dramatic artificial rock formation in the northern section of the boating lake [43]. Fundamentally, key design elements of Battersea Park's include the following:

I. Curving path network and circular nodes.
II. Irregularly shaped lake and water features, including the boating lake and cascades.

III. Exotic plantings.
IV. Densely planted tree clusters.
V. Functionally separated areas designated for walking, sports, events and rest.
VI. Architectural features such as lodges, the bandstand, Pump House and bridges.

While deeply rooted in Victorian values, Battersea Park has also embraced twentieth-century layers of meaning, as several important sculptures by twentieth-century artists are placed across the park. One of the most significant additions is the Peace Pagoda, a Buddhist stupa built in 1985, representing a cross-cultural call for global peace. Today, the park has adapted to the modern expectations of urban parks and serves as multifunctional spaces, offering a children's zoo, a sports arena, all-weather pitches, a boating lake, and various event venues.

c. **Városliget Park, Budapest, HUN**

The City Park in Budapest, originally known as Stadtwäldchen in German, was the world's first public park established by a city on its own territory and from its own resources for the free use of its citizens, with the designer selected through a public competition announced in 1813. The winning entry, designed by Heinrich Nebbien, is significant not only because it was the first of its kind in the world, but also because the City Park, as designed, is one of the most beautiful public gardens of its time [44].

According to Nebbien, the essence and foundation of a landscape garden is "the free, pure form of nature, born from the most intimate connection between economic and esthetic aspirations. (...) These two aspirations [in the case of the garden] are so closely intertwined that their results come about simultaneously and together, each arising from the other everywhere, and consequently manifesting themselves as the Beautiful as Useful and the Useful as Beautiful." [35]. The design of the Pest People's Garden was also born out of the professional and creative conviction that beauty and usefulness reinforce each other.

In terms of spatial composition, Nebbien considered classical garden design and picturesque gardens to be the defining sources of landscape gardening, viewing the subsequent sentimentalism as a mistake. The landscape architecture plan for the City Park, the "Ungarns Folks-Garten", is therefore characterized by classical principles: the long, transparent, freely formed spatial structure, the proportions, floor plans, and connections of the individual spaces, the design of the walls, the paths running along the edges, offering a comprehensive perspective of the whole, the enclosed patches of forest, the bouquet-like groups of trees formed from a few species, dividing the spaces—all these are the characteristic of the romantic transcription of the classicist picturesque garden.

Nebbien describes the separate spaces as painted images, suggesting that he intends to create paintings. The richness of design is represented by the following:

I. Various building features: buildings, the classicist gate, the restaurant, the amphitheater, the dance hall, the manor house.
II. Asymmetrical, freely formed layout, loose spatial structure, tucked away between the walls of the space and the plantings, avoiding any exposed position.
III. A regular circular central element, the Köröndi, designed as a walking and carriage area.
IV. Water mirror with a space-organizing effect.
V. In the axis of the City Park Avenue, behind the entrance gate to the park, there is a triple road system built around a circular lawn. Pedestrians strolled along the inner lane, carriages along the middle lane, and horsemen along the outer lane, each on a different surface, framed by a tree-lined avenue on the outer edge. [45].

Much of the original spatial structure and functions have been preserved to this day, in a more or less modified form.

d. Kalemegdan Park, Belgrade, SRB

Kalemegdan Park was initially designed in the landscape style, in line with the prevailing trends of its time, but over the years it has developed into a space characterized by a mixture of styles shaped by historical layers, urban growth and changing public needs. In the beginning, park consisted solely of Upper Kalemegdan, but today park includes Lower Kalemegdan, the Inner Fortress, and the Lower Town. In its early development, the park's main design elements focused on the organization of pedestrian paths and the planting of trees, reflecting the landscape style trends of the time. These early interventions emphasized circulation and visual connectivity, with pathways leading visitors toward monuments, fountains, and viewpoints overlooking the Danube and Sava rivers. Resting areas were strategically positioned along the promenades to encourage leisure and enhance the experience of movement through the park [36,46]. In its earliest form, Kalemegdan Park featured only landscape elements such as pedestrian paths, tree plantings, monuments, and fountains. However, as the park expanded and evolved into a multifunctional urban space, particularly with the development of Lower Kalemegdan, its character began to shift. New functions were introduced to accommodate broader public interests, including recreation, tourism and family activities. Lower Kalemegdan underwent the most visible transformation, especially with the addition of the Belgrade Zoo and an amusement park. As the park expanded, the integration of the fortress's historical walls, gates, towers and bridges became central to its identity. These features are preserved not only as cultural and historical landmarks but are also incorporated into the spatial logic of the park itself. However, over time, the following have consistently emerged as key elements of the design:

I. Curved and organic paths in Upper Kalemegdan, designed in the landscape garden style of the late 19th and early 20th centuries.
II. Main axial routes that connect major features.
III. Connections and stairways across steep terrain and level changes.
IV. Terraces, cliffs and elevated viewpoints.
V. Historical walls, gates, towers and bridges fully integrated into the park's spatial logic.
VI Tree-lined promenades.

4.4. Spatial Structure and Tree Canopy Coverage Analysis

a. Birkenhead Park, Liverpool, UK

Originally, the park featured lawns, gently curving paths, ornamental lakes, and framed vistas, complemented by strategic tree planting to enhance sightlines and provide spatial definition without overwhelming the landscape. In the early stage, tree canopy coverage was relatively sparse, in line with Paxton's design intent. The openness allowed for public gatherings, promenading, and uninterrupted visual connections across the park's terrain. Trees were carefully placed to frame views of key architectural elements, such as the Grand Entrance, Swiss Bridge and Boathouse, without dominating the visual or physical structure of the landscape.

By 2025, the spatial experience of Birkenhead Park has evolved significantly through the natural maturation of its vegetation. The tree canopy has become denser and more widespread; areas once defined by openness are now partially enclosed by mature tree growth. This increased canopy coverage has restructured how space is perceived. However, it is not only vegetation that has reshaped the spatial structure. Over time, the function and usage of the park have diversified, leading to the introduction of modern recreational infrastructure. Spaces have been changed to accommodate contemporary needs, including playgrounds, sports courts, fitness areas, cafés, visitor facilities and event spaces. This evolution in use has introduced a multi-zoned spatial pattern, where traditional scenic and passive-use areas now coexist with dynamic, activity-oriented zones.

b. **Battersea Park, London, UK**

The spatial structure of Battersea Park reflects a balance between formal design and organic flow, shaped by its Victorian origins and evolving urban context. The park was laid out with a clear internal hierarchy: wide tree-lined avenues and axial paths define major movement routes, while smaller curved secondary paths create a contrasting sense of intimacy and discovery.

This layered arrangement allows for seamless transitions between separated areas, open lawns, secluded woodland areas, and riverfront promenades. The design used the park's flat terrain to establish long sightlines, often ending in architectural or horticultural features. Today, these traditional spatial layers are joined by clearly zoned areas for modern facilities, including children's playgrounds, cafés, sports courts and event spaces, reflecting contemporary urban needs.

Tree canopy coverage has played a defining role in shaping both the aesthetic and ecological qualities of this spatial experience. In its original design, tree planting was formal and strategic: avenues were defined by rows of trees, and ornamental species were concentrated in designated gardens. Open lawns and recreation grounds were intentionally left unshaded, preserving visual openness and supporting active use. Over time, however, the tree canopy has matured and expanded significantly, resulting in broader shaded areas. This growth has softened the park's formal geometry, especially along meandering paths and near the water bodies, creating a different character.

c. **Városliget park, Budapest, HUN**

Nowadays Városliget's spatial structure is characterized by a strong central compositional axis, extending from Heroes' Square into the heart of the park. This axis acts as a formal spine, anchoring the park to the urban grid and establishing a clear directional logic. The rest of the park unfolds with a contrasting informal structure, a mix of curved paths, open lawns, and wooded areas that surround the central alignment. Initially, the park's vegetation was relatively sparse, characterized by simple (single-species) groupings of trees and open lawn areas. Early maps reveal these elements as the core structural components, with limited diversity in plant forms and a focus on open green spaces. The tree canopy coverage has expanded and matured over time, changing the original intent of the design. The evolving spatial structure and canopy coverage of Városliget are the product of changing social values and environmental needs, reflecting a broader European trend towards making multifunctional spaces.

d. **Kalemegdan Park, Belgrade, SRB**

Developed around the historical Belgrade Fortress, Kalemegdan Park has undergone substantial transformation since its formal landscaping in the late 19th and early 20th centuries. Kalemegdan's spatial structure is defined by its multi-level topography and fortified setting, which organize the park into a clearly articulated, three-level system. These levels are spatially and visually distinct, creating a hierarchical structure that guides movement and use. The topography plays a critical role, offering elevated viewpoints and structuring circulation along sloped paths, terraces, and stairways. The combination of its evolving landscape design, military legacy, strategic position, and varied elevation makes Kalemegdan a structurally complex park, where spatial orientation and terrain transitions are integral to the overall experience. As the oldest public park in Belgrade, Kalemegdan has preserved a notable portion of its original structure and vegetation. Many trees date back to the period between the two World Wars. While their age adds historic value, some now require replacement, not only because of their reduced visual appeal but also due to their diminished ecological function [37].

Summing up, the primary changes in the park's spatial structure can be observed through the transformation of its roads and pathways, water features, and tree canopy, as illustrated in Figure 12.

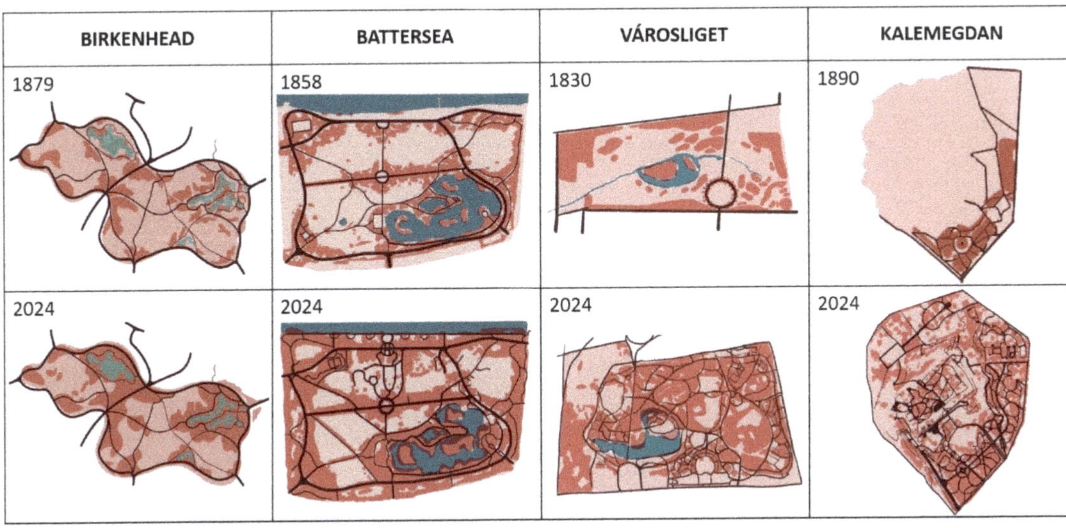

Figure 12. Space structure development comparison of the 4 selected parks. The sketches focus on the spatial structure development of the parks, highlighting the most influential elements in this aspect: the road systems, the canopy coverage, the main water features and the open spaces of the parks. Source: prepared by the Authors.

4.5. Visual Connection Analysis

a. Birkenhead Park, UK

Birkenhead Park features a carefully made arrangement of views, open spaces and architectural landmarks. Key buildings, such as the park's lodge, the main entrance and the pavilion, are carefully situated to frame vistas that guide the visitor's journey through the open space. Panoramic views across the lake and toward the pavilion remain crucial visual assets. Over time, as trees matured and the canopy expanded, some of these views became partially blocked or less visible. The park's management plan acknowledges this change and emphasizes the need to balance maintaining historic sightlines with supporting natural growth.

b. Battersea Park, UK

The River Thames serves as one of the most significant visual connections within Battersea Park, with numerous landscaped areas, elevated viewpoints and pathways oriented to frame the river's presence. The Boating Lake pathways were carefully designed to guide visitors along scenic routes, allowing them to experience a sequence of vistas where the water was beautifully framed by green lawns, trees and plantings that enhanced the picturesque quality of the landscape. The Grand Vista and the fountain lake with statues serve as focal points within the park, thoughtfully positioned to draw attention. The Battersea Power Station, a listed building located near the park, dominates the eastern skyline and creates a visual landmark when viewed from the bowling green, offering a strong visual connection between the park and its urban surroundings. While many primary axes remain clearly defined, the denser and more expansive tree cover has softened some of the original long-distance views, creating a more enclosed and intimate atmosphere in parts of the park (Figures 13 and 14).

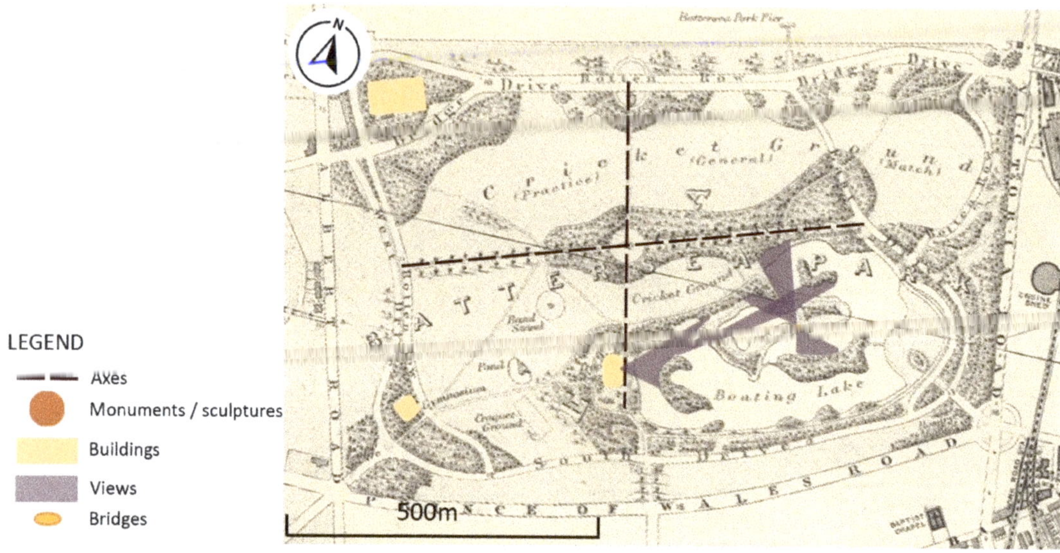

Figure 13. Visual connections in Battersea Park in 1858, organized mainly along the park's main cross axes of the park and around the water features. Source: prepared by the Authors.

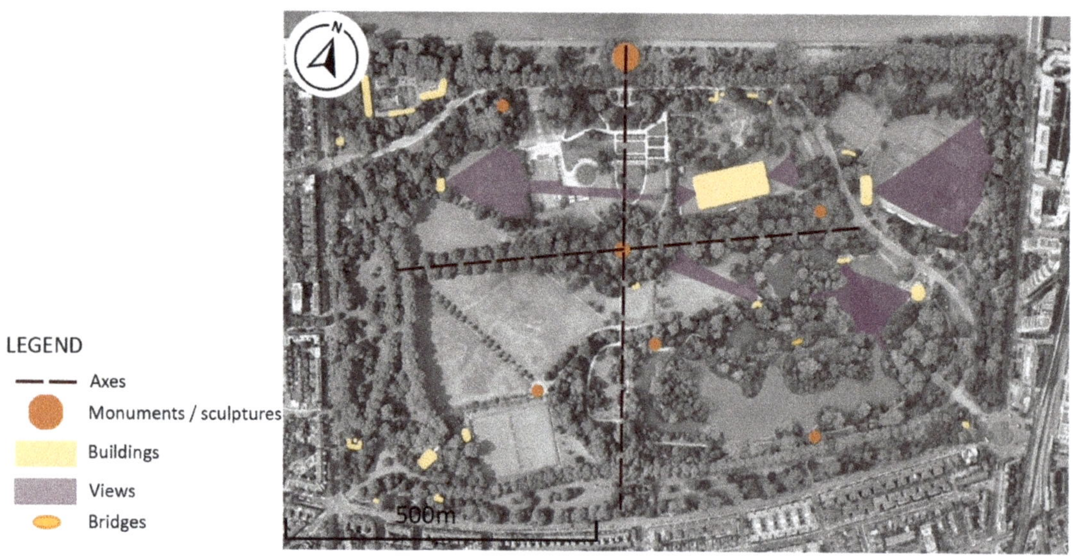

Figure 14. Visual connections in Battersea Park in 2025. The visual connections are more limited and shaped by dense vegetation developed over the centuries. Source: prepared by the Authors.

c. **Városliget Park, HUN**

When the Park first opened, its spatial composition was defined by a single primary visual axis that stretched across the park. The relatively sparse tree coverage at that time allowed for long, uninterrupted sightlines connecting key landmarks. As the park evolved over the following century, the introduction of new building elements created multiple observation points, enabling complex, multi-layered views across different planes within the park. Today, the park's dense and mature vegetation has significantly altered these sightlines. While the original long-distance views have largely been obscured, a notable exception remains—the bridge overlooking the pond still offers one of the few remaining distant vistas. Elsewhere, views are predominantly restricted, giving visitors a more enclosed and intimate spatial experience. This transition from expansive to more contained visual corridors reflects both the natural growth of the park's canopy and a shift in landscape design priorities (Figures 15 and 16).

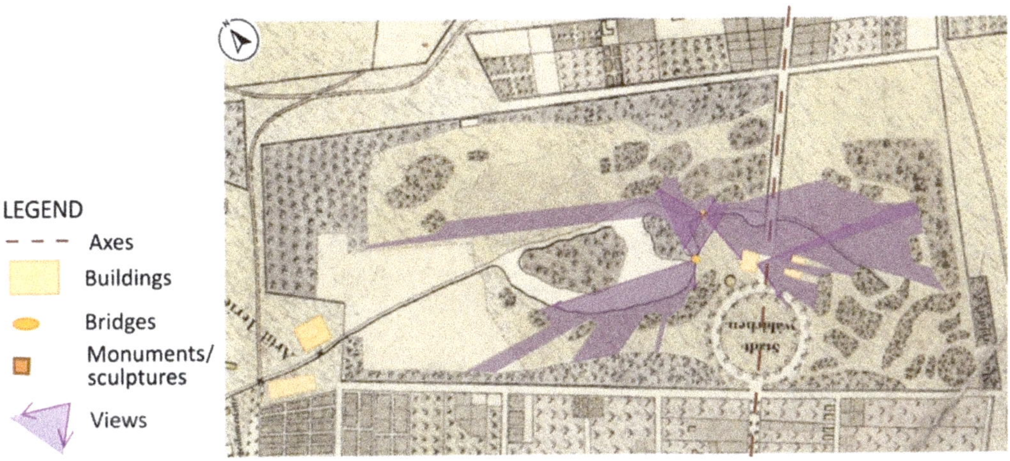

Figure 15. Visual connections in Városliget Park in 1830. Source: prepared by the Authors.

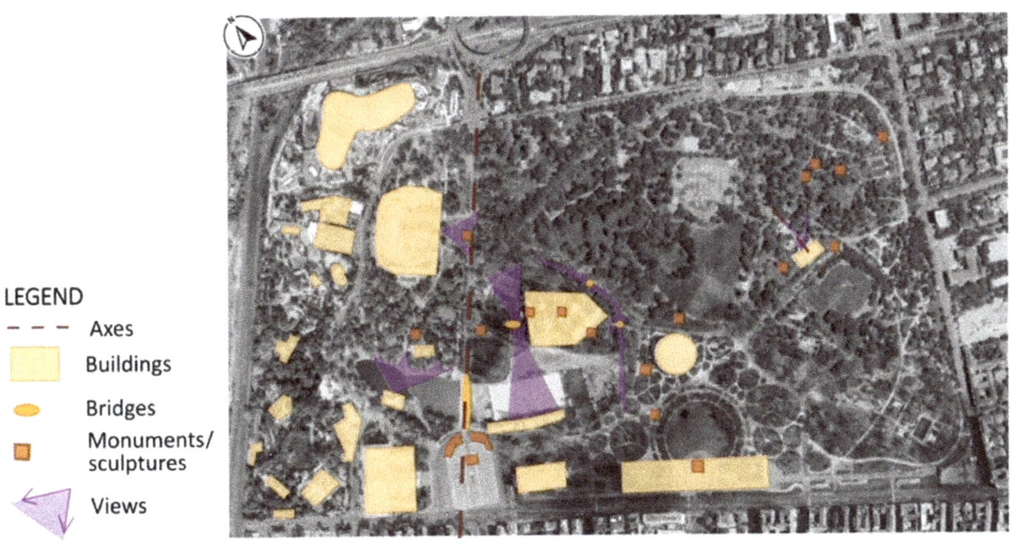

Figure 16. Visual connections in Városliget Park in 2025, led by the several building features (monuments, sculptures and buildings). Source: prepared by the Authors.

d. **Kalemegdan Park, Belgrade, SRB**

The axes in Kalemegdan can be divided into two types: those that lead to monuments or other important structures within the park and those that serve as panoramic viewpoints of Belgrade. Given the park's varied terrain and elevated position, these sightlines are a defining characteristic and a key asset of the park, offering visitors astonishing views of the confluence of the Sava and Danube rivers. Some axes, such as the one leading to the Monument of Gratitude to France, have remained largely unchanged over time, preserving their original narrative. However, some sightlines have been altered or partially obscured by the natural growth and maturation of the tree canopy. While the overall framework of the axes remains intact, the evolving vegetation has transformed the way these views are experienced (Figure 17).

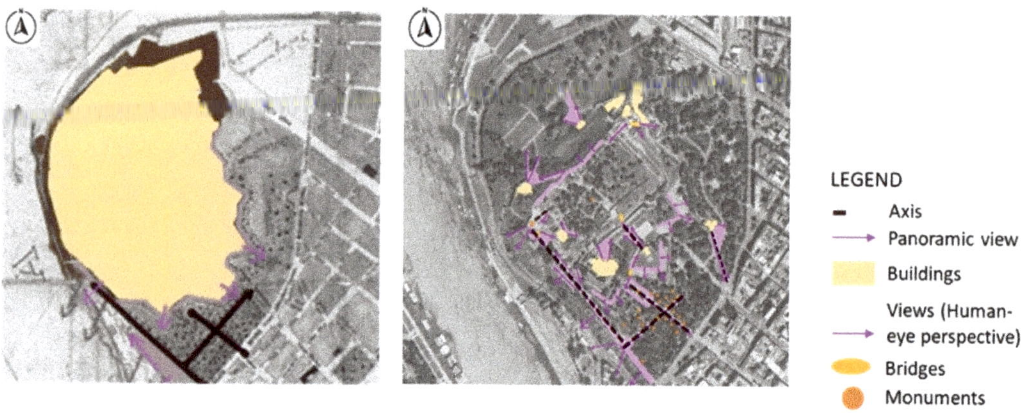

Figure 17. Kalemegdan Park visual connections in the past and nowadays. The extension of the park area creates several new visual axes and perspectives in the park. Source: prepared by the Authors.

4.6. Road and Water System Analysis

a. **Birkenhead Park, Liverpool, UK**

The water features, including lakes, streams and ornamental bridges, were strategically incorporated to add esthetic value, regulate water flow, and provide habitats for wildlife. These elements also contributed to the park's environmental sustainability and public health by controlling water flow, enhancing air quality, and providing wildlife habitats. Unlike parks situated along major rivers, Birkenhead Park's water system operates independently of large external waterways. It relies on local drainage and water retention strategies to maintain its lakes and streams.

Birkenhead Park's road system combines broad, formal avenues with winding, naturalistic paths to guide visitors through varied landscapes and key landmarks. The main routes provide clear sightlines and easy navigation, while smaller paths encourage exploration. Designed for pedestrians and carriages, the network reflects Victorian ideals of accessibility and social inclusion.

b. **Battersea Park, London, UK**

Battersea Park's waterworks are a distinctive feature of its 19th-century landscape design; the park's prominent water feature is an irregularly shaped lake, a characteristic element typical of Victorian landscape style parks. Grade II listed Pump House was the central point for park's water management. Originally built in a fine classical style using stock brick with rusticated quoins, the Pump House housed a coal-fired steam engine designed to supply water to the boating lake and the ornamental cascades [43]. These cascades, situated north of the boating lake, are an artificial rock structure made to mimic natural rock formations. The cascades and the lake were linked through a water system fed by the Pump House, which drew water from a reservoir to maintain the flow. After falling into disrepair following the war, the Pump House was restored between 1988 and 1992 and repurposed as an art gallery, preserving its historic significance. The waterworks contribute not only to the park's visual appeal but also to the micro-climate that supports the renowned Sub-Tropical Garden, enabling the cultivation of exotic plants like palms, tree ferns and bananas.

The park is divided into four quarters by two main axes: an east–west avenue and a straight north–south path. These intersect at a central circular space, creating a focal point within the park. The north–south axis extends to terminate in semi-circular paths and spaces, providing graceful transitions within the landscape. The entrances to the park are strategically positioned at each of its four corners, with an additional main entrance located at the end of the north–south axis, ensuring easy access and flow of visitors. The road system is characterized by wavy lines.

c. **Városliget Park, Budapest, HUN**

Water features have played a key role in shaping Városliget since its creation in the early 19th century. Originally, the park included a large pond with two islands connected by a canal system that added both beauty and function to the landscape. This water system was designed not only for esthetic purposes but also to support recreational activities. In summer, the pond was used for boating, while in winter, when it froze over, it became a popular ice-skating area. Over time, some parts of the water system, such as the canal, were removed, but the pond itself largely remained. Additional water elements were added throughout the park, including fountains and smaller ponds, enhancing the park's appeal and providing cooling effects during hot weather. The water bodies also help support local wildlife and contribute to the park's ecosystem.

Over the course of two centuries, the road system in Városliget has undergone significant changes in both its development and character. Initially, the road network was poorly developed and exhibited a predominantly geometric and formal layout. However, by the early twentieth century, the park experienced substantial growth, marked by a more intricate system of pedestrian paths that connected the increasingly complex park structures and architecture.

d. **Kalemegdan Park, Belgrade, SRB**

In the 19th century, Belgrade's waterscape, including areas like Kalemegdan, played a key role in the city's urban development and public health. The first modern water supply system was built on the foundations of earlier Roman, Ottoman and Austrian waterworks, channeling water to the fortress and surrounding areas [46]. Due to its topography and the presence of tunnels beneath the park, Kalemegdan does not feature many water installations. It mainly contains drainage channels, some fountains, and small water features. Historically, the fortress relied on cisterns and wells for its water supply. The park's steep slope promotes natural drainage, reducing water stagnation and supporting healthy soil conditions for the vegetation [37].

The road and path network in Kalemegdan Park reflects its military origins and later adaptation into a public space. The current circulation structure is defined by a hierarchy of paths and promenades, designed to balance accessibility, strategic views, and recreational use. Axial and radial pathways were laid out during the park's transformation in the late 19th and early 20th centuries [36]. These formal promenades offer direct visual access to key monuments, viewpoints, and the fortress. The Upper Kalemegdan promenade serves as the main circulation route, connecting the entrance near Knez Mihailova Street to the Belgrade Fortress plateau. Secondary, curvilinear paths in the lower areas and slopes follow the natural topography, offering informal, leisurely routes toward the Sava and Danube rivers.

4.7. Vegetation

Several authors describe contemporary trends in plant use in detail in comprehensive works [29,47–55]. Comparing these with the field surveys conducted related to the present study and other available surveys of the current situation of the parks reveal similarities and differences in the historical and current plant use at the sites studied.

a. **Birkenhead Park, Liverpool, UK**

The planting scheme included a wide variety of native and exotic tree species, carefully positioned to demonstrate botanical richness and serve an educational purpose. Artificial mounds created from lake excavations shaped the undulating terrain of the park and helped it to resist the prevailing winds, while also supporting plant growth and visual interest [30].

b. **Battersea Park, London, UK**

The park's layout features broad avenues lined with trees as well as more informal woodland areas and flower beds. The park's canopy coverage from the construction reveals a significant transformation, including over 4000 trees, many of which follow the original Victorian landscaping plan.

c. **Városliget Park, Budapest, HUN**

Many trees were planted during the park's initial development in the early nineteenth century and have matured into a dense canopy that defines the park's character. Currently, 7105 tree specimens compose the canopy of Városliget. There are 297 trees older than 130 years, including 53 trees older than 180 years [56].

d. **KalemegdanPark, Belgrade, SRB**

The park's green system was planned during its transformation from a military site into a public space in the late 19th and early 20th centuries, with additional planting phases occurring between the two World Wars and after World War II [37]. A large number of mature trees can be considered major park features, many of which have surpassed a century in age. The tree composition includes both native and introduced ornamental species, reflecting the horticultural trends of the 19th century in urban park design.

In summary, it can be observed that the current composition of tree species in the parks (Figure 18) has developed through similar processes. In each park, native tree species such as English oak (*Quercus robur*), European beech (*Fagus sylvatica*), poplar (*Populus* sp.), silver birch (*Betula pendula*) and silver lime (*Tilia tomentosa*) are present. A significant number of ornamental species typical of the landscape garden period, such as London plane (*Platanus x hispanica*), European ash (*Fraxinus excelsior*), bald cypress (*Taxodium distichum*) and Japanese pagoda tree (*Styphnolobium japonicum*), are also represented. In addition, numerous evergreen taxa can be found, including black pine (*Pinus nigra*), Scots pine (*Pinus sylvestris*), Himalayan (*Cedrus deodara*) and Atlas cedar. (*Cedrus atlantica*).

BIRKENHEAD (UK)	BATTERSEA (UK)	VÁROSLIGET (HU)	KALEMEGDAN (SRB)
	Acer platanoides (Norway maple)	***Acer platanoides*** (Norway maple)	***Acer platanoides*** (Norway maple)
			Acer pseudoplatanus (Sycamore maple)
	Aesculus hippocastanum (Horse chestnut)	***Aesculus hippocastanum*** (Horse chestnut)	***Aesculus hippocastanum*** (Horse chestnut)
Arbutus unedo (Strawberry tree)			
	Arbutus x andrachnoides (Hybrid strawberry tree)		
Araucaria araucana (Monkey puzzle tree)			
	Betula pendula (Silver birch)	***Betula pendula*** (Silver birch)	***Betula pendula*** (Silver birch)
			Carpinus betulus (Hornbeam)
	Castanea sativa (Sweet chestnut)		***Castanea sativa*** (Sweet chestnut)
	Cedrus atlantica (Atlas cedar)		
Cedrus deodara (Himalayan cedar)			
		Celtis occidentalis (Common hackberry)	***Celtis australis*** (European nettle tree)
Cornus kousa (Korean dogwood)			
			Corylus avellana (Common hazel)
			Corylus colurna (Turkish hazel)
Cupressus macrocarpa (Monterey cypress)			
Fagus sylvatica (European beech)			***Fagus sylvatica*** (European european beech)
Fraxinus americana **'Autumn Purple'** (Ash)			
Fraxinus excelsior (European ash)		***Fraxinus excelsior*** (European ash)	***Fraxinus excelsior*** (European ash)
			Juglans regia (Walnut)
		Larix decidua (European larch)	
Magnolia acuminata (Cucumber tree)			
	Pinus nigra (Black pine)	***Pinus nigra*** (Black pine)	***Pinus nigra*** (Black pine)
Pinus strobus (White pine)		Pinus strobus (White pine)	
	Pinus sylvestris (Scots pine)		
	Platanus x hispanica (London plane)	***Platanus x hispanica*** (London plane)	***Platanus x hispanica*** (London plane)
			Populus alba (White poplar)
		Populus canescens (Grey poplar)	
		Populus nigra (Black poplar)	
		Populus simonii (Simon's poplar)	
Prunus cerasifera (Cherry plum)	***Prunus sp*** . 'Shibayama' (Shibayama cherry)		***Prunus cerasifera*** (Cherry plum)
	Pyrus betulifolia (Birchleaf pear)		
Quercus ilex (Holm oak)			
Quercus robur (English oak)	***Quercus robur*** (English oak)	***Quercus robur*** (English oak)	***Quercus robur*** (English oak)
	Robinia pseudoacacia **'Semperflorens'** (Semperflorens black locust)	***Robinia pseudoacacia*** (Black locust)	
		Salix alba (White willow)	
Salix babylonica (Weeping willow)			
	Sorbus aucuparia (Mountain ash)		
	Styphnolobium japonicum **'Pendula'** (Weeping japanese pagoda tree)	***Styphnolobium japonicum*** (Japanese pagoda tree)	
	Taxodium distichum (Bald cypress)		
Taxus baccata **'Fastigiata'** (English & Irish yew)			
		Tilia tomentosa (Silver linden)	***Tilia tomentosa*** (Silver linden)
Tilia x europaea (Common lime)		***Tilia x europaea*** (Common lime)	
	Ulmus pumila (Siberian elm)	***Ulmus leavis*** (European white elm)	

Figure 18. Comparison of dominant tree species in the four selected parks. Source: Prepared by the Authors.

5. Conclusions

The protection and renewal of historic urban parks is a professional task that poses ongoing challenges. Due to their historical significance, their maintenance and renewal must be based on traditional values.

This study shows that historical urban public parks established in the 19th century still play an important recreational, urban structural and ecological role. They form the basic elements of urban green infrastructure, while also providing opportunities for active recreation for the population and serving as venues for social and cultural life. Their current central location in the urban structure is largely due to the time of their creation, as since the 19th century, cities have grown around these parks, further emphasizing their importance in the life of the settlements.

This analysis intended to highlight and to evaluate the relevant basic design elements of the parks which define their historical characteristics and have been preserved to this day.

From a **functional** point of view, all of the parks have been dynamically enriched since their creation: in addition to their traditional functions (to provide rest, fresh air, moral uplift to the working classes, the possibility for the aristocracy to be represented, relief from the overcrowded and polluted urban environment, etc.), numerous new functions (sports, arts, play, cultural events, catering) have also appeared over the decades. The new functions have mostly been well integrated with the old ones in the parks. The most relevant developments in this direction can be observed in the cases of Kalemegdan Park (Belgrade) and Városliget Park (Budapest). In the first case, the extension of the park area with the integration of the former fortress area contributed essentially to the enrichment of functions; in the case of Városliget, the partial transformation of the park into a cultural district of the Hungarian capital had a significant influence toward the functional development.

Functional development has also partly influenced structural changes in the parks, affecting their main **compositional** elements, such as spatial structure and visual connections. Nevertheless, it can be said that in all four parks, the structural design corresponding to the original concept is still recognizable today in all four parks. The original composition has been preserved almost entirely in the case of Birkenhead and Battersea parks. Due to the urban developments from the end 19th century, the main entrance of Városliget Park in Budapest has been moved compared to where it was initially. Despite this transformation, the original structure has been mainly preserved; only the main flows and some functions have been changed. Kalemegdan Park presents—through its already mentioned extension—the biggest changes from compositional point of view as well, still preserving some of its original compositional aspects.

The transformation of **visual connections** can be interpreted in close connection with the development of plant material. Over the centuries, as a result of aging, spreading and expanding plant populations, all four sites studied have experienced significantly higher **canopy cover** (15–25%) than originally planned. The leveling of the plant population and the strengthening of the shrub layer have resulted in a stronger separation of individual functions and a further reduction in visual connections.

Although these parks are located in different regions of Europe, there is a high degree of overlap in the main **tree species** planted. Certain tree species—such as linden (*Tilia cordata*), oak (*Quercus robur*), plane tree (*Platanus x hybrida*), birch (*Betula pendula*), horse chestnut (*Aesculus hippocastanum*), ash (*Fraxinus excelsior*) and even certain coniferous species (*Pinus strobus*)—have been present in large numbers in almost all locations in the past and now, suggesting that these species have defined the image and atmosphere of parks for centuries.

The results prove that, in light of increasingly intense urbanization processes, new environmental challenges (climate change, loss of biodiversity) and the demands for a healthy and livable urban environment, the fundamental characteristics of historic public parks can still be preserved as far as possible, as they are imprints of an era, an artist, and a social message, and their cultural and historical significance is invaluable.

Author Contributions: Conceptualization, A.F.; methodology, A.F.; software, T.P.; validation, A.T.; formal analysis, E.B.Ö.; investigation, A.T., T.P., E.B.Ö., L.B.H., D.K., A.P. and A.G.; resources, D.K. and L.B.H.; data curation, A.F.; writing—original draft preparation, A.F., A.T., A.P., E.B.Ö. and T.P.; writing—review and editing, T.P., A.G. and A.F.; visualization, A.T., A.P., E.B.Ö. and T.P.; supervision, A.G. and A.F.; project administration, D.K. and L.B.H. All authors have read and agreed to the published version of the manuscript.

Funding: The authors report financial support was provided by the Hungarian University of Agriculture and Life Sciences.

Data Availability Statement: The datasets used and/or analyzed during the current study are available from the corresponding author on reasonable request.

Conflicts of Interest: The authors declare no conflicts of interest. The authors declare that they have no known competing financial interests that could have appeared to influence the work reported in this paper.

References

1. van den Toorn, M. New functions and roles for public parks in Europe—The future relation between public space and public health. *4D J. Landsc. Archit. Gard. Art* **2020**, *55–56*, 2–27.
2. Fleming, J.; Honour, H.; Pevsner, N. *The Penguin Dictionary of Architecture and Landscape Architecture*, 5th ed.; Penguin: London, UK, 1998.
3. Cross, N. Designerly ways of knowing. *Des. Stud.* **1982**, *3*, 221–227. [CrossRef]
4. Cross, N. Science and design methodology: A review. *Res. Eng. Des.* **1993**, *5*, 63–69. [CrossRef]
5. Frayling, C. Research in art and design. *RCA Res. Pap.* **1993**, *1*, 1–5.
6. Dorst, K. The core of 'design thinking' and its application. *Des. Stud.* **2011**, *32*, 521–532. [CrossRef]
7. Deming, M.E.; Swaffield, S. *Landscape Architecture Research—Inquiry, Strategy, Design*; John Wiley and Sons: Hoboken, NJ, USA, 2011.
8. Zeisel, J. *Inquiry by Design—Environment/Behavior/Neuroscience in Architecture, Interiors, Landscape and Planning*; Norton & Co.: New York, NY, USA, 2006.
9. Francis, M. A case study method for landscape architecture. *Landsc. J.* **2001**, *19*, 15–29. [CrossRef]
10. Yin, R.K. *Case Study Research—Design and Methods*, 3rd ed.; Sage: Thousand Oaks, CA, USA; London, UK, 2003.
11. van den Toorn, M.; Bihuňová, M.; Rechner-Dika, I.; Tóth, A.; Vasilijevic, N. Research Methods in Landscape Architecture: An Analytical Framework as Basis for Research for the Redesign of 19th Century Public Parks in Central and Eastern Europe. In *Teka Komisji Urbanistyki i Architektury Oddziału Polskiej Akademii Nauk w Krakowie*; Cracŏw University of Technology: Krakow, Poland, 2019; Volume 47, pp. 189–190.
12. Nehring, D. *Stadtparkanlagen in der Ersten Hälfte des 19. Jahrhunderts. Ein Beitrag zur Kulturgeschichte des Landschaftsgartens (Geschichte des Stadtgrüns, 4)*; Patzer: Hannover/Berlin, Germany, 1979.
13. Rittel, H.W. The reasoning of designers. In Proceedings of the International Congress on Planning and Design Theory, Boston, MA, USA, 17–20 August 1987; pp. 1–12.
14. Ignatieva, M. *Design and Future of Urban Diversity Urban Biodiversity and Design*, 1st ed.; Müller, N., Werner, P., Kelcey, J.C., Eds.; Blackwell Publishing Ltd.: Oxford, UK, 2010; pp. 118–144.
15. Kruger, J. Parks Recreation, and Public Health Collaborative. *Environ. Health Insights* **2008**, 2. [CrossRef]
16. Xiaoyu, C.; Yang, J.; Zhang, Y.; Xiao, X.; Xia, J. Cooling Island Effect in Urban Parks from the Perspective of Internal Park Landscape. Humanitiees and Social Sciences Communications. 2025. Available online: https://www.researchgate.net/publication/374556563_Cooling_island_effect_in_urban_parks_from_the_perspective_of_internal_park_landscape (accessed on 21 August 2025).
17. Aktürk, G.; Dastgerdi, A.S. Cultural Landscapes under the Threat of Climate Change: A Systematic Study of Barriers to Resilience. *Sustainability* **2021**, *13*, 9974. [CrossRef]
18. Melnick, Z.R. Climate Change and Landscape Preservation: A Twenty-First-Century Conundrum. *APT Bull. J. Preserv. Technol.* **2009**, *40*, 2009.
19. Millar, C.I.; Stephenson, N.L.; Stephens, S.L. Climate Change and Forests of the Future: Managing in the Face of Uncertainty. *Ecol. Appl.* **2007**, *17*, 2145–2151. [CrossRef] [PubMed]
20. Klein, R.J.T.; Midgley, G.F.; Preston, B.L.; Alam, M.; Berkhout, F.G.H.; Dow, K.; Shaw, M.R. Adaptation opportunities, constraints, and limits. In *Climate Change 2014: Impacts, Adaptation, and Vulnerability. Part A: Global and Sectoral Aspects. Contribution of Working Group II to the Assessment Report of the Intergovernmental Panel on Climate Change*; Cambridge University Press: Cambridge, UK; IPPC: New York, NY, USA, 2014; pp. 899–943.
21. Measham, T.G.; Preston, B.L.; Smith, T.F.; Brooke, C.; Gorddard, R.; Withycombe, G.; Morrison, C. Adapting to climate change through local municipal planning: Barriers and challenges. *Mitig. Adapt. Strateg. Glob. Change* **2011**, *16*, 889–909. [CrossRef]
22. Pröbstl-Haider, U.; Kelemen-Finan, J.; Haider, W.; Schauenlehner, T.; Melzer, V.; Mostegl, N. Will climate change influence the attractiveness of cultural landscapes in Austria? In *Tourism and Leisure: Current Issues and Perspectives of Development*; Springer Science + Business Media: Berlin/Heidelberg, Germany, 2015; pp. 355–370.
23. Jellicoe, G.; Jellicoe, S. *The Landscape of Man—Shaping the Environment from Prehistory to the Present*; Viking Press: New York, NY, USA, 1975.

24. Shearer, A.W. Abduction to argument: A framework of design thinking. *Landsc. J.* **2015**, *34*, 127–138. [CrossRef]
25. Yang, Y.; He, R.; Tian, G.; Shi, Z.; Wang, X.; Fekete, A. Equity Study on Urban Park Accessibility Based on Improved 2SFCA Method in Zhengzhou, China. *Land* **2022**, *11*, 2045. [CrossRef]
26. Wallace, D.; Han, B.; Cohen, D.A.; Derose, K. The Effects of Park-Based Interventions on Health-Related Outcomes Among Youth: A Systematic Review. *Am. J. Health Promot.* **2022**, *36*, 1029–1044. [CrossRef] [PubMed]
27. Valánszki, I.; Kristensen, L.S.; Jombach, S.; Ladányi, M.; Filepné Kovács, K.; Fekete, A. Assessing Relations between Cultural Ecosystem Services, Physical Landscape Features and Accessibility in Central-Eastern Europe: A PPGIS Empirical Study from Hungary. *Sustainability* **2022**, *14*, 754. [CrossRef]
28. Gkoltsiou, A.; Paraskevopoulou, A. Landscape character assessment, perception surveys of stakeholders and SWOT analysis: A holistic approach to historical public park management. *J. Tour. Outdoor Recreat.* **2021**, *35*, 100418. [CrossRef]
29. Conway, H. *Public Parks, World of Books, Shire Garden History*; Parkwest Publications: Eatontown, NJ, USA, 1996.
30. Donald Insall Associates. Birkenhead Park Conservation Area Conservation Area Appraisal and Management Plan. 2009. Available online: https://www.wirral.gov.uk/files/birkenhead-park-appraisal.pdf (accessed on 24 May 2025).
31. Donald Insall Associates. Birkenhead Park Conservation Area Conservation Area Appraisal and Management Plan for Wirral Metropolitan Borough Council. 2025. Available online: https://www.wirral.gov.uk/files/draft-appraisal-and-management-plan-birkenhead-park-jun-2025.pdf (accessed on 20 August 2025).
32. Wandsworth Conservation & Design Group; Wandsworth Council. *Battersea Park Conservation Area Appraisal and Management Strategy*; Wandsworth Borough Council: Wandsworth, UK, 2014.
33. Smith, A. Justifying and resisting public park commercialisation: The battle for Battersea Park. *Eur. Urban Reg. Stud.* **2018**, *25*, 171–185. [CrossRef]
34. Reining, H. Nebbien, Heinrich: Ungarns Folks-Garten Der Koeniglichen Frey—Stadt Pest (1816). *SOF* **2021**, *1*, 441–442.
35. Jámbor, I. *Nebbien Városligete—A Világ Első Népkertje Pesten*; Terc Kft.: Budapest, Hungary, 2019.
36. Ćorović, D. Three Parks in Nineteenth Century Belgrade. *Serbian Stud. J. N. Am. Soc. Serbian Stud.* **2010**, *24*, 86–89. [CrossRef]
37. Anastasijević, N.; Anastasijević, V. *Funkcionalnost Zelenih Površina Beograda*; Univerzitet u Beogradu, Šumarski Fakultet: Belgrade, Serbia, 2012.
38. Aerial Photograph of Birkenhead Park. Available online: https://webbaviation.co.uk/aerial/picture.php?/14932/tags/1-aerial/2-photograph/6-foto/10-photo/11-drone/17-eye/40-uk/1573-birkenhead (accessed on 21 August 2025).
39. Aerial Photograph of Battersea Park. Available online: https://www.kalmars.com/about-kalmars/london-area-buying-guides/battersea-area-guide (accessed on 21 August 2025).
40. Aerial Photograph of Városliget Park. Available online: https://ligetbudapest.hu/en/the-project (accessed on 21 August 2025).
41. Aerial Photograph of Kalemegdan Park. Available online: https://www.facebook.com/photo.php?fbid=3468864623437573&id=1721031731554213&set=a.2456584744665571 (accessed on 21 August 2025).
42. Taylor, H.A. Urban Public Parks, 1840–1900: Design and Meaning. *Gard. Hist.* **1995**, *23*, 201–221. [CrossRef]
43. Nathaniel Lichfield & Partners Ltd. Wandsworth Borough Council Retail and Town Centre Needs Assessment. 2008. Available online: https://www.wandsworth.gov.uk/media/1964/retail_needs_assessment.pdf (accessed on 21 August 2025).
44. Nehring, D. The landscape architect, Christian Heinrich Nebbien and his design for the Municipal Park in Budapest. *J. Gard. Hist.* **1985**, *5*, 261–279. [CrossRef]
45. Butenschön, S. Der früherste Entwurfsplan Lennés. Ein Wettbewerbsbeitrag für den Stadtpark in Budapest. *Die Gartenkunst* **2015**, *29*, 1–24.
46. Ćorović, D.; Blagojević, L. *Water, Society and Urbanization in the 19th Century Belgrade*; University of Belgrade, Faculty of Architecture: Belgrade, Serbia, 2012.
47. Hobhouse, P. *Plants in Garden History*; Pavilion Books Ltd.: London, UK, 2004.
48. Hobhouse, P. *The Story of Gardening*; Dorling Kindersley Limited: London, UK, 2002.
49. Quest-Ritson, C. *The English Garden—A Social History*; Penguin Books: London, UK, 2003.
50. Smith, A. Paying for parks. Ticketed events and the commercialisation of public space. *Leis. Stud.* **2018**, *37*, 533–546. [CrossRef]
51. Rapaics, R. *Magyar Kretek*; Királyi Magyar Egyetemi Nyomda: Budapest, Hungary, 1943.
52. Sárospataki, M. Dendrológiai Kertek a 19. Századi Magyarországi Kertépítészetben. Ph.D. Dissertation, Corvinus University of Budapest, Budapest, Hungary, 2014.
53. Fekete, A. Közparkok növényalkalmazásának sajátosságai: Characteristics of plants application in public parks. 2009. In *Lippay János—Ormos Imre—Vas Károly Tudományos Ülésszak: Összefoglalók*; Sallay, Á., Várszegi, R., Eds.; Budapesti Corvinus Egyetem: Budapest, Hungary, 2009; pp. 44–45.
54. Fekete, A.; Sárospataki, M.; Takács, K. Landscape ecological and visual significance of dendrological gardens in the Carpathian Basin. *Acta Univ. Sapientiae Agric. Environ.* **2014**, *6*, 57–68. [CrossRef]

55. Fekete, A.; Tar, I.; Sárospataki, M.; Győri, P. Educational role of public parks: Windows to past, to future and to exotic worlds. *4D J. Landsc. Archit. Gard. Art* **2020**, *55–56*, 38–53. [CrossRef]
56. Liget Mérnök Konzorcium. Városliget Fakataszter. 2016. Available online: https://ligetbudapest.hu/storage/70/V%C3%A1rosliget-fakataszter.pdf (accessed on 18 August 2025).

land

MDPI

Article

Identifying Coastal Heritage Vulnerabilities: The Case of Historical Fortified Structures in Northern Portugal

Isabel Vaz Freitas [1,2], Hélder Silva Lopes [3,4] and Fátima Matos Silva [1,2,5,*]

1 Department of Tourism, Heritage and Culture, University Portucalense Infante D. Henrique, 4200-072 Porto, Portugal; immvdfbc@upt.pt
2 REMIT—Research on Economics, Management and Information Technologies, University Portucalense Infante D. Henrique, 4200-072 Porto, Portugal
3 Lab2PT—Laboratory of Landscapes, Heritage and Territory/IN2PAST—Associated Laboratory for Research and Innovation in Heritage, Arts, Sustainability and Territory, 4800-058 Guimarães, Portugal; helderlopes@geografia.uminho.pt
4 Department of Geography, Institute of Social Sciences, University of Minho, 4800-058 Guimarães, Portugal
5 CITCEM—Transdisciplinary Culture, Space and Memory Research Centre, Oporto University, 4150-564 Porto, Portugal
* [illegible]

Abstract: Landscapes and heritage sites hold significant historical, scientific, and social value but face increasing threats from climate change and human activities. Coastal and maritime heritage are at risk from sea-level rise, storms, erosion, ocean acidification, and pressures such as urbanization, construction, and industrial development. Assessing vulnerability involves considering physical, geomorphological, and socioeconomic factors, including land use, population density, tourism, and ecosystem sensitivity. Long-term monitoring, interdisciplinary research, and holistic approaches are essential for effective risk assessment and planning. This study focuses on the coastal landscapes of northern Portugal, where climate change adaptation is urgent. These areas contain important historical heritage, especially fortified military structures that reflect regional identity and maritime history shared with other coastal nations. The research highlights significant risks to these monuments because of their proximity to the sea and expanding urban areas, providing insights to guide policymakers and support localized adaptation strategies. A two-phase methodology was employed, beginning with a comprehensive literature review to identify key indicators that informed field observations, surveys, and archival research, resulting in a detailed inventory of coastal and estuarine fortifications. The second phase assessed their vulnerability to sea-level rise, coastal flooding, and shoreline retreat. The study presents a methodological approach that provides local decision-makers with strategic guidance to enhance the protection and sustainable management of coastal heritage.

Keywords: historical fortified structures; cultural heritage; northern Portugal; Atlantic coast; natural hazards; climate change; anthropogenic pressures; heritage safeguard

1. Introduction

Since the World Heritage Convention (1972) [1], cultural landscapes have been recognized and studied within historical and scientific contexts, especially through the lens of preservation and conservation, including concepts such as belonging, outstanding value, locality, universality, meaning, uniqueness, and singularity. From another perspective, the Landscape Convention (2000) [2] emphasizes the importance of landscapes for individual and social well-being, highlighting their role in improving quality of life, reflecting cultural, scientific, aesthetic, and historical values, and shaping a place's identity. Cultural landscapes are highly valuable but vulnerable, and their preservation is difficult amid natural threats and rapid global development [3]. Cultural heritage, including historic buildings, monuments, archaeological sites, traditions, and landscapes, cannot be replaced once it is lost, destroyed, damaged, altered, or erased; it is gone forever. [4]. Therefore, it is crucial to protect, maintain, and manage these assets so that future generations can experience, learn from, and benefit from them. Identifying risks to cultural heritage requires assessing factors such as historical context, topography, geology, land use, and urban development patterns. While considering geomorphological and geological processes can improve hazard assessments, most long-term vulnerability studies tend to overlook these variables.

The coastal zone between the Douro and Minho Rivers preserves an exceptional ensemble of fortifications built between the late sixteenth and eighteenth centuries, reflecting Portugal's historical response to privateer incursions, territorial disputes, and the strategic control of river mouths. These fortresses embody a distinctive interaction between defensive architecture and coastal geomorphology, adapting military design to local environmental conditions. Today, some of these sites serve cultural, institutional, or tourism functions, while others are in advanced stages of deterioration due to environmental stressors, coastal erosion, and inadequate maintenance [5,6]. Their preservation and sustainable reuse are therefore essential to safeguard this unique testimony of Portugal's maritime and military heritage, "engaging cultural heritage in climate change" as proposed by Downes (2019) [3].

Since the adoption of the World Heritage Convention (1972) [1] and the European Landscape Convention (2000) [2], cultural landscapes have been recognized as integral expressions of identity, memory, and collective well-being. However, they remain increasingly vulnerable to both natural and anthropogenic pressures. Climate change, manifested through sea-level rise, intensified storms, coastal erosion, and ocean acidification, combined with urban expansion, tourism, and industrial activity, poses a growing threat to coastal heritage worldwide [7]. Studies have shown that historic fortifications, monuments, and archaeological remains are particularly at risk from flooding, material degradation, and structural instability [8]. In Portugal and beyond, the combined impact of urbanization, erosion, and rising sea levels exacerbates the fragility of tangible and intangible heritage, calling for urgent, integrated management approaches.

Safeguarding these irreplaceable assets requires integrating geomorphological, environmental, and social dimensions into hazard and vulnerability assessments and adaptation strategies. Long-term preservation depends on interdisciplinary collaboration, adaptive policy frameworks, and active community involvement, aligned with international climate resilience and sustainability objectives. In this article, we use the term 'hazard' for potentially damaging physical processes (sea-level rise, storm surge, coastal flooding, coastal erosion), and 'risk' for the likelihood and consequences of damage to heritage resulting from the interaction between hazard, exposure, and vulnerability. These differences have been documented in the literature on risk assessment [9,10].

Accordingly, this study addresses the question: How vulnerable are the coastal fortifications of northern Portugal to the combined effects of natural and anthropogenic pressures, particularly those intensified by climate change, and how can their preservation and adaptive management be improved?

To answer this, the study aims to:

(a) Identify and classify the main natural and anthropogenic risks affecting coastal fortifications along the northern Portuguese Atlantic coast.
(b) Develop and apply a standardized observational framework with measurable indicators to systematically assess conservation status and environmental exposure.

(c) Map and analyze the vulnerability of the fortifications to support evidence-based conservation and policy decisions.

(d) Provide recommendations and adaptation strategies to support policymakers, heritage managers, and local authorities in developing sustainable conservation plans aligned with climate adaptation goals.

2. History, Architecture and Contemporary Uses

The coastal strip between the Douro and the Minho preserves a remarkable set of coastal fortifications, whose construction spans from the end of the sixteenth century to the eighteenth, a period in which Portugal faced threats from privateers, territorial disputes, and the need to control the main river bars [11]. These properties, classified mainly as Buildings of Public Interest and, in some cases, as National Monuments, constitute a material testament to the evolution of military engineering, influenced by Italian and French traditions and adapted to the geographical specificities of the Portuguese Atlantic coast [12]. Strategically positioned along the northern Portuguese Atlantic coast, the forts and fortresses extend from the Porto metropolitan area to Caminha, near the Spanish border, reflecting a historical defensive system designed to control and protect the coastline [12]

Twelve heritage sites are identified, representing key elements of the region's historical coastal defensive network (Figure 1). These structures, dating mainly from the 16th to 18th centuries, were tactically built near estuaries, harbors, and coastal promontories to protect maritime routes and settlements from naval attacks and piracy. The spatial distribution shows a clear concentration around major coastal urban centers such as Porto, Viana do Castelo, and Caminha, areas that historically held high military and commercial importance. Their proximity to the shoreline highlights their dependence on maritime visibility and control but also exposes them to contemporary threats such as sea-level rise, storm surges, and coastal erosion. The linear alignment of these fortifications along the Atlantic fringe demonstrates both the historic defensive logic and the modern environmental vulnerability of this coastal cultural heritage [13].

In Porto, the fortifications of the Douro Atlantic Front stand out, such as the Fort of S. João Baptista da Foz or Castle of S. João da Foz, which began construction in 1570 with bastioned walls designed by Simão de Ruão. Between 1646 and 1653, under the direction of engineer Charles Lassart, it was expanded, sacrificing the original church and Benedictine residence. The fortress, with an irregular, rectangular plan, features three bastions and a half-bastion. It has a neoclassical portal from 1796, designed by Reinaldo Oudinot, which includes a drawbridge, a corridor with a casemate, and a guardhouse [14]. Today, it houses the National Defense Institute, which is open for visits, events, and training.

Also in Porto, the Fort of São Francisco Xavier (or Castelo do Queijo) was first designed by Charles Lassart in 1561. However, construction was not completed until 1661, based on a plan by Miguel de L'École. The fort features a trapezoidal shape surrounded by a moat, with cannons, domed sentry boxes, and a portal displaying the royal coat of arms. After several management changes, it was handed over to the Commandos Association in 1978 and now serves as a cultural and museum space [12].

In Matosinhos, the Fort of Nossa Senhora das Neves began construction in 1638 to bolster the defense of the Douro bar, along with the forts of Porto, and was completed in 1720. It features a star-shaped layout with sentry boxes connected by sloping curtain walls. In the nineteenth century, it served as the Porto Customs House, and in 1962, it was the subject of a landscaping project by Ilídio de Araújo. Today, it functions as the headquarters of the Port of Leixões Captaincy [12,14].

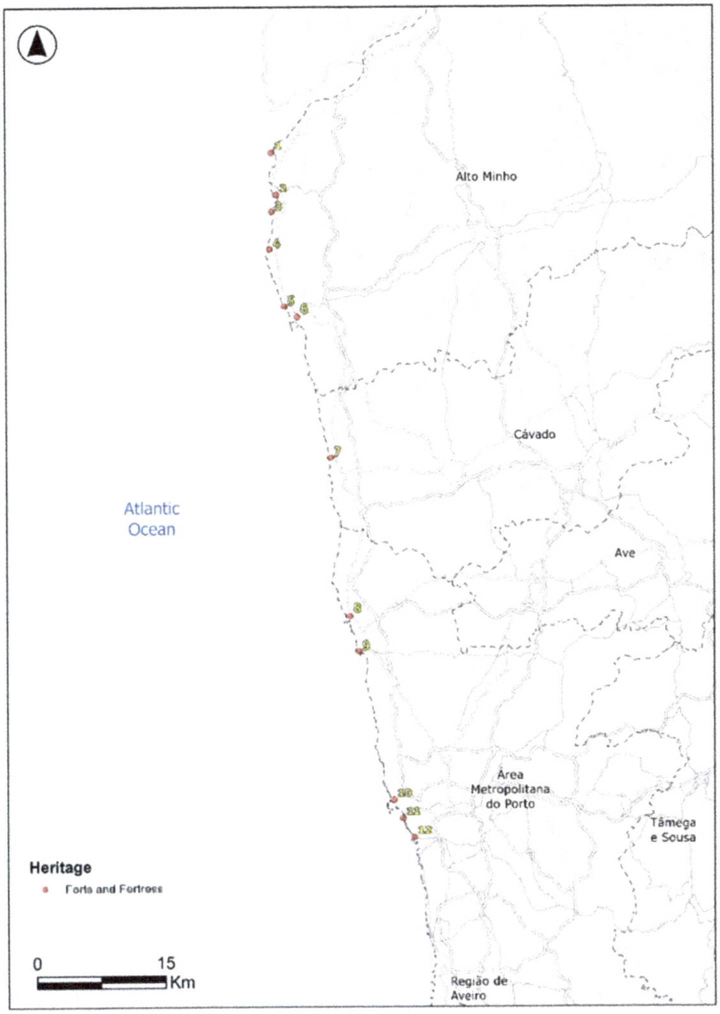

Figure 1. Forts and Fortresses in northern Portugal (between Porto and Caminha). Source: Own elaboration, based on fieldwork analysis. 1—Forte do Ínsua (Moledo, Caminha); 2—Fortaleza or Forte da Largateira, Forte de Âncora, Caminha); 3—Forte do Cão, Forte da Gelfa (V. P. de Âncora, Caminha); 4—Forte or Fortim de Montedor, Forte do Paço (Carreço, Viana do Castelo); 5—Forte da Areosa, Fortim da Areosa, Forte ou Fortim da Vinha, Castelo or Forte Velho (Areosa, Viana do Castelo); 6—Forte de Santiago da Barra, Castelo de Santiago da Barra (Monteserrate, Viana do Castelo); 7—Forte de S. João Baptista (Esposende); 8—Fortaleza de N.ª Sr.ª da Conceição, Castelo da Póvoa (Póvoa de Varzim); 9—Forte de S. João Baptista (Vila do Conde); 10—Forte da N.ª Sr.ª das Neves, Forte de Leça da Palmeira, Castelo de Matosinhos (Leça da Palmeira); 11—Forte de S. Francisco Xavier do Queijo, Castelo do Queijo (Nevogilde, Porto); 12—Forte de S. João Baptista da Foz or Castelo de S. João da Foz (Porto, Foz do Douro).

Continuing north, we encounter fortifications along the Ave River, including the Fort of S. João Baptista de Vila do Conde, designed by Filippo Terzi. Built from the late 1500s to the early 1600s under António de Vila Lobos' leadership, it features a polygonal shape with five bastions and only three guardhouses. Inside are casemates, the governor's house, the kitchen, and the chapel. Deactivated after 1834, it was used for vessel registration in the

Ave bar area. In the 1990s, it was converted into a hotel, which has since been deactivated and now serves as an event space [12,14].

The Fortress of Nossa Senhora da Conceição in Póvoa de Varzim was constructed between 1701 and 1740 on the site of the old Fort of Torreão, by order of D. Pedro II. It features a polygonal shape with four bastions, a vaulted entrance topped with a royal shield, and a bell tower. The main square features several buildings, including a chapel. Recently renovated, it has been transformed into a space for recreation and dining [14].

In Esposende, the Fort of S. João Baptista, built between 1699 and 1702 according to a project by the engineer Manuel Pinto Vila Lobos and under the direction of master Pedro da Rocha Vale, had a starry plan with bastions at each angle. In 1866, it was modified with the installation of a metallic lighthouse—a rare structure in Portugal—and an annex building for the lighthouse keeper [15].

Regarding the defensive system of the Minho Litoral in the Viana do Castelo region, the Fort of Santiago da Barra is notable for being the successor to structures initiated in 1552. Designed by Filippo Terzi, it was constructed between 1589 and 1596, remodeled between 1652 and 1654, and a moat was added in 1700. The fort's polygonal layout includes trapezoidal walls, triangular bastions, and circular guardhouses [16]. Inside, key structures include the Chapel of Santiago and the Roqueta Tower; today, it functions as the headquarters of Turismo do Porto e Norte de Portugal [17,18].

Complementing the coastal defenses, smaller structures that are no longer in use include the Fort of Areosa (1640–1668), which features a star-shaped layout with four bastions, possibly designed by the same architect as the forts of Montedor and Cão; the Fort of Montedor, also from the mid-17th century, showcases four uneven bastions and several tourism development plans that never materialized [19]; and the Forte do Cão (1640–1668), with four bastions—two facing the sea and connected by a curved face, and two larger ones facing inland [14].

In Vila Praia de Âncora, the Lagarteira Fortress features a star-shaped plan with four bastions, faceted guardhouses, a portal topped by a royal shield, as in many other fortresses, and three barracks buildings. Maintains the local Maritime Delegation integrated in the Captaincy of the Port of Caminha [19].

Lastly, the unique Fort of Ínsua in Moledo, a National Monument, combines the Franciscan Convent of the Observance, completed in 1392, with the fortress, finished in 1653. Currently, it sits on a small island due to siltation caused by sediment from the Minho River entering the Atlantic Ocean, among other factors, as access is only by boat. Its irregular star-shaped layout features five bastions and ravelins, three of which serve as breakwaters. Notably, the presence of a freshwater well in the middle of the islet is one of only three known worldwide [20]. This was included in a requalification project by architect Fernando Távora as part of the government's Revive program to enhance cultural heritage [14]. However, the project was not executed.

These fortifications demonstrate the spread of bastioned military architecture in Portugal, marked by Italian (Filippo Terzi) and French (Charles Lassart) influences. Its chronology coincides with periods of conflict—the Iberian Union and the Restoration Wars—and with the need to control strategic river bars. They also reveal the diffusion of military architecture of Renaissance and Baroque origin, as well as the ability to adapt to the topographical characteristics and defensive needs of the Portuguese Atlantic coast [11,21]. Currently, there is a diversity of uses, ranging from residual military and museum functions to cultural, tourist, or unaffected spaces.

These fortresses on the Douro and Minho coasts are therefore unique testimonies of Portuguese military and maritime history, reflecting technical developments and defensive strategies over three centuries.

The preservation and adaptation of these properties reinforce cultural identity and offer potential for the development of heritage tourism and historical education activities, which are essential for the transmission of this heritage to future generations.

In the context of preservation and adaptation, it is imperative to undertake a comprehensive analysis of the subject's integration over the past decades, with particular emphasis on population growth and land use (Figures 2 and 3). The analysis of coastal population and land-cover maps reveals a markedly heterogeneous, yet clearly anthropized, narrow strip along the northwestern Portuguese coast. The population is concentrated in areas surrounding the primary coastal towns and estuaries. The highest population density is observed in areas with continuous built-up areas, transportation infrastructure, and adjacent agricultural or pastureland. Conversely, less populated sectors are distinguished by a more fragmented mosaic of forests, shrubland, and, in some cases, wetlands and water bodies, suggesting the presence of areas where natural or semi-natural cover still predominates. Forts and fortresses are distributed along nearly the entire coastline. However, many of these cultural heritage assets are in or immediately adjacent to the most urbanized and densely populated segments. This observation highlights the cumulative exposure of these heritage assets to both intensive human pressure and coastal hazards.

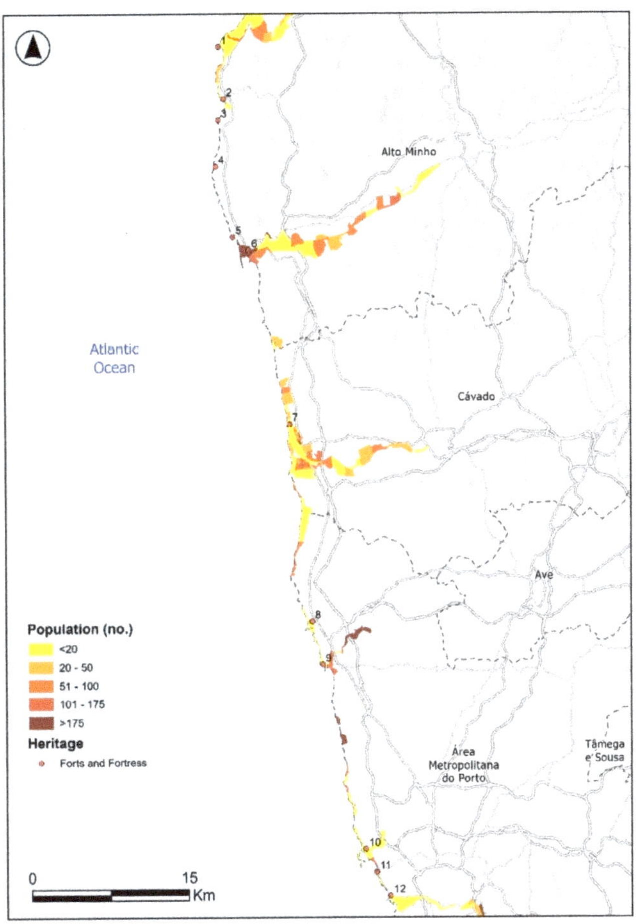

Figure 2. Coastal population distribution and location of forts and fortresses along the northwestern Portuguese coast. Source: Own elaboration, based on INE (2022) [22].

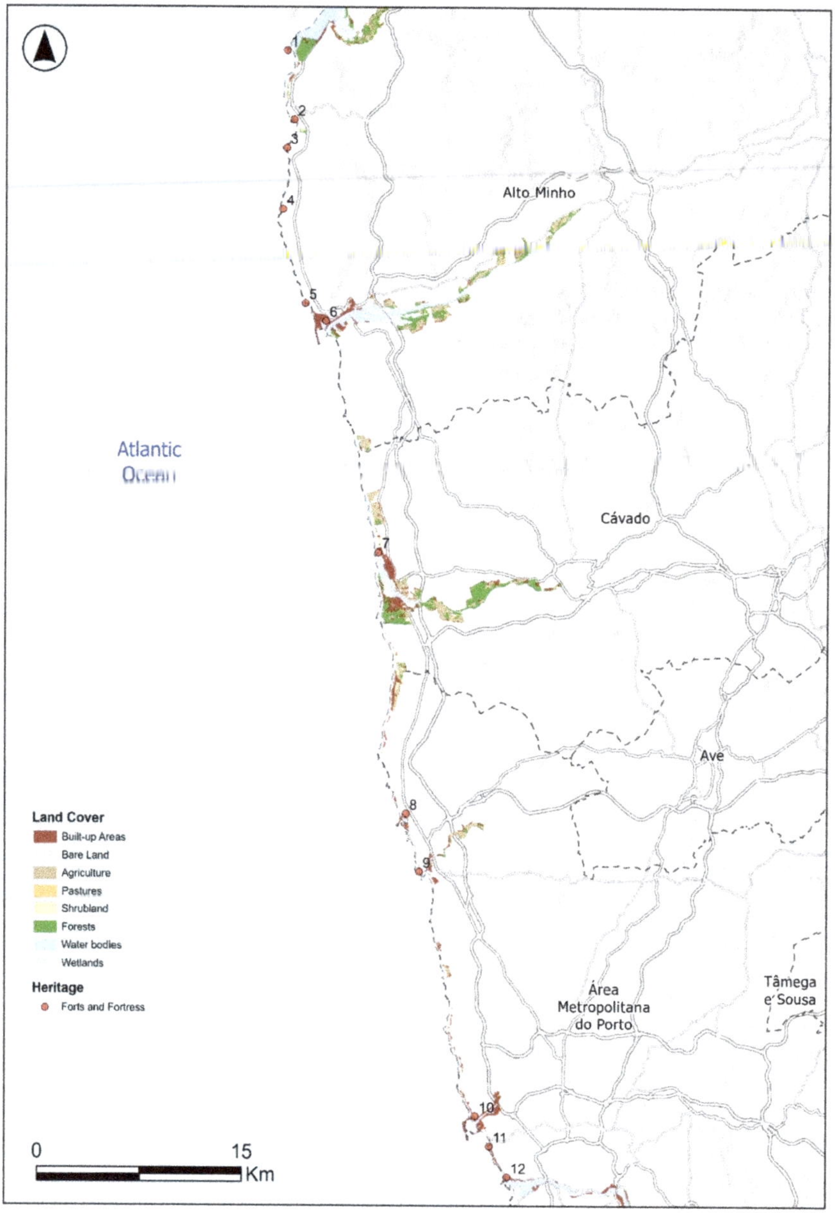

Figure 3. Land cover classes and locations of forts and fortresses along the same coastal stretch, highlighting the spatial overlap between built-up areas, sensitive habitats, and fortress sites. Source: Own elaboration, based on INE (2022) [22].

3. Coastal Cultural Heritage Under Climate and Human Pressure

Coastal vulnerability depends on both physical characteristics, such as geomorphology, slope, shoreline changes, tides, wave height, and sea-level rise, and human activities, including population density, tourism, infrastructure, land use, and ecosystem sensitivity. In fact, socioeconomic factors can change more rapidly than physical hazards and significantly affect coastal vulnerability, underscoring the need to incorporate cultural, economic, and environmental values into assessments [23]. Alexopoulou, Karathanassi, and Kremezi (2024) [24] emphasize that coastal archeological sites and monuments are increasingly threatened by shoreline retreat, accelerated erosion, and human activity.

Research conducted in recent years has confirmed the existence of real risks posed by climate-driven natural threats, particularly to the preservation of cultural coastal heritage. Factors driven by climate change, such as sea-level rise, ocean acidification, stronger storms, rising temperatures, and coastal erosion, threaten the stability and preservation of coastal cultural heritage. Although coastal heritage is crucial for understanding past cultures and history, long-term monitoring data on the environmental conditions affecting these sites remain limited [25].

It is widely recognized that climate change threatens cultural heritage sites worldwide. This emerging threat highlights the urgent need to develop targeted policies to protect historical sites, particularly those vulnerable not only to sea-level rise—one of the most significant natural hazards in coastal areas [26,27]—but also to other climate-related risks.

Extreme weather, sea-level rise, acidification, and temperature fluctuations damage structures through flooding, erosion, material deformation, and cracking, with coastal buildings being especially vulnerable [28]. The study by Arns, A. et al. (2017) [29] shows that sea-level rise interacts with tides, waves, and storm surges, thereby amplifying their combined impacts, particularly in shallow coastal regions. Natural defenses like continental shelves and wetlands currently reduce these impacts, but if they are lost or fail to adapt, coastlines will face much harsher ocean conditions.

This means that relying only on sea-level rise projections can underestimate future threats, and more advanced approaches are needed to design reliable coastal defenses. Water significantly accelerates the deterioration of cultural heritage through flooding, heavy rainfall, and sea-level rise. In particular, saltwater intrusion adds threats such as salt crystallization, placing many coastal monuments, UNESCO sites, and archeological areas at serious risk of damage [30].

Recent studies provide numerous examples of the climate change-related threats to historical fortifications along the coast. This is not only a concern on the Portuguese coast but also observed internationally, where many such examples can be found. Reimann, L. et al. (2018) [31] studied 49 cultural World Heritage Sites (WHS) located in low-lying coastal areas of the Mediterranean. They found that 37 of these sites are currently vulnerable to a 100-year flood (a flood so severe it occurs once every 100 years) and 42 are threatened by coastal erosion. Looking ahead to 2100, they predict that the risk of flooding could increase by 50% and the risk of erosion by 13% across the region. Some individual sites could face much higher increases in risk than these regional averages. In the Atlantic, Sánchez, Sánchez, and Ribalaygua (2020) [32] emphasize coastal fortifications as significant cultural heritage in the Canary Islands, highlighting that the castles of San Juan Bautista in Tenerife and San Cristóbal in Gran Canaria are especially vulnerable to the impacts of sea-level rise in the coming years. Given that heritage is a non-renewable resource, it is essential to take proactive measures to safeguard and preserve these assets from the impacts of climate change. Gomez Pretel and Carvajal Díaz (2025) [33] note that since the late 20th century, Cartagena de Indias has used groynes, barriers, mangrove restoration, and coastal management projects to curb erosion and protect Manzanillo Fort. However, the study reveals that climate change, sea-level rise, and human activity are increasing risks, underscoring the need for stronger protection of heritage and archeological sites. Further research is needed to assess vulnerable sites in line with the UN Sustainable Development Goal on Climate Action.

In Portugal, the expansion of urban areas along the coast, combined with climate change, has heightened coastal populations' vulnerability to hazards such as sea-level rise and coastal erosion, with erosion rates exceeding 8 m per year in some locations [34]. As highlighted by Zanchettin et al. (2021) [35], heritage risks are not solely caused by natural factors but are often driven by human activities such as urbanization, groundwater extraction, and construction, among others, underscoring the need for effective management and regulation to control and mitigate anthropogenic factors. Population growth in coastal zones is driving rapid development, intensifying or modifying both short- and long-term natural sedimentary processes. Trakadas and Karra (2022) [36] highlight that the most immediate threat to the area's cultural and natural heritage comes from human activities, particularly urban development and construction, which intensify pressure on the coast and risk damaging habitats and historical sites.

In parallel with the impact of urbanization, Ferreira et al. (2023) [37] argue that tourism is viewed as a double-edged sword, stimulating economic growth and revitalizing heritage, while also creating social, environmental, and cultural pressures, with impacts varying by destination maturity, and placing intangible heritage at risk of dilution and misrepresentation.

In summary, the combined effects of climate change and human activity pose a pressing challenge to the preservation of coastal cultural heritage. Rising seas, intensifying storms, and environmental degradation converge with urbanization and tourism pressures, threatening both tangible and intangible heritage in ways that are often underestimated. Safeguarding these irreplaceable assets and landscapes requires moving beyond traditional preservationist approaches toward integrated strategies that combine climate adaptation, sustainable development, and community engagement. Only through proactive management, informed by interdisciplinary research and aligned with global frameworks such as the UN Sustainable Development Goals, can coastal cultural heritage be protected for future generations.

4. Materials and Methods

This study employed an initial qualitative research approach that integrated documentary analysis, a literature review, and systematic field observation. Historical and typological data on the coastal fortifications were consolidated from bibliographic sources and existing inventories, particularly those of the Directorate-General for Cultural Heritage, to establish a historical and contextual foundation for subsequent field analysis. In this study, we also used the data available in SIPA (the Sistema de Informação para o Património Arquitectónico) [14] as the primary national source of cultural-heritage information. Portugal also maintains official lists of sites classified as "Buildings of Public Interest" and "National Monuments", which we incorporated. These registries provide descriptive, historical, and legal-status information, but they vary in completeness, offer inconsistent or missing geospatial attributes, and focus mainly on architectural heritage. As a result, many landscape-scale or locally recognized heritage elements relevant to our analysis are not included. To address these gaps, we added standardized spatial data, harmonized metadata, landscape-level attributes, and additional heritage elements obtained from regional and local inventories. This enriched dataset serves as the basis for the analysis presented in the paper.

Based on the literature review, the principal drivers of risks affecting coastal heritage were identified by collecting and analyzing indicators most frequently reported by authors. These risks were categorized into anthropogenic pressures, including urbanization, groundwater extraction, construction works, tourism pressure, and population growth, and natural pressures, such as sea-level rise, ocean acidification, intensified storms, temperature increase, coastal erosion, meteorological variability, saline intrusion, salt crystallization, flooding, and heavy rainfall.

A structured observation matrix was then developed, comprising the two main categories of pressures—anthropogenic and natural—derived from the literature (Table 1). For each pressure, specific observable indicators were defined to ensure consistency and reproducibility across all surveyed sites (Tables 2 and 3). For each fortification, the observable indicators listed in Tables 2 and 3 were recorded in a simple binary way (0 = not observed; 1 = observed). This binary coding allowed us to systematically compare the presence or absence of each risk manifestation across sites.

Table 1. Main risks affecting coastal heritage sites. According to the authors, based on the literature review.

Anthropogenic Risks	Natural Hazards
• Urbanization and Coastal Development—Intensify pressure on heritage sites and disrupt natural sedimentary processes. • Groundwater Extraction—Can lead to land subsidence and structural instability. • Construction Activities—Cause vibrations, pollution, and landscape alteration. • Tourism and social pressure—Lead to wear and tear, pollution, and potential vandalism. • Population Growth in Coastal Zones—Drives infrastructure expansion and environmental stress. • Heritage abandonment (Van Nguyen & Duy, 2024 [38]; Pedretti, L. et al., 2024 [39]; Erban, Gorelick & Zebker, 2014 [40]; Azam, Riaz, & Javaid, 2024 [41]; Silva, Arreiol & Fragata, 2024 [42]; Wu & Barrett, 2022 [43]; Freitas, Koskowski, 2021 [44])	• Sea-Level Rise—Increases the risk of flooding and submersion of coastal structures. • Ocean Acidification—Affects building materials, especially those containing calcium carbonate. • Stronger Storms—Lead to physical damage and accelerated erosion. • Rising Temperatures—Cause material deformation and thermal stress. • Coastal Erosion—Undermines structural foundations and leads to loss of land. • Weather and Temperature Fluctuations—Contribute to cracking and material fatigue. • Saltwater Intrusion and Salt Crystallization—Accelerate material decay, especially in porous stones. • Flooding and Heavy Rainfall—Increase moisture-related deterioration. (Reimann et al., 2018 [31]; Spiridon & Sandu, 2023 [23]; Alexopoulou, Karathanassi & Kremezi, 2024 [24]; Turiashvili, 2024 [28]).

For example, urbanization and coastal development were assessed through proximity to modern buildings or infrastructure. At the same time, natural hazards such as erosion or salt crystallization were evaluated by observing material degradation, structural instability, or alterations in the surrounding landscape. This standardized framework enabled a systematic and comparative assessment of both human-induced and environmental factors affecting the fortifications. Table 1 presents the key authors whose research informed the development of Tables 1–3. Their studies provided the theoretical and empirical foundation for identifying and selecting the risk factors and for constructing the indicators used to assess the vulnerability of coastal fortifications in this study.

Fieldwork involved direct observation and site surveys of coastal fortifications located between Porto and Caminha, encompassing both open-ocean fronts and estuarine environments influenced by the Douro, Leça, Ave, Cávado, Lima, and Minho Rivers. Each site was analyzed using the standardized observational framework, systematically documenting its physical condition, environmental exposure, and the presence of anthropogenic and natural pressures identified during field inspection. Based on this set of indicators, a comprehensive inventory of 16th–18th century coastal and estuarine fortifications was compiled, prioritizing officially listed heritage assets. For each site, data were recorded on chronology, typology, conservation status, current function, and observable evidence of human-induced (e.g., tourism, urbanization, coastal development) and natural stressors (e.g., marine humidity, erosion, storm impact).

Table 2. Anthropogenic pressures and observational indicators.

Anthropogenic Pressures	Observable Indicators
Urbanization and Coastal Development	Nearby contemporary structures or transportation routes close to the fortress Installation of coastal engineering works, such as seawalls, jetties, or piers Heritage abandonment
Groundwater Extraction	Cracks or fissures in walls and structural foundations Uneven, tilted, or sagging floors Evidence of ground settlement, such as sunken areas, leaning trees, or shifted terrain
Construction Activities	Active construction or operation of heavy equipment in the vicinity Accumulation of dust, debris, or pollutants on walls and surfaces Temporary barriers that restrict or alter access to the site
Tourism and social pressure	Markings, graffiti, or other forms of deliberate damage on walls, vandalism Litter accumulation in and around the fortress Signs of unauthorized access
Population Growth	New road construction, utility networks, or residential developments close to the fortress Higher levels of traffic or noise in the area Noticeable alterations in land use or constructions around the site
Lack of maintenance or visible neglect	Accumulated debris, dirt, or vegetation is overtaking the structure. Deteriorated structural elements, including collapsed walls, missing roof sections, or exposed reinforcements. Absence of protective measures, like missing fences, signage, or monitoring systems. No visible conservation or monitoring activity, suggesting institutional or administrative abandonment.

In the second phase, all architectural structures were georeferenced using GPS, enabling the creation of a spatially and visually accessible database integrating geographic and observational data. Google Earth imagery was employed to complement field observations, with aerial images collected in 2020 and 2025 to detect and illustrate macroscale environmental and morphological changes over time. In a second phase, this study assessed the vulnerability of coastal fortifications to coastal flooding and coastline retreat resulting from sea-level rise. The methodology combined scientific data on climate change projections with the geographic location and constructive typology of each fortification, following a semi-quantitative approach inspired by previous frameworks such as the risk assessment in the fortifications of the Canary Islands [27] and Historic Environment Scotland (HES) scorecard and the Forino et al. [28] model for archeological heritage in Australia. The study then presents a quantitative analysis of the Sea Level Rise Hazard Index (SLRHI) and the Coastal Vulnerability Index (CVI).

4.1. Sea Level Rise Hazard Index (SLRHI)

The Sea Level Rise Hazard Index (SLRHI) Scenarios were calculated based on the Mean Sea Level Rise projection under the Intermediate Hazard Scenario, in accordance with Directive 2007/60/EC on flood risk assessment and management.

Table 3. Natural hazards and observational indicators.

Natural Pressures: Sea-Related Indicators	**Observable Indicators**
Sea-Level Rise	Water marks, staining, or mineral deposits on walls indicate recent inundation Coastline differences Erosion or scouring around foundations caused by repeated tidal exposure Areas of the site regularly submerged or waterlogged
Ocean Acidification	Powdery or granular surfaces indicate chemical degradation
Coastal Erosion	Undermined walls or foundations Loss of land at the base of cliffs or shoreline adjacent to the fortress
Saltwater Intrusion and Salt Crystallization	Granular disintegration or flaking of porous stones Pitting or surface roughness linked to repeated salt crystallization
Natural Pressures: Weather-Related Indicators	**Observable Indicators**
Stronger Storms	Fallen or displaced masonry, roof tiles, or defensive structures Broken or bent gates, doors, or railings Accelerated erosion of the surrounding soil or cliffs
Rising Temperatures	Thermal cracking in masonry or plaster Deformation of wooden elements Discoloration or surface flaking caused by repeated heating
Weather and Temperature Fluctuations	Cracks, spalling, or fatigue patterns in stone or mortar
Flooding and Heavy Rainfall	Mold, moss, or algae growth indicates prolonged dampness Saturation of water in soil

Sea level rise hazard and coastal floods occur because of the interaction between three main processes:

- Abnormally high-tide events, which take place during new or complete moon phases when the Moon is at its closest point to Earth, generating what are known as spring tides.
- Storm surges occur when the sea level rises temporarily due to low atmospheric pressure and strong onshore winds associated with storms.
- Persistent winds and waves, which produce repeated wave action and wind-driven water accumulation along the coast.

When combined, these processes generate extreme scenarios that significantly intensify the impacts on populations, infrastructure, and heritage sites located near the shoreline.

SLHRI represents extreme coastal forcing conditions. This index includes five confidence levels, corresponding to conditional probabilities of flooding, each separated by 20% probability intervals relative to the central estimate of sea level rise projection.

The values were then mapped along the study coastline using the five classes defined in Table 4, producing a Sea Level Rise Hazard map on which the locations of the forts and fortresses were overlaid (see Figure 2 in Section 5.2). This analysis provides the spatial framework for interpreting each site's exposure before comparing it with the field-based indicators listed for each fortification.

Table 4. Sea Level Rise Hazard Index (SLRHI): class definitions and decision rules.

Confidence Level	Description	Probability Interval	Values
1	Very High	>80%	0.00–2.65 m
2	High	60–80%	2.65–2.75 m
3	Medium	40–60%	2.75–2.80 m
4	Low	20–40%	2.80–2.90 m
5	Very Low	<20%	2.90–3.25 m

Source: Authors' own elaboration, based on the requirements of Directive 2007/60/EC.

4.2. Coastal Vulnerability Index

In parallel, coastline retreat—defined as the inland movement of the coastline due to erosion and rising sea levels—was analyzed to assess its potential effects on the stability of fortifications near the coast. The Coastal Vulnerability Index (CVI) was used to represent each site's physical susceptibility to flooding and shoreline retreat.

The CVI is a composite index with values ranging from 1 (Very Low) to 5 (Extreme). For each fortification, six criteria were considered: (1) flooded area under the extreme sea-level rise scenario; (2) distance to the coastline; (3) coastal typology (rocky, sandy/dune or estuarine shore); (4) geology; (5) land use; and (6) hydrographic network (proximity to and density of drainage channels).

Data for these criteria were derived from standard national spatial datasets. The flooded area was mapped by intersecting the extreme sea-level rise scenarios defined by Directive 2007/60/EC with the national digital elevation model in a GIS environment. Distance to the coastline and the hydrographic network were calculated from the official Portuguese topographic mapping at a 1:25,000 scale. Coastal typology was defined by geomorphological interpretation of national orthophotos and the regional coastal management plan. Geological information was extracted from the national geological map, and land-use patterns were obtained from the official national land-use/land-cover map (COS 2018, available in DGT—Directorate General of Territory [45]).

Each criterion was classified on a five-point ordinal scale from 1 (very low contribution to vulnerability) to 5 (very high contribution), and the criteria were combined using predefined weights (w_i) based on expert judgement and previous CVI applications. The CVI for site j was computed as a weighted sum of the scores of all criteria (Equation (1))

$$CVI_j = \Sigma\ (w_i \times s_{ij}), \text{ with } \Sigma\ w_i = 1 \quad (1)$$

where s_{ij} is the score assigned to criterion i at site j. The resulting CVI values were finally reclassified into five vulnerability levels (Very Low, Low, Medium, High and Very High).

Informed by reference studies and standard Analytic Hierarchy Process (AHP) practices applied to coastal vulnerability assessment [46–49], and by expert judgement on the specific geomorphological and management context of the northern Portuguese coast, we proposed the hierarchy of importance shown in Table 5, which reflects the relative weight of each parameter in determining coastal flooding susceptibility and shoreline retreat. These studies are not site-specific for our area, but they provide widely used examples of AHP-based CVI applications.

This multi-parameter assessment enabled the classification of each fortification according to its degree of exposure and vulnerability to the combined effects of coastal flooding and coastline retreat, thereby supporting the definition of priority levels for adaptation and conservation measures.

Table 5. Criteria used in the Coastal Vulnerability Index (CVI), with their description, relative importance and weights.

Criterion	Operational Definition	Direction of Effect on Vulnerability	Scoring (1–5) *	Relative Weight (w_i)
Flooded area	Proportion of the fortification footprint inundated under the extreme sea-level rise scenario	Larger flooded area corresponds to higher coastal vulnerability	1 = no inundation; 2 = very small proportion inundated; 3 = limited partial inundation; 4 = extensive partial inundation; 5 = almost total/full inundation	0.30
Distance to the coastline	Shortest planimetric distance between the fortification footprint and the present-day shoreline	Shorter distance corresponds to higher coastal vulnerability	1 = very far from the coastline; 2 = far; 3 = intermediate distance; 4 = near; 5 = immediately adjacent to the shoreline	0.20
Coastal typology	Coastal setting where the fortification is located (rocky coast, sandy/dune coast, estuarine lowland, islet)	Less resistant/lower-lying settings correspond to higher coastal vulnerability	1 = high rocky shore or cliff; 2 = rocky shore with limited beach; 3 = mixed rocky/sandy pocket beach; 4 = open sandy/dune coast; 5 = estuarine or low-lying alluvial flat	0.15
Geology	Dominant lithology and substrate properties in the vicinity of the fortification (buffer zone—ZEP-determined by the heritage classification 50 m)	Less consolidated/softer materials correspond to higher coastal vulnerability	1 = very resistant crystalline rocks; 2 = resistant sedimentary rocks; 3 = moderately resistant formations; 4 = weakly consolidated deposits; 5 = unconsolidated sands, silts or alluvium	0.15
Land use	Predominant land cover within a buffer surrounding the fortification (e.g., natural vegetation, agriculture, urban areas)	More artificial/sealed surfaces correspond to higher coastal vulnerability	1 = predominantly natural vegetation; 2 = semi-natural/agricultural mosaic; 3 = mixed built-up and vegetated; 4 = predominantly built-up; 5 = highly urbanized or artificial surfaces	0.10
Hydrographic network	Proximity to main river channels and local drainage density	Closer and more connected fluvial network corresponds to a higher flood-propagation risk	1 = far from channels, very low drainage density; 2 = far, low density; 3 = moderate distance/density; 4 = near channels and/or high density; 5 = located on/adjacent to main channel or confluence	0.10

* All criteria are ranked from 1 (very low contribution to vulnerability) to 5 (very high contribution), and the CVI is computed as a weighted combination of these scores. Source: Authors' own elaboration based on previous coastal-vulnerability studies and standard AHP procedures.

5. Results

5.1. Field Work Analysis: Observation Based on the Set of Indicators

The analysis was based on a direct comparison between the list of risks and indicators in the bibliography (Tables 1–3) and the risks recorded for each fortification in the provided set of sites (Table 6). Recurrent patterns, local factors such as the presence of rivers like the Douro, Leça, Ave, Cávado, Lima, or Minho, and management factors including maintenance, vandalism, and incomplete rehabilitation were considered. For better understanding, we compiled a table listing the georeferenced locations of each fort, along with the aerial images collected in 2020 and 2025, using Google Earth.

The fieldwork and observations presented in Table 6 demonstrate strong alignment between the real coastal hazards affecting the northern Portuguese fortifications and those identified in the literature.

Almost all sites show exposure to sea-level rise, coastal erosion, stronger storms, meteorological and temperature fluctuations, saltwater intrusion, and excessive maritime humidity associated with the Atlantic influence, as well as, where relevant, fluvial dynamics from the Douro, Leça, Ave, Cávado, Lima, and Minho Rivers. The frequent occurrence of flooding, heavy rainfall, and salt crystallization across nearly all forts confirms the significance of these degradation drivers in marine and estuarine environments, as highlighted in previous studies.

Similarly, the anthropogenic risks identified in the literature—namely, urbanization and coastal development, tourism pressure, and population growth in coastal zones—are consistently reflected in the observed sites. These pressures are particularly pronounced in urban settings such as Porto, Matosinhos, and Póvoa de Varzim, where fortifications are encircled by dense infrastructure and tourism activity. Poor maintenance and neglect, observed in several cases, further amplify vulnerability, reinforcing the notion of management deficits as risk multipliers. At specific sites, such as Areosa, Montedor, Forte do Cão/Gelfa, and Ínsua, issues such as vandalism, uncontrolled vegetation growth, and inadequate maintenance emerge as active degradation factors. These problems, though indirectly referenced in the literature under broader categories like tourism pressure or neglect, appear locally as more critical threats, revealing deficiencies in protection, monitoring, and cultural governance. Uncontrolled vegetation, for example, facilitates water infiltration and structural fracturing, complicating conservation efforts. While often overlooked in broader theoretical discussions, this factor is a tangible and recurrent issue in several of the studied forts.

Hydrographic conditions also accentuate the vulnerability profile. The interaction of marine and riverine forces, especially in the Douro, Leça, Ave, Cávado, Lima, and Minho River systems, creates hybrid risks (tidal surge combined with river flooding) that intensify moisture infiltration and capillary humidity, producing site-specific degradation patterns that exceed generalized bibliographic predictions.

Additionally, incomplete or inadequate rehabilitation efforts, such as at Forte do Ínsua (Moledo), pose operational risks that compromise structural integrity and accelerate decay—an issue that is less emphasized in theoretical frameworks, which tend to focus on external environmental factors.

Overall, the results reveal consistent vulnerability patterns across the coastline. The dominant anthropogenic pressures, such as urban expansion, tourism, and maintenance neglect, combine with persistent natural hazards, such as salt crystallization, temperature variation, and flooding, to accelerate deterioration. In contrast, forts located in less urbanized or more isolated settings face greater environmental exposure but fewer direct human alterations. At the same time, those forts integrated into urban contexts degrade more rapidly, linked to anthropogenic intensity.

Table 6. Analysis of the anthropogenic pressures and natural hazards affecting forts.

Designation and Location	Anthropogenic Pressures	Natural Hazards	Images—2020 and 2025
Forte de S. João Baptista da Foz, Castelo de S. João da Foz—Porto, Foz do Douro **41.149082, −8.674492**	Urbanization and coastal development; Tourism and social pressure; Population growth in coastal areas; Lack of maintenance or visible neglect.	Increased temperature; Meteorological and temperature fluctuations; Saltwater intrusion and salt crystallization; Floods and heavy rains.	 **2020**  **2025**
Forte de S. Francisco Xavier do Queijo, Castelo do Queijo—Nevogilde, Porto **41.168696, −8.689917**	Urbanization and coastal development; Tourism and social pressure; Population growth in coastal areas; Lack of maintenance or visible neglect.	Meteorological and temperature fluctuations; Saltwater intrusion and salt crystallization; Floods and heavy rains.	 **2020**  **2025**

Table 6. *Cont.*

Designation and Location	Anthropogenic Pressures	Natural Hazards	Images—2020 and 2025
Forte de N. S. das Neves, Forte de Leça da Palmeira, Castelo de Matosinhos—Matosinhos, Leça da Palmeira **41.187676, −8.702222**	Urbanization and coastal development; Tourism and social pressure; Population growth in coastal areas; Lack of maintenance or visible neglect.	Coastal erosion; Meteorological and temperature fluctuations; Saltwater intrusion and salt crystallization; Floods and heavy rains.	 **2020**  **2025**
Forte de S. João Baptista—Vila do Conde **41.341731, −8.751788**	Urbanization and coastal development; Tourism and social pressure; Population growth in coastal areas; Lack of maintenance or visible neglect.	Coastal erosion; Meteorological and temperature fluctuations; Saltwater intrusion and salt crystallization; Floods and heavy rains.	 **2020**  **2025**

Table 6. *Cont.*

Designation and Location	Anthropogenic Pressures	Natural Hazards	Images—2020 and 2025
Fortaleza de N. S. da Conceição, Castelo da Póvoa—Póvoa de Varzim **41.378177, −8.764017**	Urbanization and coastal development; Tourism and social pressure; Population growth in coastal areas; Poor maintenance.	Meteorological and temperature fluctuations; Saltwater intrusion and salt crystallization; Floods and heavy rains.	 **2020**  **2025**
Forte S. João Batista—Esposende **41.543032, −8.790527**	Urbanization and coastal development; Tourism and social pressure; Population growth in coastal areas; Lack of maintenance or visible neglect.	Meteorological and temperature fluctuations; Saltwater intrusion and salt crystallization; Floods and heavy rains.	 **2020**  **2025**

Table 6. *Cont.*

Designation and Location	Anthropogenic Pressures	Natural Hazards	Images—2020 and 2025
Forte de Santiago da Barra, Castelo de Santiago da Barra—Monserrate, Viana do Castelo **41.688862, −8.838142**	Urbanization and coastal development; Tourism and social pressure; Population growth in coastal areas; Lack of maintenance or visible neglect.	Meteorological and temperature fluctuations; Saltwater intrusion and salt crystallization; Floods and heavy rains.	 **2020**  **2025**
Forte da Areosa, Fortim da Areosa, Forte, Fortim da Vinha, Castelo, Forte Velho—Areosa, Viana do Castelo **41.699723, −8.856206**	Tourist and social pressure (Vandalism); Lack of maintenance or visible neglect (uncontrolled growth of vegetation).	See level rise; Meteorological and temperature fluctuations; Saltwater intrusion and salt crystallization; Floods and heavy rains.	 **2020**  **2025**

Table 6. *Cont.*

Designation and Location	Anthropogenic Pressures	Natural Hazards	Images—2020 and 2025
Forte or Fortim de Montedor, Forte de Paçô—Carreço, Viana do Castelo **41.758863, −8.876477**	Tourist and social pressure (vandalism); Lack of maintenance or visible neglect (uncontrolled growth of vegetation).	See level rise; Coastal erosion; Meteorological and temperature fluctuations; Saltwater intrusion and salt crystallization; Floods and heavy rains.	 **2020**  **2025**
Forte do Cão, Forte da Gelfa—Vila Praia de Âncora, Caminha **41.797679, −8.873928**	Tourist and social pressure (vandalism); Lack of maintenance or visible neglect (uncontrolled growth of vegetation).	See level rise; Coastal erosion; Meteorological and temperature fluctuations; Saltwater intrusion and salt crystallization; Floods and heavy rains.	 **2020**  **2025**

Table 6. *Cont.*

Designation and Location	Anthropogenic Pressures	Natural Hazards	Images—2020 and 2025
Fortaleza or Forte da Lagarteira, Forte de Âncora—Vila Praia de Ancora **41.815555, −8.867823**	Urbanization and coastal development; Tourism and social pressure; Population growth in coastal areas; Lack of maintenance or visible neglect.	Coastal erosion; Meteorological and temperature fluctuations; Saltwater intrusion and salt crystallization; Floods and heavy rains.	 **2020**  **2025**
Forte da Ínsua—Moledo, Caminha (Franciscan Fort and Convent) **41.8593126, −8.874496**	Tourism and social pressure (vandalism); Lack of maintenance or visible neglect (uncontrolled growth of vegetation and incomplete architectural rehabilitation).	See level rise; Coastal erosion; Meteorological and temperature fluctuations; Saltwater intrusion and salt crystallization; Floods and heavy rains.	 **2020**  **2025**

Source: Own elaboration, based on fieldwork analysis and Google Maps, 2020 and 2025.

We began investigating the Forts of the Northern Coastline, which follow the maritime border, in 2020; therefore, we collected aerial imagery on Google Maps at that time. Since we decided to publish part of the study in 2025, we revisited this collection. We concluded that in some forts, the aerial photographs did not show any significant visible changes at the macro level. In the Forts or Small Forts of Areosa and Montedor/Paçô, it was observed that the vegetation cover has advanced considerably, showing less impact on the Forts of Cão/Gelfa and Ínsua. These are the monuments where the lack of maintenance is most evident. The comparative analysis of 2020 and 2025 imagery also shows ongoing deterioration, including the vegetation encroachment referred to, moisture staining, and visible terrain loss, especially at northern sites exposed to stronger marine conditions. Collectively, these findings underscore that the interaction between neglect, urban encroachment, and intensified marine dynamics severely compromises the long-term preservation of these coastal fortifications.

This highlights the urgent need for integrated coastal management, periodic monitoring, and adaptive conservation strategies that address both climatic and socio-economic factors to enhance the resilience of Portugal's maritime and military heritage.

5.2. Future Scenarios and Mapping of Hazard and Vulnerability Patterns

The spatial distribution of the Sea Level Rise Hazard Index (SLRHI) and the Coastal Vulnerability Index (CVI) (Figures 2 and 3) highlights two main exposure contexts along the northern Portuguese coast. The first corresponds to open-ocean sectors on rocky or sandy/dune shore, such as Castelo do Queijo, Areosa, Montedor, Cão/Gelfa, and Ínsu, where direct wave action and persistent swell generate very high to extreme hazard levels (SLRHI 4–5) (Figure 4). The second encompasses estuarine areas, including Foz do Douro, Leça da Palmeira, Vila do Conde, Esposende, Santiago da Barra, and Lagarteira, which register high hazard levels (SLRHI 4).

Figure 4. Sea Level Rise Hazard Index (SLRHI) for forts and fortresses, based on classification. Source: Authors' own elaboration.

The mapping indicates that very high sea-level rise (class 5) is concentrated in low-lying or isolated coastal features, such as Forte da Ínsua and the Areosa–Montedor–Cão/Gelfa coastal arc. In contrast, high level (class 4) dominates in the estuarine lowlands of the Douro, Leça, Ave, Cávado, Lima, and Âncora Rivers.

Figure 5 shows in detail the SLRHI levels for 2050 and 2100, considering the characteristics of the various locations.

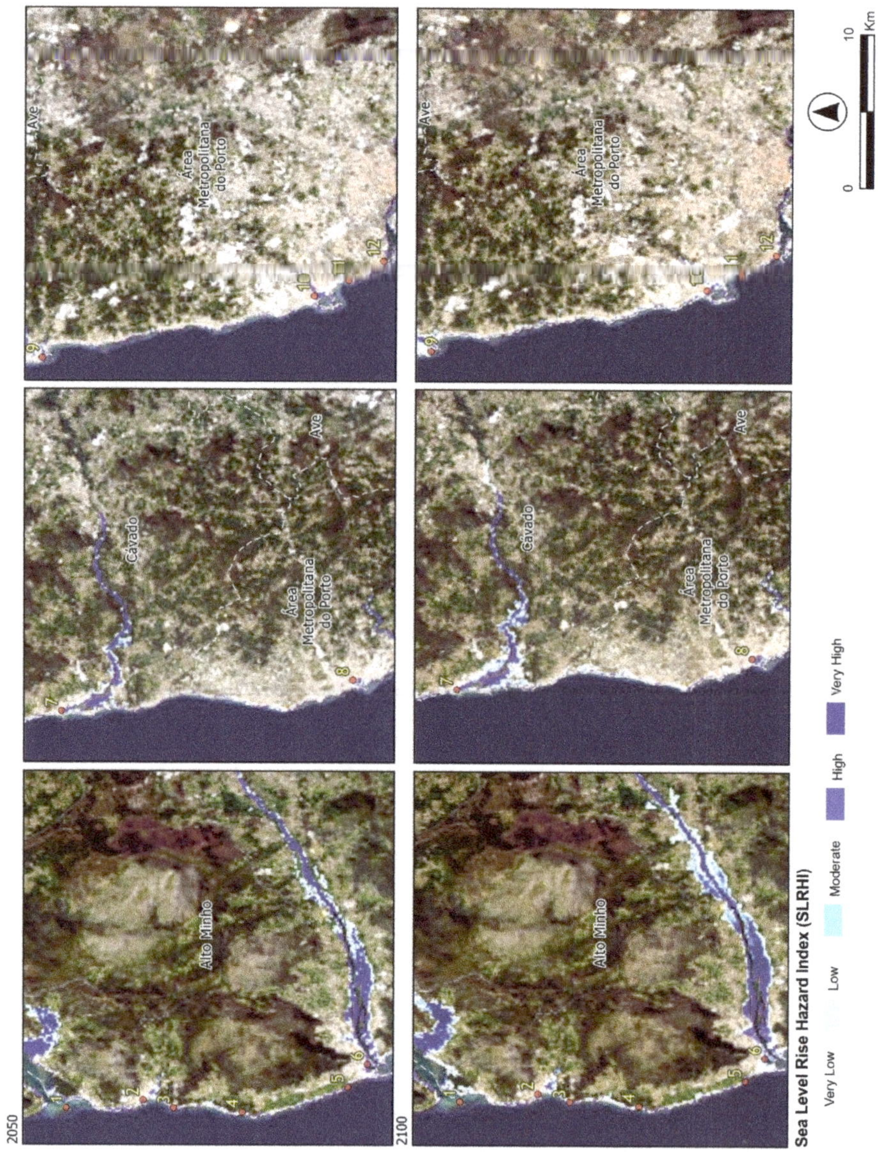

Figure 5. Sea Level Rise Hazard Index (SLRHI) for forts and fortresses, with three areas of analysis. Source: Authors' own elaboration.

The CVI results are, as expected, consistent with the SLRHI pattern, since sea-level-rise-driven flooding is one of the main drivers of site vulnerability (Figure 6). The highest vulnerability (CVI High and Very High) corresponds to segments with limited natural buffering capacity, whereas medium to high vulnerability occurs mainly in stretches with low topography and intense fluvial–marine interaction.

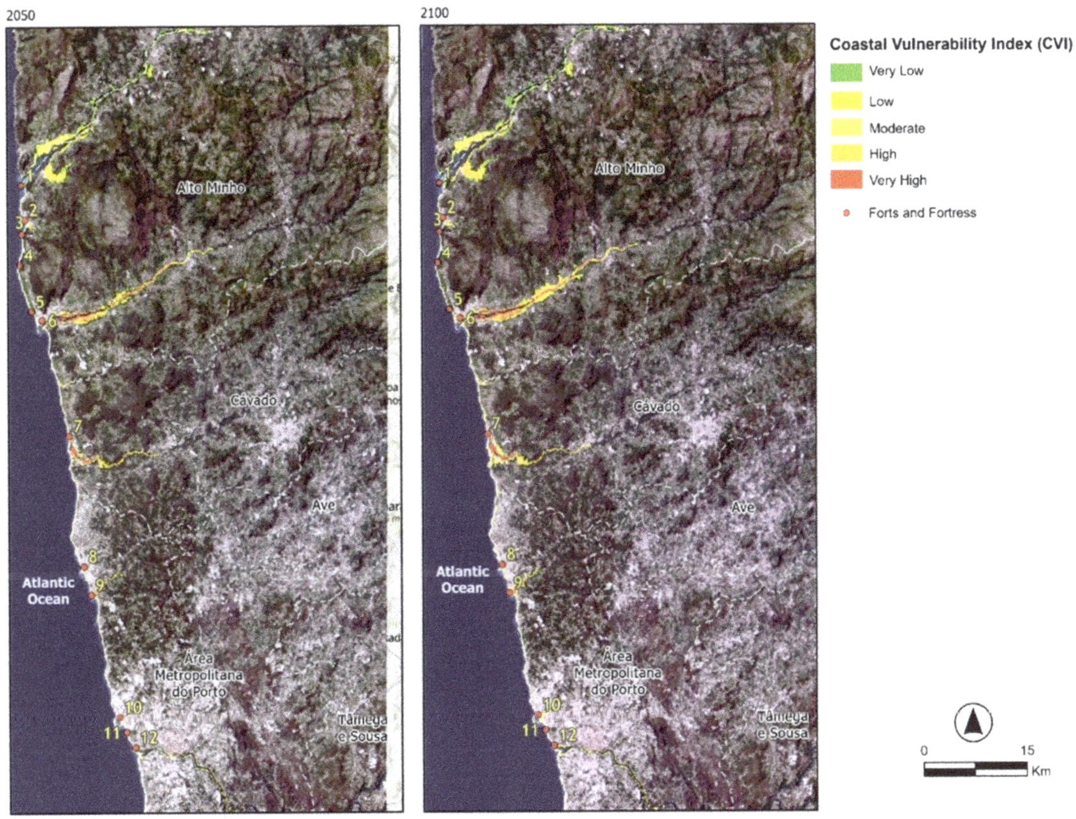

Figure 6. Coastal Vulnerability Index (CVI) for forts and fortresses, based on classification. Source: Authors' own elaboration.

Compared with the previous fieldwork analysis, these mapped results reveal a spatially coherent pattern. The same coastal stretches identified as subject to flooding, humidity, and saltwater influence are also among the most sensitive to sea-level rise and shoreline retreat. Thus, the mapping reinforces the notion that northern open-ocean sectors and estuarine transitions are the primary zones of future inundation and morphological change under projected sea-level rise scenarios.

Figure 7 provides a comprehensive analysis of the CVI levels across 2050 and 2100 for three distinct regions of the northwest, with a focus on the characteristics inherent to each location.

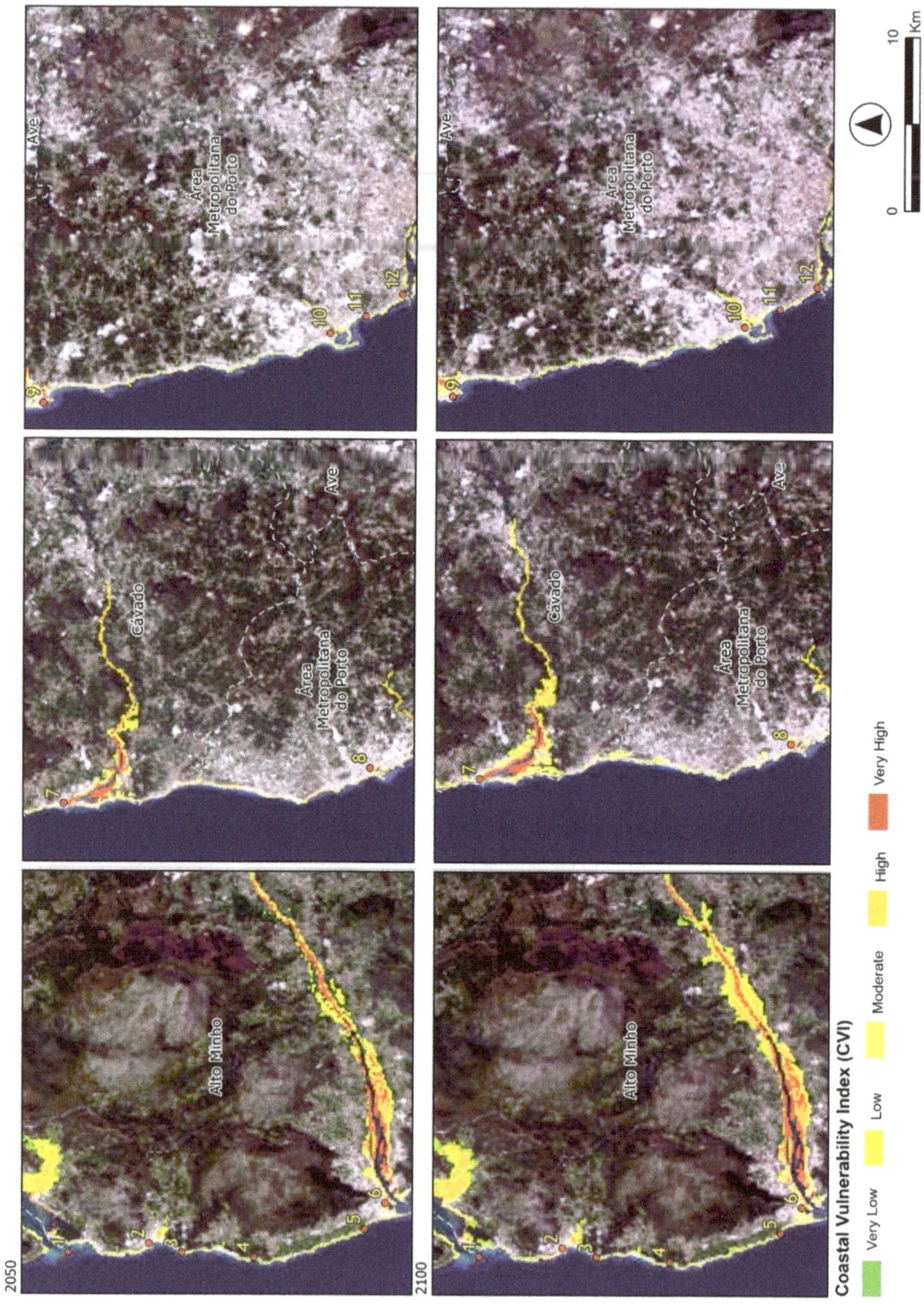

Figure 7. Coastal Vulnerability Index (CVI) for forts and fortresses, with three areas of analysis. Source: Authors' own elaboration.

6. Discussion and Conclusions

The comparison between the risks identified in the literature and those observed in the field at the coastal fortifications of northern Portugal (between Porto and Caminha) reveals a significant similarity between natural factors and a wider range of local anthropogenic risks, primarily related to maintenance and management.

The natural hazards described in the bibliography, such as sea-level rise, coastal erosion, marine humidity and flooding, were confirmed in practically all the structures analyzed (presence/absence of indicators as listed in Tables 2 and 3), although our field notes documented clear variations in their intensity, from incipient and localized manifestations to extensive, visibly damaging effects, particularly in low-lying or estuarine settings and in sites located near major river mouths (Douro, Leça, Ave, Cávado, Lima, and Minho).

Anthropogenic risks, also identified in the literature, are exacerbated by tourist pressure, urbanization, and, in particular, by the lack of or poor maintenance, often linked to vandalism and uncontrolled vegetation growth.

The results, therefore, demonstrate strong alignment between the risks identified in the literature and the empirical evidence, confirming the importance of coastal processes and climate change in the degradation of historical site structures. However, specific local factors emerge that intensify vulnerability. These local factors are related to the lack of maintenance and abandonment of several fortifications (Areosa, Montedor, Cão, Ínsua), which accelerates the impact of natural agents; vandalism and misappropriation, often associated with a lack of vigilance and the absence of active cultural use; the uncontrolled growth of vegetation, which aggravates infiltrations and cracking; incomplete or inadequate interventions, which leave structures partially exposed and fluvial influence combined with tides, which amplifies humidity and salinization processes in estuarine fortifications.

Overall, the combined influence of natural and anthropogenic pressures accelerates deterioration, with urban forts showing faster decay due to human-induced stress, and remote oceanfront forts exposed to more severe environmental forces.

Across the study coast, the conjunction of sea-level rise (SLR), storminess, and shoreline change implies a structural shift in the hazard baseline for heritage by 2100. In practical terms, what used to be exceptionally high-water events will occur more frequently, compressing return periods and expanding the geographic footprint of flooding. Within our SLHI/CVI screening, this translates into upward pressure on hazard classes (flood probability bands) at exposed oceanfront sites (e.g., Castelo do Queijo, Areosa–Montedor–Cão/Gelfa, Ínsua) and into more frequent compound flooding in estuaries (Foz/Douro, Leça, Ave, Cávado, Lima, Minho), where tidal set-up, storm surge and river discharge interact. Even without site-specific hydrodynamic modeling, the mapped patterns and the literature converge on more frequent overtopping and spray, longer wetting cycles, and broader areas of transient inundation as credible futures that erode safety margins built into historical siting and construction.

The integrated SLRHI–CVI reading supports a phased prioritization:

(i) Urgent safeguarding and continuous monitoring for Ínsua and the Areosa–Montedor–Cão/Gelfa arc (maximum exposure/vulnerability; access constraints);
(ii) Estuarine management plans combining salt-barrier measures, drainage, vegetation management, and preventive maintenance for Foz/Leça/Vila do Conde/Esposende/Santiago da Barra/Lagarteira;
(iii) Cross-cutting reinforcement of governance and routine maintenance, mitigating vandalism and incompatible uses, and integrating nature-based solutions (dune restoration, halophyte vegetation) wherever spatially feasible.

Hybrid marine–fluvial risk deserves special emphasis until 2100. The estuarine forts display a dual sensitivity: modest SLR increments expand storm-episode water coverage, while higher antecedent base levels reduce freeboard against river discharges. The mapped corridors of preferential inundation and persistent humidity around Foz, Leça, Ave, Cávado, Lima and Âncora are consistent with this mechanism. This hybridization challenges traditional single-hazard measures (e.g., local crest raising) and favors integrated drainage/salinity management at the site scale and along the connected waterfront.

These analyses highlight that vulnerability is not only a function of natural forces but also results from the interaction between environmental processes and factors of human management and use.

The overall vulnerability of coastal fortifications in northern Portugal is high and requires a future, systemic, interdisciplinary approach. From a prioritization standpoint, the SLRHI–CVI synthesis already points to three near-term trajectories aligned with 2100 conditions: (1) Oceanfront critical—Ínsua and the Areosa–Montedor–Cão/Gelfa arc—require urgent safeguarding and continuous condition monitoring, because exposure and vulnerability will rise fastest here with limited scope for nature-based dissipation on small headlands or islets; (2) Estuarine compounds—Foz/Leça/Vila do Conde/Esposende/Santiago da Barra/Lagarteira—call for drainage, salinity control and vegetation management combined with conservative fabric care to slow moisture-driven decay; (3) Governance fixes—regularized maintenance, access control, and deterring vandalism—are cross-cutting accelerators of resilience under the same physical forcing.

The results emphasize the urgent need for integrated coastal management, adaptive conservation, and systematic monitoring that account for both climatic threats and socio-economic pressures to ensure the long-term resilience of northern Portugal's maritime military heritage. In this way, the Programa da Orla Costeira (POC) Caminha–Espinho identifies this coastal stretch as one of the most vulnerable in northern Portugal due to its strong erosive dynamics, high exposure to overtopping and flooding, and the concentration of human occupation in at-risk zones. Anticipated climate change impacts—such as sea-level rise, altered wave regimes, and more frequent and intense storms—are expected to intensify these hazards. In response, the POC proposes a comprehensive adaptation framework that combines protection, accommodation, and planned retreat, grounded in the precautionary principle and integrated territorial planning. Key measures include prioritizing coastal defense in urbanized areas, restoring degraded ecosystems, controlling human activities that destabilize the coastline, and identifying high-risk zones to safeguard populations, natural systems, and cultural heritage. The POC also establishes conditions for the phased relocation of vulnerable infrastructure, the progressive adaptation of built environments, and the incorporation of coastal risk mitigation into the management of classified areas. Importantly, territorial plans must align with the Critical Areas for Protection, Accommodation, and Planned Retreat defined in the POC's Territorial Model, ensuring that site-specific strategies address both current and future coastal risks through proactive, long-term planning.

The ENAAC 2030 strategy highlights the vulnerability of Portugal's coastline, which is among the country's most exposed areas to climate hazards. Climate change is expected to amplify these risks through sea-level rise, altered wave regimes, more frequent and intense storms, and ongoing sediment deficits. Dense coastal occupation further increases the potential socio-economic impacts, with approximately 14% of the population living along the coast, underscoring the urgency of coordinated adaptation measures.

Priority measures should include the implementation of continuous environmental monitoring systems, reinforcement of preventive maintenance and vegetation control, integrated management plans that consider the interplay between culture, environment, and tourism, heritage education, and community surveillance in abandoned structures, as well as the incorporation of these fortifications into local climate change adaptation plans.

Additional measures can enhance the resilience of northern Portugal's coastal fortifications by using structural reinforcement with minimally invasive, compatible materials that strengthen walls and foundations without compromising historical authenticity. Similarly, targeted desalination treatments or poultices can address salt crystallization in masonry, mitigating one of the most common degradation processes identified in field observations. Early warning systems and predictive modeling tools could provide anticipatory measures for extreme events, improving preparedness and timely intervention. Simultaneously, digital preservation techniques, including 3D scanning, BIM modeling, and virtual reality documentation, can create permanent records of structural conditions and inform intervention planning without physically impacting the sites.

Controlling visitor access through designated pathways and guided tours is essential to minimize physical wear and erosion caused by tourism. Moreover, integrating cultural and ecological functions—such as research, educational activities, and eco-tourism programs—can sustain a positive human presence that discourages vandalism while enhancing public awareness and engagement in heritage preservation.

The adequate protection of the coastal fortified heritage depends, therefore, on a management model that combines technical knowledge, permanent monitoring, and active participation of local communities, transforming these structures not only into historical testimonies but also into living and resilient elements of the Portuguese Atlantic landscape.

Coastal adaptation strategies have advanced, but most remain small-scale and fragmented. Effective long-term adaptation to sea-level rise requires integrated planning, strong monitoring systems, and actionable knowledge. Placing communities at the center, engaging stakeholders, incorporating local knowledge, and co-developing context-specific solutions are essential. However, current efforts remain small and fragmented, underscoring the need for more inclusive participation and long-term strategies to achieve equitable and sustainable outcomes [32]. Engaging communities, raising awareness of cultural landscapes, and integrating nature-based and traditional solutions, along with regenerative thinking that considers carrying capacity and risk, are essential for promoting sustainable, harmonious, and long-term preservation.

Adapting cultural heritage to climate change is complex, as measures such as shelters or relocation can threaten both physical and intangible values, including authenticity, memory, and a sense of place, underscoring the need for more substantial research, collaboration, and international knowledge exchange [50]. Effective adaptation, therefore, requires the involvement of both decision-makers and local communities, fostering an awareness of heritage values and ensuring that responses respect cultural significance. Turiashvili (2024) [28] emphasizes that assessing the impacts of climate change on cultural heritage requires a holistic, interdisciplinary approach. This involves integrating climate modeling, historical data analysis, site monitoring, risk and vulnerability assessments, hydrological and geological studies, cultural landscape research, and archeological documentation. Experts in an interdisciplinary effort can more precisely determine the threats confronting heritage sites.

Strengthening policies and embracing hybrid and transformative approaches are key to meeting future challenges. Better monitoring helps assess how adaptation strategies reduce damage and vulnerability across ecosystems, economies, and communities.

Despite improvements, climate change risk assessments for cultural heritage still face conceptual and operational gaps, underscoring the need for robust, socially informed risk evaluation to support effective adaptation and resilience [51].

Recognizing cultural heritage assets at risk from sea-level rise or broader climate-related hazards, such as air pollution, extreme temperature shifts, strong winds, and heavy rainfall, is a vital first step toward discussing adaptation and mitigation options that can strengthen the resilience of these monuments. To enhance resilience against climate change, specific measures can be applied to heritage structures in a way that harmonizes with the surrounding landscape [30]. These interventions aim to safeguard coastal fortifications from sea-level rise and erosion while preserving the site's visual character and historical significance.

Assessing and identifying possible adaptation strategies is a crucial initial task for policymakers. These strategies lay the groundwork for robust regulations and action plans to protect coastal cultural heritage from the impacts of climate change.

Building on these foundations, the assessment and design of potential adaptation strategies offer policymakers an initial framework for safeguarding coastal heritage, particularly historical fortifications near the sea, which were the focus of this study, and face increasing threats from sea-level rise. Proposed adaptation measures focus on protecting communities and ecosystems from sea-level rise, emphasizing the need for proactive planning and management to mitigate climate-related impacts [30].

Possible solutions include adopting a long-term approach to spatial planning, relocating vulnerable infrastructure to reduce exposure, establishing preventive measures such as designating high-risk zones for future urban development, and adapting urban areas to sea-level rise through a combination of infrastructure modifications and nature-based strategies [34].

Future Directions

Methodologically, the index approach used here is appropriate for screening and prioritization under uncertainty through 2100, and it is consistent with international heritage-risk practice referenced in the manuscript. However, the next step toward investment-grade adaptation is to couple these indices with site-scale hydrodynamic and groundwater modeling (for compound flood pathways), material-specific deterioration models (for salt/thermal cycling), and monitoring (Unmanned Aerial System—UAS) photogrammetry, ground-based imaging, moisture/salinity sensing to quantify trend slopes and test measures. This layered approach (screening, targeted modeling, pilot measures, iterative monitoring) fits both the diverse sitting contexts and the management realities documented in the results.

Author Contributions: Conceptualization, I.V.F., H.S.L. and F.M.S.; methodology, I.V.F., H.S.L. and F.M.S.; validation, I.V.F., H.S.L. and F.M.S.; formal analysis, I.V.F., H.S.L. and F.M.S.; investigation, I.V.F., H.S.L. and F.M.S.; resources, I.V.F., H.S.L. and F.M.S.; data curation, I.V.F., H.S.L. and F.M.S.; writing—original draft preparation, I.V.F., H.S.L. and F.M.S.; writing—review and editing, I.V.F., H.S.L. and F.M.S. All authors have read and agreed to the published version of the manuscript.

Funding: This research received no external funding.

Data Availability Statement: The original contributions presented in the study are included in the article; further inquiries can be directed to the corresponding author.

Conflicts of Interest: The authors declare that they have no conflicts of interest.

References

1. UNESCO: Convention Concerning the Protection of the World Cultural and Natural Heritage, 16 November 1972. Available online: https://whc.unesco.org/en/conventiontext/ (accessed on 27 November 2025).
2. European Landscape Convention, Florence, 20 October 2000. Available online: https://www.coe.int/en/web/landscape/the-european-landscape-convention (accessed on 27 November 2025).
3. Downes, J. *Future of Our Pasts: Engaging Cultural Heritage in Climate Action*; ICOMOS: Paris, France, 2019.
4. Spiridon, P.; Sandu, I.; Stratulat, L. The conscious deterioration and degradation of the cultural heritage. *Int. J. Conserv. Sci.* **2017**, *8*, 81–88.

5. Figueiredo, A.; Martinville, S. O património arqueológico costeiro em áreas de galgamento e erosão sedimentar em Portugal: Impacto ambiental na preservação do património. *Finisterra* **2024**, *59*, e36569. [CrossRef]
6. Caeiro-Gonçalves, F.; Bio, A.; Iglesias, I.; Avilez-Valente, P. Sea Level Rise Effects on the Sedimentary Dynamics of the Douro Estuary Sandspit. *Water* **2023**, *15*, 2841. (In Portugal) [CrossRef]
7. Gkaifyllia, A.; Hasiotis, T.; Tzoraki, O. Vulnerability of Coastal Heritage in the Context of Climate Adaptation: A Review. *Environ. Earth Sci. Proc.* **2025**, *32*, 14. [CrossRef]
8. Hollesen, J. The impact of climate change on archaeological sites. *UNESCO Cour.* **2025**, *2025*, 18–19. [CrossRef]
9. Lopes, H.S.; Silva, P.F.; Pinto, B.G.; Cuevas, P.D.; Ribeiro, V.; Remoaldo, P. Urban heat stress in the context of socio-economic and environmental challenges: Heat risk analysis and online surveys in northwestern Portugal. *Int. J. Disaster Risk Reduct.* **2025**, *120*, 105384. [CrossRef]
10. Lopes, H.; Remoaldo, P.; Ribeiro, V.; Martín-Vide, J. The impacts of climate change on human wellbeing in the municipality of Porto—An analysis based on remote sensing. In *Climate Change and Health Hazards: Addressing Hazards to Human and Environmental Health from a Changing Climate*; Springer: Cham, Switzerland, 2023; pp. 135–172.
11. Callixto, C. As Fortificações Marítimas do Tempo da Restauração. In *História das Fortificações Portuguesas no Mundo de Rafael Moreira*; Publicações Alfa: Lisbon, Portugal, 1989; pp. 207–220.
12. Barroca, M. *As Fortificações do Litoral Portuense*; Edições Inapa: Lisbon, Portugal, 2001.
13. Ribeiro, R. A topografia da província de Entre-Douro e Minho em 1758: A sua análise. In *Atas do VI Simpósio Luso-Brasileiro de Cartografia Histórica*; U.Porto Press: Braga, Portugal, 2015; p. 11.
14. SIPA-DGPC, Direção-Geral do Património Cultural, Sistema de Informação para o Património Arquitetónico, Ministério da Cultura de Portugal. Available online: http://monumentos.gov.pt/site/APP_PagesUser/Default.aspx (accessed on 16 April 2025).
15. Amândio, B. *O Forte de São João Baptista e o Farol de Esposende*; S. Bartolomeu do Mar; Centro Social da Juventude de Mar: Lisbon, Portugal, 1995.
16. Antunes, J.M.V. Obras Militares do Alto Minho: A Costa Atlântica e a Raia ao Serviço das Guerras da Restauração. Master's Thesis, Faculdade de Letras da Universidade do Porto, Porto, Portugal, 1996.
17. Abreu, A.A. *O Castelo de, S. Tiago da Barra*; Câmara Municipal de Viana do Castelo; Centro de Estudos Regionais: Viana do Castelo, Portugal, 1986.
18. Ermelinda Almeida—CAF. *Alto Minho*; Ermelinda Almeida—CAF: Lisbon, Portugal, 1987.
19. Peixoto, A.M. *O Litoral de Viana e a Sua Arquitectura Militar*; Arquivo Municipal de Viana do Castelo: Viana do Castelo, Portugal, 2001.
20. Gil, J.; Cabrita, A. *Os Mais Belos Castelos e Fortalezas de Portugal*, 4th ed.; Editorial Verbo: Lisbon, Portugal; São Paulo, Brazil, 1996; ISBN 972-22-1135-8.
21. Almeida, J. *Roteiro dos Monumentos Militares Portugueses*; LIVRARIA: Lisbon, Portugal, 1947.
22. INE. Censos 2021. 2022. Available online: https://censos.ine.pt/xportal/xmain?xpgid=censos21_main&xpid=CENSOS21&xlang=pt (accessed on 27 November 2025).
23. Spiridon, P.; Sandu, I. Conservation of the Cultural Heritage: From participation to collaboration. *ENCATC J. Cult. Manag. Policy* **2015**, *5*, 43–53. [CrossRef]
24. Alexopoulou, C.; Karathanassi, V.; Kremezi, M. Coastal Erosion Assessment for Protecting Cultural Heritage. In Proceedings of the IGARSS 2024—2024 IEEE International Geoscience and Remote Sensing Symposium, Athens, Greece, 7–12 July 2024; IEEE: Piscataway, NJ, USA, 2024; pp. 2203–2207.
25. Pavlopoulos, L.; Michalis, P.; Vlachos, M.; Georgakopoulos, A.; Tsiakos, C.; Amditis, A. Integrated sensing solutions for monitoring heritage risks. In Proceedings of the IGARSS 2024—2024 IEEE International Geoscience and Remote Sensing Symposium, Athens, Greece, 7–12 July 2024; IEEE: Piscataway, NJ, USA, 2024; pp. 3352–3355.
26. Fatorić, S.; Seekamp, E. Are cultural heritage and resources threatened by climate change? A systematic literature review. *Clim. Change* **2017**, *142*, 227–254. [CrossRef]
27. Sesana, E.; Gagnon, A.S.; Ciantelli, C.; Cassar, J.; Hughes, J.J. Climate change impacts on cultural heritage: A literature review. *WIREs Clim. Change* **2021**, *12*, e710. [CrossRef]
28. Turiashvili, L. Impact of climate change on Immovable Monuments of Cultural Heritage. *Sci.-Tech. J. Build.* **2024**, *2*, 21–25. [CrossRef]
29. Arns, A.; Dangendorf, S.; Jensen, J.; Talke, S.; Bender, J.; Pattiaratchi, C. Sea-level rise-induced amplification of coastal protection design heights. *Sci. Rep.* **2017**, *7*, 40171. [CrossRef]
30. Chalkidou, S.; Georgiadis, C.; Roustanis, T.; Patias, P. A methodology for identifying coastal cultural heritage assets exposed to future sea level rise scenarios. *Appl. Sci.* **2024**, *14*, 7210. [CrossRef]
31. Reimann, L.; Vafeidis, A.T.; Brown, S.; Hinkel, J.; Tol, R.S. Mediterranean UNESCO World Heritage at risk from coastal flooding and erosion due to sea-level rise. *Nat. Commun.* **2018**, *9*, 4161. [CrossRef]

32. Sánchez, F.G.; Sánchez, H.G.; Ribalaygua, C. Cultural heritage and sea level rise threat: Risk assessment of coastal fortifications in the Canary Islands. *J. Cult. Herit* **2020**, *44*, 211–217. [CrossRef]
33. Gomez Pretel, W.; Carvajal Díaz, A. Climate Change Impacts on UNESCO Heritage: Assessing Sea-Level Rise and Coastal Erosion at Fort San Juan de Manzanillo (1854–2023), Cartagena de Indias. *Conserv. Manag. Archaeol. Sites* **2025**, *27*, 1–24. [CrossRef]
34. Silva, A.A.; Trindade, J.; Rocha, J. Análise da inundação em zonas urbanas expostas ao aumento do nível do mar em Portugal Continental. In *10th Congresso Nacional de Geomorfologia*; Dinâmicas geomorfológicas no espaço e no tempo; Atas; Associação Portuguesa de Geomorfólogos: Porto, Portugal, 2024; pp. 211–215.
35. Zanchettin, D.; Bruni, S.; Raicich, F.; Lionello, P.; Adloff, F.; Androsov, A.; Antonioli, F.; Artale, V.; Carminati, E.; Ferrarin, C.; et al. Sea-level rise in Venice: Historic and future trends. *Nat. Hazards Earth Syst. Sci.* **2021**, *21*, 2643–2678. [CrossRef]
36. Trakadas, A.; Karra, A. Coastal landscapes, environmental change, and maritime cultural heritage resources in Morocco: The case study of Essaouira. *J. Marit. Archaeol.* **2022**, *17*, 401–420. [CrossRef]
37. Ferreira, T.C.; Romão, X.; Freitas, P.M.; Mendonça, H. Risk Assessment and Vulnerability Analysis of a Coastal Concrete Heritage Structure. *Heritage* **2023**, *6*, 6153–6171. [CrossRef]
38. Van Nguyen, M.; Duy, K.H. Prioritizing barriers to the conservation of cultural heritage buildings in adaptation to urbanization and climate change. *J. Clean. Prod.* **2024**, *473*, 143529. [CrossRef]
39. Pedretti, L.; Giarola, A.; Korff, M.; Lambert, J.; Meisina, C. Comprehensive database of land subsidence in 143 major coastal cities around the world: Overview of issues, causes, and future challenges. *Front. Earth Sci.* **2024**, *12*, 1351581. [CrossRef]
40. Erban, L.E.; Gorelick, S.M.; Zebker, H.A. Groundwater extraction, land subsidence, and sea-level rise in the Mekong Delta, Vietnam. *Environ. Res. Lett.* **2014**, *9*, 084010. [CrossRef]
41. Azam, R.; Riaz, M.R.; Javaid, M.F. Vibration vulnerability assessment of heritage structures in Lahore: A step towards sustainable conservation. *Built Herit* **2024**, *8*, 51. [CrossRef]
42. Silva, F.M.; Arreiol, M.; Fragata, A. The Impact of Pollution on Cultural Heritage in the Historic Centre of Porto, Portugal. *Urban Sci.* **2024**, *8*, 31. [CrossRef]
43. Wu, T.; Barrett, J. Coastal land use management methodologies under pressure from climate change and population growth. *Environ. Manag.* **2022**, *70*, 827–839. [CrossRef] [PubMed]
44. Freitas, I.V.; Koskowski, M.R. Heritage and Sustainable Development: Capacity Building Through Tourism. In *Capacity Building Through Heritage Tourism*; Srivastava, S., Ed.; Apple Academic Press: Palm Bay, FL, USA, 2021; pp. 113–131.
45. DGT. Carta de Uso e Ocupação do Solo. 2018. Available online: https://inspire-geoportal.ec.europa.eu/srv/api/records/b498e89c-1093-4793-ad22-63516062891b (accessed on 27 November 2025).
46. Forino, G.; MacKee, J.; von Meding, J. A proposed assessment index for climate change-related risk for cultural heritage protection in Newcastle (Australia). *Int. J. Disaster Risk Reduct.* **2016**, *19*, 235–248. [CrossRef]
47. Ramieri, E.; Hartley, A.; Barbanti, A.; Santos, F.D.; Gomes, A.; Hilden, M.; Santini, M. *Methods for Assessing Coastal Vulnerability to Climate Change*; ETC CCA Technical Paper 1; European Environment Agency: Copenhagen, Denmark, 2011; pp. 1–93.
48. Torresan, S.; Critto, A.; Rizzi, J.; Marcomini, A. Assessment of coastal vulnerability to climate change hazards at regional scale: The case study of the North Adriatic Sea. *Nat. Hazards Earth Syst. Sci.* **2012**, *12*, 2347–2368. [CrossRef]
49. Bongarts Lebbe, T.; Rey-Valette, H.; Chaumillon, É.; Camus, G.; Almar, R.; Cazenave, A.; Claudet, J.; Rocle, N.; Meur-Férec, C.; Viard, F.; et al. Designing coastal adaptation strategies to tackle sea level rise. *Front. Mar. Sci.* **2021**, *8*, 740602. [CrossRef]
50. Dimabayao, J.J.; Lara, J.L.; Canoura, L.G.; Solheim, S. Integrating Climate Risk in Cultural Heritage: A Critical Review of Assessment Frameworks. *Heritage* **2025**, *8*, 312. [CrossRef]
51. Ferreira, V.L.D.; Pereira, E.S.; de Mello, L.P.S.; Silva, R.A.G.; Dias, F.F. Analysis of Coastal Environmental Vulnerabilities in the Municipality of Niterói, Rio de Janeiro, Brazil, in the Face of Sea Level Rise Projections. *Coasts* **2025**, *5*, 11. [CrossRef]

land

MDPI

Article

Understanding the Dynamics of Urban Heritage: A Quantitative Analysis of Historic Districts Across China's Rapid Urbanization

Honghao Zhang, Xingming Li * and Xiang Zhang

College of Urban and Environmental Sciences, Central China Normal University, Wuhan 430079, China; honghao@mails.ccnu.edu.cn (H.Z.); znnyt@163.com (X.Z.)
* Correspondence: xingming@mail.ccnu.edu.cn

Abstract

Understanding the spatial distribution and associated contextual factors of historic and cultural districts is essential for heritage conservation and land-based spatial planning. Previous studies have largely focused on site-level authenticity and development models, while national-scale quantitative analyses remain limited, particularly in the Chinese context. This study constructs a nationwide dataset of 1212 historic and cultural districts across 31 provincial-level regions in China and applies a GIS-based framework integrating Global Moran's I, kernel density estimation, and standard deviation ellipse analysis. The results reveal significant spatial clustering broadly aligned with the Hu Huanyong Line, with high-density concentrations in major urban agglomerations such as the Yangtze River Delta and an overall orientation along an east–west axis at the national scale. Further analysis suggests that these spatial patterns are broadly consistent with natural geographic conditions and long-term historical–cultural contexts; river systems, dynastic capital construction, economic development, and population migration provide important interpretive backgrounds for understanding regional differentiation. By providing a national-scale quantitative characterization of designated historic and cultural districts, this study offers spatial evidence to support more differentiated, land-oriented conservation and spatial planning strategies and to facilitate comparable spatial-heritage analytics in other regions where designation lists exist.

Keywords: historic and cultural districts; spatial distribution patterns; historical evolution; contextual factors; China

1. Introduction

Across the world, historic areas and heritage districts are increasingly exposed to development pressures driven by urban renewal, real-estate investment, infrastructure upgrading, and tourism-led transformation, which may undermine their spatial integrity and long-term sustainability [1,2]. In response, international heritage governance has progressively emphasized more integrative frameworks—most notably the UNESCO "Historic Urban Landscape" (HUL) approach—which advocates embedding heritage conservation into comprehensive planning and multi-stakeholder governance to balance preservation with urban regeneration and sustainable development [3]. These challenges are particularly pronounced in China, where rapid urbanization and economic growth have accelerated land-use change and reshaped traditional urban spaces, making it increasingly important to understand the spatial organization of historic and cultural districts and the factors that associated with their distribution at broader geographic scales.

Historic and cultural districts typically refer to areas with a rich concentration of preserved cultural relics, clusters of historic buildings, which can comprehensively and authentically reflect traditional layouts and historical features on a significant scale [4]. These districts hold irreplaceable value in terms of cultural preservation, architectural aesthetics, cultural heritage, and commercial development. In many European cities, historic city centers have been analyzed as key heritage spaces under intense tourism and real-estate pressure, with research examining how touristification and visitor flows reshape urban form and threaten the long-term conservation of historic urban landscapes [1]. Such understanding is crucial for formulating more scientifically sound and effective policies and measures. Therefore, studying the spatial distribution patterns and influencing factors of historic and cultural districts is of great significance, especially in the context of China.

Previous research on historic and cultural districts has primarily focused on four aspects: authenticity [1,2], development models [4,5], cultural inheritance [6,7], and heritage governance and planning operationalization [8,9]. Regarding the authenticity of historic and cultural districts, scholars emphasize the importance of preserving the spatial structure of these areas. This involves maintaining and protecting their unique regional characteristics and historical features while ensuring the orderly and rational use of their spatial environment [6]. Additionally, the preservation of the spatial environment itself is critical [10]. On one hand, this entails maintaining the original layout of the natural environment, including distinctive surrounding residential buildings. On the other hand, when restoring damaged areas, efforts should be made to recreate their original appearance and maintain harmony within the human environment, aligning preservation with modern urban development [11].

In terms of development models, some scholars have noted that many Chinese cities, during urban renewal processes, have developed historic and cultural districts without adequate planning, altering their original features and being associated with their decline or even disappearance [12]. Given the current state of historic and cultural districts in China, some researchers have proposed a classification-based protection approach. Well-preserved districts can maintain their status and functions, while those in poor condition should undergo restoration to their original appearance, integrating as seamlessly as possible into modern urban renewal efforts to maximize their value [13].

Regarding cultural inheritance, past studies have often explored aspects of heritage preservation. The original inhabitants of historic and cultural districts are the bearers and transmitters of cultural heritage, forming the soul of these districts. However, the mass relocation of these residents has led to a gradual decline in the cultural richness of some historic districts [14]. Some scholars suggest that to reduce resident displacement, it is possible to restore the cultural scenes of these districts. For example, utilizing public spaces in historic and cultural districts to regularly host unique folk activities such as dragon and lion dances during the Spring Festival, temple fairs and lantern festivals, or Dragon Boat Festival races along historic waterways can enrich the lives of residents and provide a space for broader cultural exchange [15]. Other scholars believe that appropriate modifications to the internal architecture of these districts can protect them without compromising their authenticity [16].

Historic and cultural districts also function as land-based policy instruments that operationalize heritage values through planning and governance [8]. In practice, designation typically triggers conservation planning requirements, development and design controls (e.g., restrictions on land use, building intensity, facade and streetscape guidelines), and review procedures that influence redevelopment decisions and infrastructure upgrading

https://doi.org/10.3390/land15020239

within and around designated areas [9]. In addition, contemporary heritage governance increasingly emphasizes integrating conservation objectives into urban planning and multi-stakeholder coordination so as to manage change in historic areas under rapid urbanization. From this governance-and-planning perspective, observed spatial patterns of designated districts may reflect not only historical–cultural endowments but also differences in policy implementation capacity and local institutional resources, which should be acknowledged when interpreting national-scale clustering and regional disparities [17].

In terms of research methods for historic and cultural districts, traditional approaches have included field surveys, observation, and the collection of historical documents and literature [18]. Different researchers have employed various methods. Field surveys and observation are fundamental for studying historic and cultural districts, allowing researchers to record information on location, architectural style, structure, layout, historical development, community environment, and cultural activities [5]. The collection of historical documents involves reviewing archives, local records, and other materials to understand the historical evolution, construction processes, and related events and figures of these districts [19]. Additionally, semi-qualitative and semi-quantitative methods, such as Landscape Character Assessment (LCA) and Historic Landscape Characterization (HLC), have been used to describe landscapes and their features [20]. It should be noted that LCA/HLC studies may also rely on GIS as a mapping and visualization platform. However, their typical application is often case- or region-specific and focuses on landscape classification and descriptive characterization, rather than conducting reproducible, national-scale spatial statistical assessments of clustering and directional structure. In this context, Geographic Information Systems (GIS), already used in landscape design, offer a promising approach for mapping and measuring spatial patterns at broader geographic scales. However, spatial-analytic outputs are sensitive to methodological choices and data representation. The modifiable areal unit problem (MAUP) suggests that detected patterns may vary with analysis scale and aggregation units [21,22]. In addition, measures of clustering and spatial dependence depend on how spatial relationships are specified (e.g., distance-based or k-nearest-neighbor spatial weights [23,24]), and hotspot surfaces produced by kernel density estimation (KDE) can change noticeably with bandwidth/search-radius settings [25,26]. Therefore, spatial-pattern studies should explicitly report key parameters (analysis scale, weights specification, bandwidth settings) and, where feasible, conduct sensitivity checks across plausible specifications to communicate uncertainty in pattern identification and interpretation [27].

Building on the above literature, we propose a conceptual framework in which the national-scale geography of designated historic and cultural districts may reflect the combined context of heritage endowments, urbanization pressures, and governance/planning capacity. Natural settings (e.g., topography and river corridors) are linked to long-term settlement suitability and provide the environmental context for historic urban development, while historical–cultural processes (e.g., political centers, economic cores, and population movements) offer interpretive context for the emergence of heritage-rich regions over time. Under rapid urbanization, development intensity and redevelopment demand tend to concentrate in economically dynamic regions, creating uneven pressures on historic areas. Meanwhile, governance structures—including nomination incentives, planning resources, and the ability to implement conservation controls—affect which areas are more likely to be designated and effectively managed. Consequently, observed core–periphery clustering and directional structures can be interpreted as outcomes of interactions among environmental constraints, historical trajectories, urbanization dynamics, and governance-mediated designation and management.

https://doi.org/10.3390/land15020239

In conclusion, while previous scholars have conducted extensive research on the authenticity, development models, and cultural inheritance of historic and cultural districts, yielding important results, there are still two main gaps. First, much of the research has focused on single districts [5,11,19], lacking a macro-scale study of all historic and cultural districts across the country. Second, existing research in China has largely emphasized qualitative analysis [28,29], with insufficient reproducible, national-scale spatial-statistical characterization of clustering/cores/directionality. Third, contextual factors associated with the observed national-scale patterns remain under-discussed. Key questions remain: (1) What is the national-scale distribution of China's designated historic and cultural districts based on a consolidated dataset? (2) How can clustering, density cores, and directional structure be characterized reproducibly using spatial statistics? (3) How can the observed patterns be interpreted in relation to contextual factors such as topography, river corridors, and historical–cultural processes? Beyond China, this research offers a transferable national-scale spatial-analytic framework for mapping and quantifying heritage-district patterns, which can support cross-country comparison where similar designation lists exist. China provides a critical case due to rapid urbanization and large territorial heterogeneity, allowing us to observe how governance and development pressures are reflected in core–periphery and directional structures. The explicit reporting of methodological sensitivity and uncertainty further strengthens international reproducibility and comparability of heritage spatial analytics.

To address these gaps, we first assembled and systematized a nationwide dataset of 1212 historic and cultural districts from the National Cultural Heritage Administration (NCHA), where records are dispersed across multiple releases and formats, and supplemented these with historical sources. The general goal of this study is to compile a consolidated nationwide dataset of China's officially designated historic and cultural districts and to map and quantify their spatial distribution patterns, while exploring the contextual factors associated with these patterns. Building on this consolidated dataset, this study applies Global Moran's I, kernel density estimation (KDE), and the standard deviation ellipse (SDE) to examine spatial distribution and its determinants. The objectives are to: (1) provide a systematic census of historic and cultural districts in China; (2) analyze their spatial distribution patterns; and (3) explore the factors shaping these patterns. The findings aim to offer new perspectives for the protection and development of such districts and to inform related studies in other countries or regions.

This paper is organized as follows. Section 2 introduces the study area and describes data sources and processing. Section 3 presents the research methods, including Global Moran's I, kernel density analysis, and the standard deviation ellipse. Section 4 reports and discusses the results on the number of historic and cultural districts, their spatial distribution characteristics (patterns, direction, and centroid), and the contextual factors associated with pattern formation (natural and historical–cultural). Section 5 concludes with key findings, policy implications, and limitations with directions for future research.

2. Study Area and Data

2.1. Study Area

China is in Southeast Asia, along the western coast of the Pacific Ocean, spanning from 73°40′ E to 135°05′ E and from 53°30′ N to 4° N (Figure 1). The country's total land area is approximately 9.6 million square kilometers (Figure 1). Figure 1 presents the study area map of China, showing provincial boundaries, major river systems, and administrative centers over a DEM-based relief background; the color gradient represents elevation and thus visually differentiates major landform types, broadly indicating high-elevation mountains and plateaus, mid-elevation hills and basins/valleys, and low-elevation plains

https://doi.org/10.3390/land15020239

and lowlands. China's terrain is highly diverse—mountains, plateaus, basins, plains, and river valleys—with a general west-to-east decline and a commonly recognized three-step topographic structure [30]. The climate types across China are equally varied, including subtropical, temperate, plateau, mountain, and tropical rainforest zones. As of 2022, China had a Gross Domestic Product (GDP) of 17.88 trillion USD and a population of approximately 1.41 billion people (World Bank, 2023) [31]. China is one of the world's four ancient civilizations, with a history spanning over 4000 years and a rich cultural heritage accumulated through dynasties such as the Shang, Zhou, Qin, Han, Tang, Song, Yuan, Ming, and Qing [32]. This heritage encompasses various domains, including architecture, art, literature, philosophy, and science and technology. The dominant cultural influences in China include Confucianism, Daoism, Buddhism, literary arts, and traditional medicine, among others [33]. These cultural elements have developed and evolved over different historical periods, contributing to the formation of unique thematic cultures. Historic and cultural districts in China are distributed across all provinces, preserving ancient architecture, traditional crafts, and local culinary culture. These districts serve as important sites for the display and transmission of local historical and cultural heritage. Therefore, studying the national-scale spatial distribution of China's historic and cultural districts and the contextual factors associated with it is both academically and practically significant.

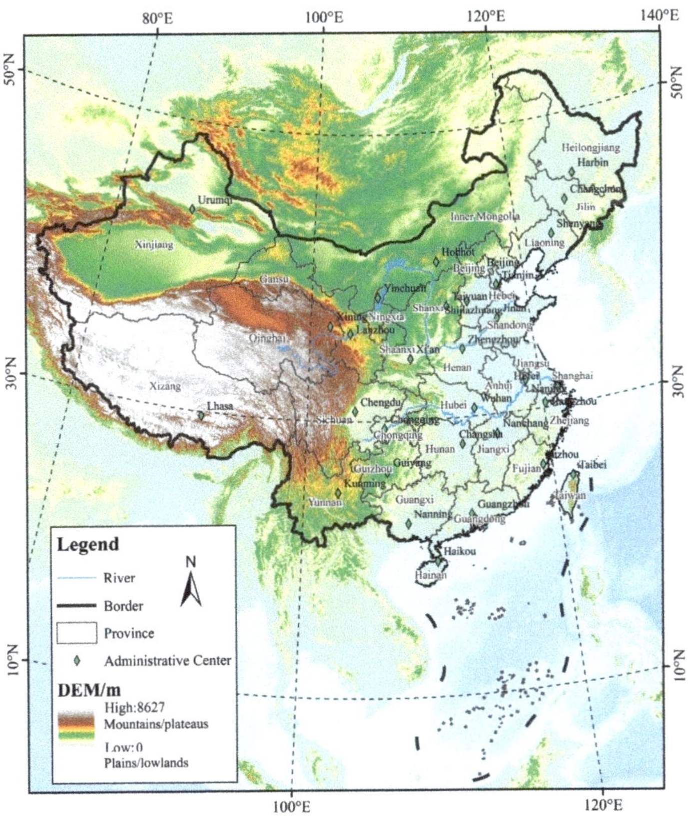

Figure 1. Overview of the Study Area.

2.2. Data Sources and Processing

In 2015, the Ministry of Housing and Urban-Rural Development and the National Cultural Heritage Administration announced the first list of 30 historic and cultural districts

https://doi.org/10.3390/land15020239

in China. Since then, additional lists have been released by relevant authorities, and by 2023, over 1200 historic and cultural districts had been officially recognized. This study obtained a total of 1212 officially recognized historic and cultural districts (as of 2023) from authoritative sources, including the NCHA and provincial government websites, and cross-checked entries to ensure consistency. Accordingly, the primary data used in this study consist of (i) the official designation list of historic and cultural districts and (ii) the geocoded point locations for each district, supplemented by national-scale spatial layers (DEM and river networks) for contextual overlay analyses. It should be noted that the dataset represents officially designated historic and cultural districts rather than an exhaustive census of all historically significant urban areas. In China, designation is typically implemented through a multi-level process involving local nomination, documentation, and compliance with planning/management requirements, followed by recognition through staged official releases. Accordingly, the observed distribution may reflect not only heritage endowments but also differences in governance capacity and nomination incentives, implying potential selection bias; therefore, results are interpreted as describing the spatial patterns of designated districts.

The geographic coordinates (latitude and longitude) of each district were collected using the Baidu Maps API coordinate picker (BD-09). To ensure coordinate consistency for national-scale analysis, all records were batch-transformed using the Xingchen Coordinate Batch Conversion Software (V2.0) following a standardized conversion workflow (BD-09 → WGS84). The converted coordinates were then imported into ArcGIS 10.8 as a geographic coordinate dataset in WGS 1984 (WGS84) and projected to a China-wide projected coordinate system, Asia North Albers Equal Area Conic, to support national-scale mapping and spatial statistics. Spatial analyses were conducted in ArcGIS, including Global Moran's I, Kernel Density Estimation (KDE), and Standard Deviational Ellipse (SDE), to characterize spatial clustering, density cores, and overall directional trends of historic and cultural districts across China.

For the analysis of spatial distribution and influencing factors, the study categorizes the districts into five historical periods: the Han-Tang period (including the Spring and Autumn Period, Warring States, Qin, Han, Three Kingdoms, Jin, Northern and Southern Dynasties, Sui, and Tang Dynasties), the Song period (including the Northern and Southern Song, Liao, Jin, Western Xia, and Dali), the Yuan-Ming period, the Qing period, and the modern era. Of the 1212 historically documented districts, 146 date back to before the Han-Tang period, 379 to the Song period, 285 to the Yuan-Ming period, 292 to the Qing period, and 110 to the modern era. Regarding the influencing factors, this study references a variety of sources, including brief histories of certain dynasties, historical records of population migration, economic development histories, and research findings from other scholars.

3. Research Methods

To address the above research gaps and ensure a transparent and reproducible analytical process, this study develops an integrated workflow that links data compilation, spatial-pattern measurement, and contextual interpretation. Figure 2 presents the overall analytical workflow of this study. It begins with a synthesis of the literature review and identified research gaps, followed by the compilation and processing of a consolidated nationwide dataset of historic and cultural districts (including geocoding, coordinate verification, and projection). The framework then applies a set of GIS-based spatial statistical methods—spatial distribution mapping, Global Moran's I, kernel density estimation (KDE), and the standard deviational ellipse (SDE)—using ArcGIS software (version 10.8; Environmental Systems Research Institute (Esri) Inc., Redlands, CA, USA) to quantify national-scale spatial patterns. Finally, the observed patterns are interpreted through contextual asso-

 https://doi.org/10.3390/land15020239

ciation analyses using DEM and river-proximity overlays, leading to conclusions, policy implications, and directions for future research.

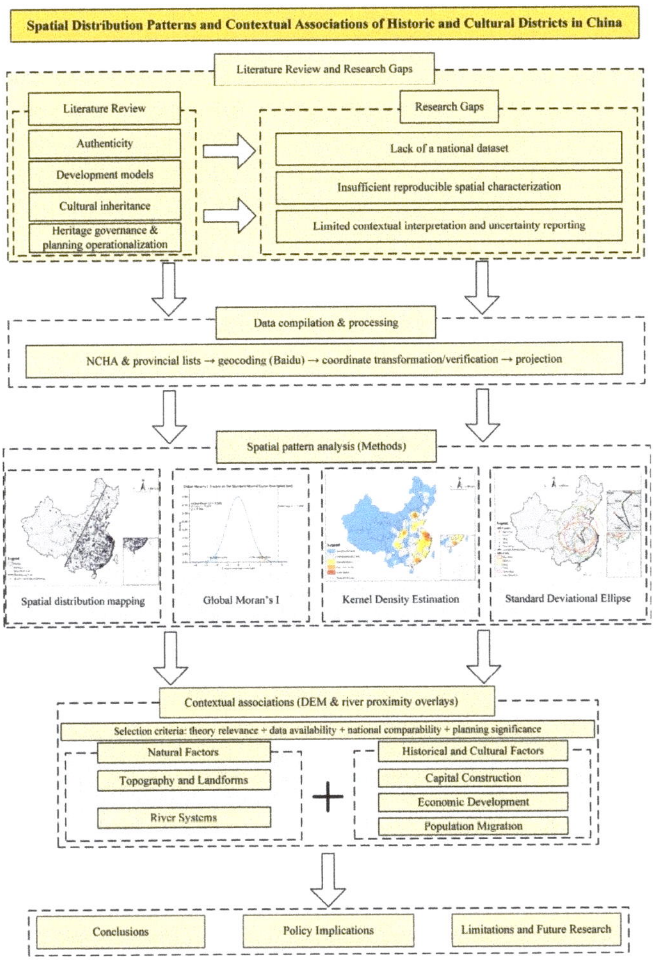

Figure 2. Methodological framework and analytical workflow of the study.

3.1. Global Moran's I

In this study, Global Moran's I is utilized to assess the spatial distribution pattern of historic and cultural districts in China [34]. Global Moran's I is a statistical measure used to detect spatial autocorrelation, indicating whether the attributes of a specific area exhibit a clustered distribution pattern in space. The formula is as follows:

$$I = \frac{n}{S_0} \times \frac{\sum_{i=1}^{n} \sum_{j=1}^{n} W_{ij}\left(y_{i-\overline{y}}\right)\left(y_{j-\overline{y}}\right)}{\sum_{i=1}^{n}\left(y_{i-\overline{y}}\right)^2} \tag{1}$$

where I represents the Global Moran's I index, with a value range of $[-1, 1]$. n is the total number of spatial units, $S_0 = \sum_{i=1}^{n} \sum_{j=1}^{n} W_{ij}$, W_{ij} is the spatial weight, for which this study adopts the adjacency weight method. y_i and y_j represent the attribute values of spatial units i and j, respectively, and $\overline{y}$ is the mean of all spatial unit attribute values.

https://doi.org/10.3390/land15020239

In this study, Global Moran's I was implemented in ArcGIS 10.8 using an inverse-distance conceptualization of spatial relationships (INVERSE_DISTANCE), Euclidean distance (EUCLIDEAN_DISTANCE), and no standardization (NONE). No threshold distance band or external weights matrix file was specified.

3.2. Kernel Density Analysis

Kernel Density Analysis is employed in this study to identify the dense areas of historic and cultural districts in China, reflecting their specific clustering patterns. Kernel Density Analysis is a method used to estimate the density distribution of spatial data points [35]. The formula is as follows:

$$f(x_0) = \frac{1}{nh}\sum_{i=1}^{n} K\left(\frac{X_i - x_0}{h}\right) \tag{2}$$

Kernel density estimation (KDE) was performed in ArcGIS 10.8 using an output cell size of 1000 m and a search radius of 200 km (projected units). To assess bandwidth sensitivity, we tested alternative search radii of 150 km, 200 km, and 250 km; the main density cores and overall spatial pattern were broadly stable across specifications, and 200 km was selected as the final setting to balance pattern stability and interpretability at the national scale. For map visualization, the KDE surface was classified into five classes using Natural Breaks (Jenks). All analyses were conducted under the China Albers Equal Area Conic projected coordinate system (meters).

3.3. Standard Deviation Ellipse

The Standard Deviation Ellipse (SDE) method is used to quantitatively analyze the directional characteristics and extent of dispersion in the distribution of historic and cultural districts in China [36]. This method is widely used in spatial analysis. The key parameters of SDE include the center, area, and the lengths of the major and minor axes of the ellipse. The center of the ellipse represents the centroid of the spatial distribution of historic and cultural districts in China. The major axis describes the distribution direction, expressed as the azimuthal angle relative to true north, while the minor axis measures the degree of dispersion of the historic districts. The eccentricity, or the ratio of the lengths of the major and minor axes, evaluates the directional tendency of the distribution. The steps involved are as follows:

(1) Center of the Ellipse

$$SDE_X = \sqrt{\frac{\sum_{i=1}^{n}(x_i - \overline{X})^2}{n}} \tag{3}$$

$$SDE_y = \sqrt{\frac{\sum_{i=1}^{n}(y_i - \overline{Y})^2}{n}} \tag{4}$$

(2) Orientation of the Ellipse

$$\tan\theta = \frac{A + B}{C} \tag{5}$$

$$A = \left(\sum_{i=1}^{n}\tilde{x}_i^2 - \sum_{i=1}^{n}\tilde{y}_i^2\right) \tag{6}$$

$$B = \sqrt{\left(\sum_{i=1}^{n}\tilde{x}_i^2 - \sum_{i=1}^{n}\tilde{y}_i^2\right) + 4\left(\sum_{i=1}^{n}\tilde{x}_i\tilde{y}_i\right)^2} \tag{7}$$

https://doi.org/10.3390/land15020239

$$C = 2\left(\sum_{i=1}^{n} \tilde{x}_i \tilde{y}_i\right)^2 \tag{8}$$

In Formulas (5)–(8), θ represents the azimuthal angle of the ellipse, i.e., the angle from the true north to the major axis in a clockwise direction. $\tilde{x}_i$ and $\tilde{y}_i$ are the differences between the coordinates of each study object and the mean center.

(3) Standard Deviation Distances for the Ellipse Axes

$$\sigma_x = \sqrt{2}\sqrt{\frac{\sum_{i=1}^{n}\left(\tilde{x}_i \cos\theta - \tilde{y}_i \sin\theta\right)^2}{n}} \tag{9}$$

$$\sigma_y = \sqrt{2}\sqrt{\frac{\sum_{i=1}^{n}\left(\tilde{x}_i \sin\theta + \tilde{y}_i \cos\theta\right)^2}{n}} \tag{10}$$

In Formulas (9) and (10), σ_x and σ_y represent the lengths of the X and Y axes, respectively, and n indicates the total number of historic and cultural districts.

The standard deviational ellipse was generated in ArcGIS 10.8 and reported at the 1-standard-deviation (1 SD) level for the full dataset and for each historical period to summarize directional trend and dispersion. Calculations were based on projected coordinates in China Albers Equal Area Conic (meters).

3.4. Qualitative Synthesis of Historical–Cultural Factors

Historical–cultural factors (capital construction, economic development, and population migration) were derived through a structured qualitative synthesis of secondary sources. We compiled evidence from peer-reviewed studies and authoritative historical references that document (i) major political centers and state-building processes, (ii) long-term economic cores and trade development, and (iii) demographic change and migration dynamics in different historical periods (e.g., [37–41]). Information relevant to the five historical periods defined in this study was extracted and organized into period-specific summaries and spatial cues (e.g., locations of major political centers, economic hubs, and migration-related contexts). These qualitative summaries were then used as contextual lenses to interpret the spatial-statistical outputs (KDE cores and SDE centroid migration/orientation), rather than as direct causal determinants.

4. Results and Discussion

4.1. National Inventory (Systematic Census) of Historic and Cultural Districts in China

A systematic national inventory based on the compiled dataset (n = 1212) reveals substantial provincial heterogeneity in the distribution of historic and cultural districts. Provinces along the eastern and southeastern seaboard account for a large share of designated districts, with Jiangsu (n = 128, 10.56%), Guangdong (n = 114, 9.41%), and Zhejiang (n = 107, 8.83%) ranking among the highest. By contrast, several inland and frontier regions contribute only a small proportion of the national total, underscoring a pronounced large-scale spatial imbalance.

To interpret this national-scale differentiation, we further use the Hu Line (Heihe–Tengchong Line), proposed in 1935 by Chinese geographer Hu Huanyong as a demographic dividing line in China [40]. It separates the country into two distinct parts: the southeast, which covers 36% of the land area and is home to 96% of the population, and the northwest, which contains only 4% of the population. Given its long-standing relevance for describing China's population and settlement geography, this study uses the Hu Line as a reference to

https://doi.org/10.3390/land15020239

interpret the large-scale spatial differentiation of historic and cultural districts. As shown in Figure 3, the overall distribution of historic and cultural districts roughly aligns with the Hu Line, with a much higher concentration in the southeast than in the northwest. Specifically, the southeast region hosts 1120 districts, accounting for 93.9% of the total, while the northwest region contains only 73 districts, representing 6.1%. This pattern indicates a strong spatial correspondence between the distribution of historic and cultural districts and the demographic contrast represented by the Hu Line.

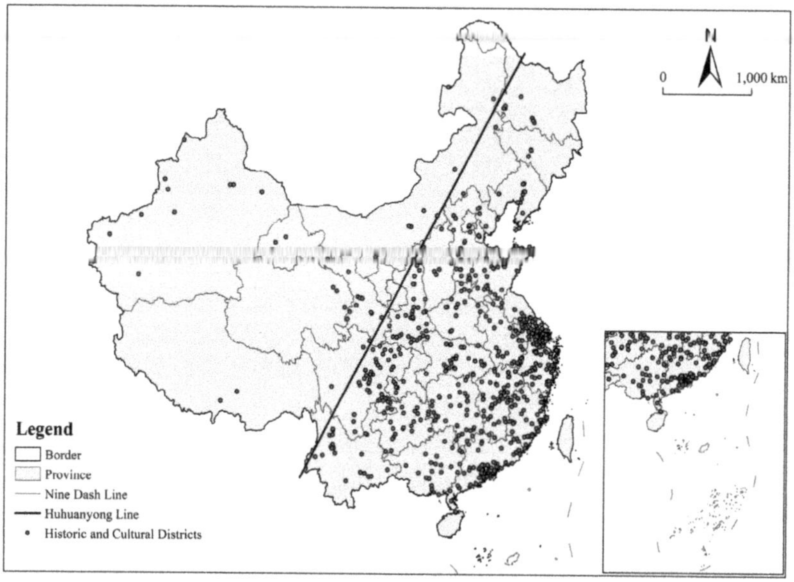

Figure 3. Spatial distribution of historical and cultural districts in China.

4.2. *Spatial Distribution Characteristics of Historic and Cultural Districts in China*

4.2.1. Spatial Distribution Patterns

Using ArcGIS 10.8, the Global Moran's I index for the spatial distribution of historic and cultural districts in China was calculated. The results yielded a Moran's I value of 0.150, a Z-score of 3.669, and a p-value of 0.000, indicating statistical significance at the 0.01 level. This suggests that, at the provincial scale, the distribution of historic and cultural districts exhibits positive spatial autocorrelation—provinces with a higher (or lower) number of districts tend to cluster spatially.

Further analysis of the spatial distribution structure across provinces (Figure 4) allows us to categorize the spatial distribution of historic and cultural districts into core areas, sub-core areas, general areas, sub-peripheral areas, and peripheral areas. The core areas are primarily concentrated in the Yangtze River Delta. The sub-core areas are distributed around the periphery of the Yangtze River Delta, the Pearl River Delta, the Beijing-Tianjin-Hebei region, and the Guanzhong Plain. Regions including the middle reaches of the Yangtze River, the Chengdu-Chongqing area, and the northern part of the Central Plains represent a transition from general areas to sub-peripheral areas. The peripheral areas are mainly located in the Northwest and Northeast, encompassing most parts of Inner Mongolia, Ningxia, Qinghai, Tibet, Xinjiang, Gansu, Sichuan, Yunnan, Heilongjiang, and Jilin.

The Yangtze River Delta region, which includes Shanghai, Jiangsu, and Zhejiang, stands out as the core area for the distribution of historic and cultural districts in China. This concentration can be interpreted in relation to a combination of contextual conditions,

https://doi.org/10.3390/land15020239

including transportation networks, climate and land resources, long-term economic activity, trade connections, and cultural traditions. The Yangtze River Delta is located in the lower reaches of the Yangtze River, with abundant water resources and extensive water-transport corridors. The region is characterized by a subtropical monsoon climate, featuring warm and humid conditions, ample rainfall, and fertile soils, which have historically supported intensive agricultural production. Since the Tang and Song dynasties, the Yangtze River Delta has been widely documented as one of China's major economic centers; cities such as Hangzhou and Suzhou were flourishing commercial hubs during the Song dynasty [42]. During the Ming and Qing dynasties, with the expansion of maritime trade routes, ports such as Ningbo and Shanghai functioned as important gateways for foreign trade, coinciding with regional economic prosperity [4]. The region is also rich in cultural heritage, including the Wu–Yue cultural tradition and the dissemination of Confucianism; cultural assets such as classical gardens, academies, and ancient towns constitute key elements of the region's cultural landscape [35].

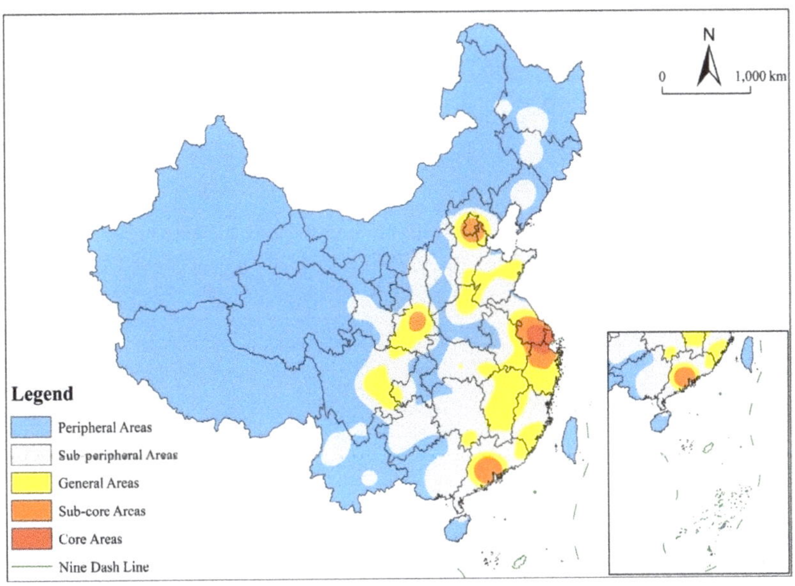

Figure 4. Spatial distribution of kernel density of Chinese historical and cultural districts.

The areas on the periphery of the Yangtze River Delta, along with the Pearl River Delta and the Beijing–Tianjin–Hebei region, can be regarded as secondary core areas. The concentration of historic and cultural districts in these regions can be discussed in relation to historical trajectories, economic development, cultural diversity, and geographic settings. Historically, these regions have long been important centers of economy, culture, and politics. In the process of modernization, these regions have experienced rapid development, and existing studies often note the role of geographic advantages and policy support in shaping conservation capacity and resource allocation for heritage preservation and restoration. Culturally, the diversity and inclusiveness of these regions are commonly associated with rich intangible cultural heritage that continues to be maintained and adapted within contemporary urban development. Geographically, these regions are predominantly alluvial plains with flat terrain, fertile land, and convenient transportation, providing a context in which dense settlement and commercial activity have been historically documented. Representative cultural districts are therefore frequently observed in these areas over time.

https://doi.org/10.3390/land15020239

4.2.2. Spatial Distribution Direction and Centroid

The calculation results of the Standard Deviation Ellipse reveal that the area of the standard deviation ellipse for China's historic and cultural districts is 254.03 km^2, with the centroid located in Dawu County, Xiaogan City, Hubei Province. The overall distribution tends to be along an east–west axis and has undergone an evolution characterized by initial contraction followed by expansion, with significant differences across historical periods (Figure 5). The centroid migration exhibits a distinct "southeast–northwest–northeast" trajectory, with the migration area primarily concentrated in the four provinces of Hubei, Jiangxi, Henan, and Jiangsu. The directional changes in the ellipse tend to flatten over time.

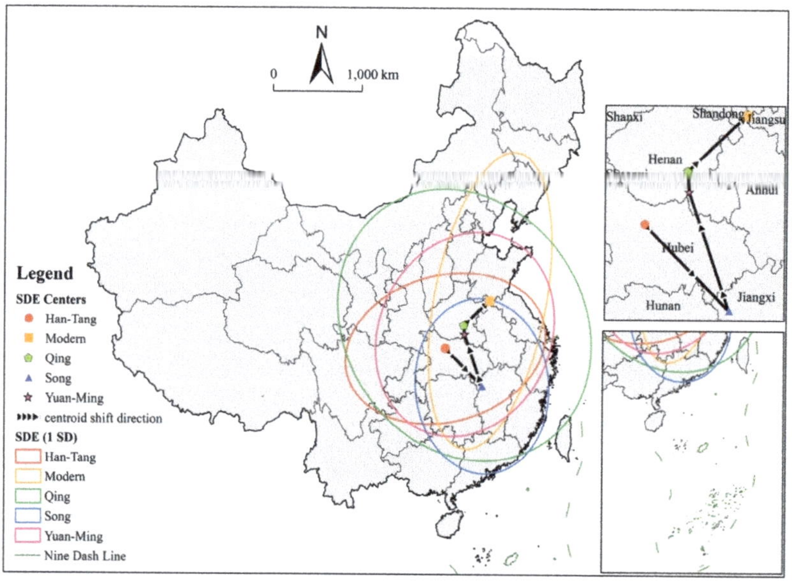

Figure 5. Elliptic distribution of standard deviation of historical and cultural districts in China.

Specifically, during the Han-Tang period, the centroid of historic and cultural districts was located in Nanzhang County, Xiangyang City, Hubei Province, with an overall east–west distribution. In the Song dynasty, the centroid shifted southeast to Tonggu County, Yichun City, Jiangxi Province, and the distribution orientation changed to a north–south axis. Compared to the Han-Tang period, the area of the ellipse decreased by 22.74%. During the Yuan-Ming period, the centroid migrated northwest to Xinyang City, Henan Province, and the area expanded by 52.08% compared to the Song period, with a spatial distribution orientation along a northeast-southwest axis. In the Qing dynasty, the centroid further moved northwest to Queshan County, Zhumadian City, Henan Province, with the ellipse area expanding by 108.54% compared to the Yuan-Ming period, and the distribution orientation shifted to a northwest-southeast axis. In the modern era, the centroid moved northeast to Xuzhou City, Jiangsu Province, with the spatial distribution orientation returning to a northeast-southwest axis, and the ellipse area decreased by 52.06%.

During the Han–Tang period, China's political and economic centers were primarily located in the Central Plains, particularly around Chang'an (modern-day Xi'an) and Luoyang. Against this background, the subsequent centroid shift to the southeast in the Song period can be interpreted in relation to widely discussed southward changes in economic gravity and the relocation of the political center to Lin'an (modern-day Hangzhou) during the Southern Song. The contraction of the ellipse relative to the Han–Tang period (−22.74%)

https://doi.org/10.3390/land15020239

may also be discussed in the context of documented warfare and urban decline in parts of northern China during the late Northern Song and Southern Song period [43]

In the Yuan–Ming period, the establishment of capitals in Dadu (modern-day Beijing) and Nanjing coincided with a northwestward shift in the centroid and an expansion of the ellipse (+52.08% compared to the Song period). The observed changes in directional structure (e.g., the northeast–southwest orientation) can be contextualized alongside historical processes of state consolidation and regional integration. For example, governance arrangements and frontier management policies during the Yuan and Ming dynasties (e.g., the Tusi system, the Weiso system, and the Gaitu Guiliu policy) are frequently discussed in relation to strengthening territorial control and administrative reach [44]. During the Qing dynasty, with the capital remaining in Beijing, historical accounts emphasize further consolidation of frontier governance and territorial integration, which provides a broader context for the continued expansion of the distribution area (+108.54% compared to the Yuan–Ming period) and the associated directional adjustments [45].

In the modern era, the centroid migrated northeastward and the ellipse area decreased substantially (−52.06%). This shift can be interpreted in relation to the widely documented acceleration of industrialization and capitalist development in the eastern coastal regions and parts of Northeast China [46], which coincided with intensified urban growth and heritage-related designation and conservation activities in these regions.

4.3. Contextual Factors Related to the Spatial Pattern Formation of Historic and Cultural Districts in China

This section provides an exploratory, descriptive interpretation of the observed spatial patterns by examining their contextual associations with natural and historical–cultural conditions. Because the analyses are based on spatial overlays and descriptive statistics rather than explanatory causal modeling, we avoid causal language and interpret the findings as associations. We selected these factors because they represent core natural constraints and historical–cultural processes repeatedly highlighted in the literature on Chinese settlement evolution and heritage formation, and they can be examined at the national scale with available data. Accordingly, we consider topography/landforms and river systems (DEM and river-proximity overlays) together with capital construction, economic development, and population migration (historical records) as contextual lenses for interpreting the observed spatial patterns.

4.3.1. Natural Factors

(1) Topography and Landforms

This study used ArcGIS 10.8 to overlay the point locations of historic and cultural districts with a digital elevation model (DEM), in order to characterize their topographic context (elevation-based distribution) across China (Figure 6). Based on the elevation standards jointly set by the International Organization for Standardization (ISO) and the International Civil Aviation Organization (ICAO), and according to the study's needs, the historic and cultural districts were categorized into three elevation zones: below 200 m, 200–500 m, and above 500 m.

The results indicate that 70.38% of historic and cultural districts are located in plains (below 200 m, 853 districts), 16.99% in hilly areas (200–500 m, 206 districts), and 12.63% in mountainous regions (above 500 m, 153 districts) (Table 1). The majority of historic and cultural districts are concentrated in plains. Plains are widely recognized as areas with favorable geographic conditions for human habitation and development, and ancient cities were commonly established in these settings, where historic and cultural districts are now frequently observed. These regions are also typically characterized by relatively mild

https://doi.org/10.3390/land15020239

climatic conditions and long-standing commercial activity, which form an important contextual background for interpreting the current spatial pattern. In addition, contemporary urban development and heritage designation practices are more commonly observed in low-altitude and highly accessible areas, which may help explain the continued concentration of historic and cultural districts in plains.

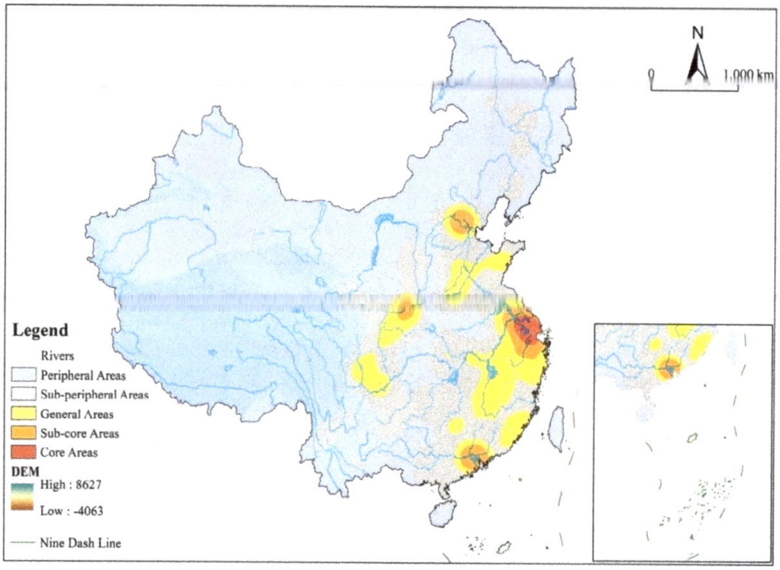

Figure 6. Geographical spatial distribution and hydrological characteristics of Chinese historical and cultural districts.

Table 1. Elevation- and landform-based distribution of historic and cultural districts in China.

Elevation	Landform	Number of Districts	Percentage (%)
Below 200 m	Predominantly plains	853	70.38
200–500 m	Hilly areas, including Liaodong, Shandong, and Jiangnan hills	206	16.99
Above 500 m	Mainly mountainous and plateau regions on the third step	153	12.63

(2) River Systems

China's major river systems, from north to south, include the Songhua River, Liao River, Hai River, Yellow River, Huai River, Yangtze River, and Pearl River basins [47]. River systems were included in this study because major rivers and their tributaries have historically structured settlement corridors, transport and trade connectivity, agricultural productivity, and the location of early cities in China; therefore, proximity to rivers provides a relevant natural-context lens for interpreting the observed spatial pattern of designated historic and cultural districts. Using ArcGIS 10.8, this study mapped the distribution of these rivers and employed buffer analysis to explore the spatial relationship between historic and cultural districts and river systems, thereby quantifying the proximity of these districts to water bodies more accurately. As shown in Table 2, a significant proportion of historic and cultural districts are clustered within 5000 m of a river, with 73.02% of these districts located within this range, and only 26.92% situated beyond 5000 m from a river. This highlights the critical role that rivers have played in the development and evolution of historic and cultural districts.

https://doi.org/10.3390/land15020239

The distribution and development of historic and cultural districts are closely linked to rivers, reflecting economic, social, and cultural connections. Economically, rivers have often been key drivers of regional development [48]. In ancient times, rivers provided more convenient transportation routes compared to overland routes, facilitating the flow of goods, people, and cultural exchange. This ease of transportation supported the prosperity of commerce and trade, ultimately laying the economic foundation for the development of historic and cultural districts and ensuring robust transport links. Additionally, the fertile lands along rivers were conducive to agricultural production, supporting the formation and growth of settlements, and providing a population base essential for the establishment of historic and cultural districts.

Table 2. Distance of Historic and Cultural Districts from Rivers in China.

Distance from River (m)	Number of Districts	Percentage (%)
0–500	657	19.06
500–1000	172	14.19
1000–5000	482	39.77
5000–10,000	197	16.25
Above 10,000	130	10.73

(3) Integrated overlay summary

To provide an integrated, zone-based summary of natural-context associations, we further overlaid the KDE-derived spatial zones (core, sub-core, general, sub-peripheral, and peripheral) with the DEM and the distance-to-river layer, and calculated mean elevation and mean distance to rivers for each zone (Table 3).

Table 3. Mean elevation and distance to rivers by KDE zone.

Regional Level	Number of Points	Mean Elevation (m)	Average Distance to River (km)	Spatial Characteristics and Implications
Peripheral areas	62	1287.52	84.06	High elevation, peripheral mountainous areas; far from major rivers.
Sub-peripheral areas	353	416.87	42.74	Transition from mountains to hills; partially distributed along river valleys and basins.
General areas	403	181.06	61.28	Transition from hills to plains; generally, far from rivers.
Sub-core areas	253	82.26	14.36	Low-elevation plains; significantly closer to the river network.
Core areas	141	8.04	8.65	Extremely low elevation; coastal/estuary/core city areas; closest to rivers.

As shown in Table 3, the overlay statistics indicate a clear natural-context gradient from peripheral to core areas. Peripheral and sub-peripheral zones are associated with higher elevations and greater distances to rivers, whereas sub-core and core zones coincide with low-elevation plains and markedly shorter river distances. This pattern is also visually supported by the integrated overlay map (Figure 6), which shows that KDE-derived core zones largely align with major river corridors and low-relief plains. The "general" zone displays a larger mean river distance, likely reflecting within-class heterogeneity and potential outlier influence. Importantly, these findings are interpreted as contextual associations based on descriptive overlays rather than causal determinants.

4.3.2. Historical and Cultural Factors

(1) Capital Construction

https://doi.org/10.3390/land15020239

Consistent with the dynasty-specific centroid migration and directional shifts identified by the SDE analysis (Section 4.2.2), the following synthesis summarizes how changes in political centers may provide historical context for interpreting the observed spatial pattern. Capital construction is included as a historical–cultural contextual factor because the establishment and relocation of major political centers historically concentrated administrative resources, population, and cultural production, thereby shaping long-term urban development trajectories. In this study, it is not treated as a causal determinant, but as a contextual lens to interpret the period-based spatial-statistical results (KDE cores and SDE centroid migration/orientation).

Changes in dynasties and shifts in political centers across Chinese history provide an important historical context for interpreting the spatial distribution of historic and cultural districts. During the Qin Dynasty, following the unification of China under Qin Shi Huang, the capital was established in Xianyang, and the construction of a nationwide road network greatly enhanced territorial connectivity. Within this historical context, the spatial pattern of early urban development and subsequent historic districts can be interpreted in relation to these institutional and infrastructural arrangements [49].

In the Han–Tang period, Chang'an adopted the "li-fang" system, characterized by a planned urban layout that clearly distinguished residential, commercial, and artisanal spaces. This spatial structure represents an important morphological reference in Chinese urban history, within which many characteristics of later historic districts can be understood [16].

During the Song Dynasty, commercial activities became increasingly prominent, and the strict spatial separation between residential and market areas gradually weakened. Historical records describe flourishing night and morning markets in the capital city of Bianjing (modern-day Kaifeng), with commercial activities distributed throughout the city. Such urban phenomena reflect a more open and mixed spatial structure, which provides a useful historical background for understanding the diversity of district forms observed in later periods [29].

In the Yuan Dynasty, the capital was established in Dadu (modern-day Beijing), and the expansion of the empire coincided with intensified ethnic interaction. Historical accounts indicate increasing cultural exchange during this period, and architectural forms in urban areas exhibit diverse stylistic elements, which are commonly reflected in the cultural characteristics of historic districts [37].

During the early Ming Dynasty, the capital shifted from Nanjing to Beijing, and large-scale urban construction activities were documented, particularly in the construction of city walls. The layout of Nanjing's city wall and surrounding districts has been widely discussed as an expression of political, military, and cultural considerations of the time, providing important historical references for interpreting the spatial organization of historic urban areas. The maritime prohibition policy in the Ming Dynasty was also closely associated with functional changes in some coastal cities, as documented in existing historical studies [7].

In the Qing Dynasty, with Beijing remaining the political center, the integration of Manchu and Han cultural traditions became evident in urban spaces. For example, hutongs and siheyuan in Beijing retained traditional Han architectural forms while incorporating Manchu cultural elements, a combination that is widely recognized as a distinctive characteristic of the city's historic districts [50]. These historical references are used to contextualize the SDE-based centroid trajectory and ellipse-orientation changes across periods, rather than to claim a direct causal relationship.

(2) Economic Development

To interpret the KDE-identified core areas and period differences, we summarize key economic contexts documented in the literature that coincide with long-term urban

 https://doi.org/10.3390/land15020239

prosperity in the major concentration regions. Economic development across Chinese history provides an important historical context for interpreting the emergence and characteristics of historic and cultural districts. Periods of sustained agricultural productivity, craft specialization, commercial activity, and urban expansion are frequently discussed in the literature as broader conditions under which distinctive urban spaces and district forms were documented and later preserved as heritage landscapes.

During the Qin and Han periods, historical accounts describe economic policies such as land privatization and currency unification, alongside the diffusion of iron tools, new farming techniques, and water conservancy projects (e.g., the "Jingqu Law"), which are commonly associated with improvements in agricultural production and irrigation conditions. Contemporary descriptions also note the rapid development of handicrafts, particularly iron smelting and silk weaving, and such developments coincided with increasingly differentiated urban functions and spatial organization [32]. In the Tang Dynasty, agricultural and handicraft industries continued to expand, and the Jiangnan region's land development and rice production, together with advances in silk weaving and ceramics, were accompanied by intensified commercial activities and the growth of urban space. The East and West Markets in Chang'an (the Tang capital) functioned as major commercial nodes and reflected the social and cultural characteristics of the time, providing historical references for understanding the district-like urban spaces discussed in existing studies [28].

During the Song–Yuan period, the spread of advanced agricultural practices (e.g., double-cropping rice and wheat), increased cultivation of cash crops, and achievements in porcelain production corresponded with a broader picture of commercial prosperity. The emergence of paper currency ("Jiaozi") and the expansion of overseas trade—often illustrated by the historical prominence of Quanzhou as a major port—are widely interpreted as part of the economic and cultural context in which commercial areas developed distinctive architectural forms and commercial cultures. These characteristics are frequently highlighted as components of later-recognized historic and cultural districts, as exemplified by the markets of Kaifeng and Hangzhou [43].

In the Ming–Qing period, historical narratives emphasize the diffusion of high-yield crops (e.g., double-cropping rice, corn, and sweet potatoes), the expansion of private handicrafts, and advances in porcelain craftsmanship (such as blue-and-white and polychrome porcelains), alongside long-distance trade and the activities of merchant guilds. Urban prosperity extended to small and medium-sized cities and towns [38]. Guild halls, shops, and residential areas established by merchant guilds are documented as important elements of urban space with distinctive regional cultural features, and such built environments provide a historical basis for interpreting the preservation of valuable urban layouts, streets, and buildings that are now represented within historic and cultural districts. This synthesis is presented as contextual evidence to support interpretation of the KDE cores and regional concentrations, not as statistical proof of causation.

(3) Population Migration

We compiled major, well-documented migration episodes and their dominant directions to provide context for the observed centroid shifts and regional pattern changes across periods. Several well-documented population migrations in Chinese history provide an important historical context for interpreting the spatial evolution of historic and cultural districts, with major movements frequently described as oriented toward the south, the northeast, and the southwest (Figure 7). The Yongjia Rebellion (307–317 AD) in the late Western Jin Dynasty, associated with the War of the Eight Princes and northern incursions, was followed by large-scale southward migration of northern aristocrats and commoners. Historical accounts report that approximately one million people moved into the Jiangnan

 https://doi.org/10.3390/land15020239

region, and existing studies discuss this demographic shift in relation to regional economic and cultural changes that are often considered relevant to later heritage landscapes in the area [51]. The An Lushan Rebellion (755–763 AD) in the mid-Tang Dynasty was also accompanied by substantial southward migration, with estimates of around one million migrants. During this period, the population distribution between north and south became more balanced, and migration flows were concentrated in Hubei, Hunan, the Taihu Basin, and along the Han River—regions that subsequently feature prominently in historical narratives of socio-economic and cultural development [41].

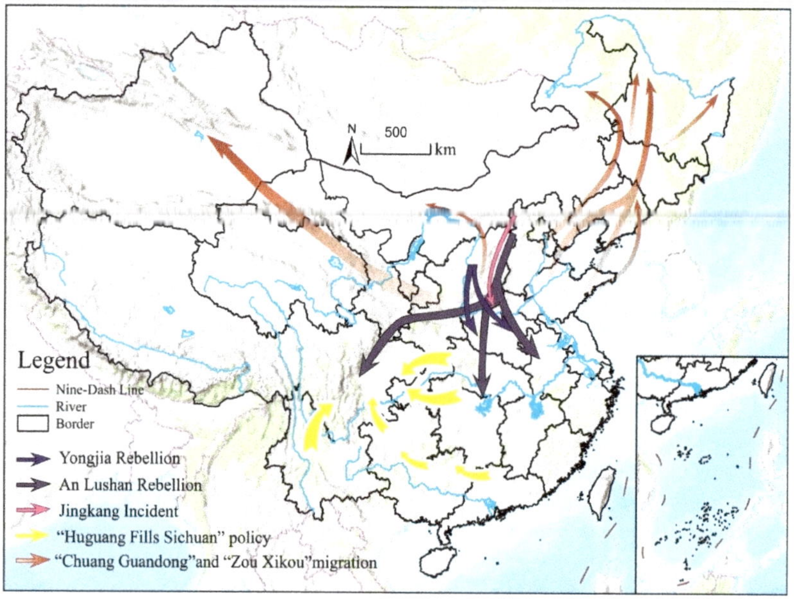

Figure 7. Map of several important historical population migrations in China.

The Jingkang Incident (1126–1127 AD), marking the fall of the Northern Song Dynasty, was followed by the southward relocation of the Song court and large-scale population movement, with historical estimates of about five million migrants mainly to Zhejiang, Jiangsu, Hubei, and Sichuan. Southern Jiangsu and Zhejiang, in particular, are frequently described as emerging cultural centers in subsequent periods, and such historical trajectories provide a useful background for interpreting the concentration of historic and cultural districts in the south [41]. In the Qing Dynasty, the policy known as "Huguang Fills Sichuan" (initiated in 1654) encouraged migration from Huguang (modern Hunan and Hubei) to Sichuan; existing records suggest that roughly six million people responded. Studies commonly discuss this migration in relation to population recovery and socio-economic reconstruction in Sichuan, offering contextual explanations for later cultural landscapes in the region [39].

In the mid-19th to early 20th centuries, "Chuang Guandong" involved residents from Shandong, Hebei, and Henan moving to northeast China. The movement is often discussed in connection with push factors such as disasters in the lower Yellow River region and pull factors such as fertile land and resources in the northeast, alongside broader regional development processes [52]. Similarly, the "Zou Xikou" migration—beginning in the late Ming and early Qing dynasties and peaking during the late Qing and early Republican periods—involved movement from Shanxi, Shaanxi, and other provinces to Inner Mongolia, Ningxia, and other northwestern areas. Historical narratives frequently link this migration

https://doi.org/10.3390/land15020239

to the expansion of livestock production, trade, and frontier commerce, which provide an interpretive background for the cultural exchange documented in these regions [53].

Overall, these migration episodes corresponded with notable demographic restructuring and the relocation of skilled artisans and other groups. Rather than implying direct causality, the evidence summarized here indicates that population mobility formed part of broader historical processes—together with institutional change, economic activity, and urban development—that can help interpret the regional differentiation and long-term evolution of historic and cultural districts. These migration contexts are discussed alongside the SDE/KDE outputs to interpret regional differentiation in the mapped patterns.

5. Conclusions and Implications

5.1. Conclusions

5.1.1. Current Spatial Distribution Patterns

Overall, the designated historic and cultural districts in China exhibit a clear macro-scale spatial differentiation, with a strong concentration in the southeast side of the Hu Line. At the national scale, the pattern can be summarized as a multi-core structure, with the most prominent concentration in the Yangtze River Delta and additional secondary concentrations in the Pearl River Delta and the Beijing–Tianjin–Hebei region.

5.1.2. Evolution of Spatial Distribution Patterns

The period-based SDE analysis indicates a consistent directional structure at the national scale and a distinct trajectory of centroid migration across historical periods. The centroid shift follows a "southeast–northwest–northeast" pathway, suggesting that the spatial focus of designated districts has changed over time in ways that correspond to broader historical regional dynamics. These results provide a compact, comparable description of how the distribution pattern evolved at the national scale.

5.1.3. Contextual Associations

The observed spatial patterns can be interpreted in relation to both natural constraints and historical–cultural contexts. Districts are more frequently located in low-elevation plains and tend to occur closer to major river corridors, consistent with long-term settlement suitability and accessibility. Historical–cultural contexts—including shifts in political centers, long-term economic cores, and major population movements—provide complementary lenses for interpreting the regional differentiation and temporal evolution of the spatial pattern. These associations are discussed as contextual interpretations rather than as causal determinants.

5.2. Policy Implications

5.2.1. Recognizing Historic and Cultural Districts as Spatially Clustered Land Units

Historic and cultural districts should be understood not only as isolated heritage sites, but as spatially clustered land units embedded within broader territorial systems. From a land and spatial planning perspective, conservation should move from site-based protection toward overlay-based zoning that can be directly embedded into statutory plans. Specifically, KDE-derived core areas can be translated into heritage cluster overlay zones with clear management objectives (e.g., core protection, buffer, and compatible development sub-zones). In these overlay zones, coordinated land-use controls—such as limits on redevelopment intensity, height and form regulations, and protection of key view corridors—should be implemented to manage cumulative development pressure. In addition, Heritage Impact Assessment (HIA) triggers can be established: major construction, infrastructure expansion, or land-use conversion projects within core and buffer overlays

https://doi.org/10.3390/land15020239

should be required to submit HIA documentation before approval, ensuring that planning decisions address cluster-level integrity rather than treating districts in isolation.

5.2.2. Integrating Natural Geographic Constraints into Long-Term Land-Use Planning

Planning in heritage-dense areas should integrate topographic and river-corridor constraints into long-term land-use planning through risk-management overlay zones. The concentration of districts along major river corridors indicates that rivers have historically structured settlement and cultural development, and therefore must be incorporated into contemporary planning controls. In operational terms, planning interventions in areas with dense historic districts should adopt river-corridor risk management overlays, integrating heritage protection with flood mitigation and ecological safeguards. For example, river-adjacent clusters can be assigned development setback requirements, restrictions on shoreline hardening, and conservation of riparian buffers to reduce exposure to flooding and land degradation. Where redevelopment is unavoidable, risk-sensitive development controls (e.g., ground-floor elevation requirements, drainage capacity upgrades, and restrictions on high-risk land uses) should be introduced as part of local development regulations. Aligning heritage conservation with river basin management can improve long-term sustainability and reduce hazard-driven heritage loss.

5.2.3. Advancing Regionally Differentiated Land-Based Conservation Strategies

Planning strategies should be regionally differentiated by adopting stricter heritage-cluster controls in core areas and preventive, low-intensity protection measures in peripheral areas. This pattern indicates that uniform planning solutions are unlikely to be effective. Instead, spatial analysis results should guide differentiated planning solutions for central core areas and outer districts. In core regions with high heritage concentration and intensive urbanization, stricter redevelopment controls and zoning overlays are needed, including tighter limits on construction intensity, mandatory HIA triggers for key projects, and active management of land-use conflicts within heritage cluster zones. In contrast, in peripheral regions where districts are fewer but historical environments remain relatively intact, policy priorities should emphasize preventive protection: controlling new land conversion, avoiding large-scale infrastructure fragmentation, and using low-intensity, landscape-compatible development guidelines. Such regionally differentiated strategies allow planning tools to match local heritage density and development pressure, thereby improving the effectiveness of land-based conservation.

5.3. Limitations and Future Research

Despite the contributions of this study, several limitations should be acknowledged. First, spatial uncertainty may exist due to the geocoding process and the use of point-based representations for historic and cultural districts, which may not fully capture their actual spatial extent. In addition, although all spatial analyses were conducted under a projected coordinate system, minor positional deviations and scale effects may still influence the accuracy of distance and density measurements. Second, the designation of historic and cultural districts is subject to administrative selection and policy priorities, which may introduce designation bias and affect the completeness and representativeness of the dataset. Moreover, the use of provincial-level aggregation may obscure finer-scale spatial heterogeneity.

Future research could further improve this framework in two directions. On the one hand, explanatory spatial modeling techniques (e.g., spatial regression or geographically weighted models) could be applied to quantitatively test the determinants of spatial distribution patterns. On the other hand, temporal designation analysis based on multi-period datasets would allow for a more dynamic understanding of how historic and cultural

https://doi.org/10.3390/land15020239

districts emerge, evolve, and decline over time. Such efforts would help strengthen the causal interpretation and deepen the policy relevance of spatial heritage studies.

Author Contributions: Conceptualization, H.Z. and X.L.; Methodology, H.Z., X.L. and X.Z.; Software, H.Z. and X.L.; Validation, H.Z., X.L. and X.Z.; Formal analysis, H.Z., X.L. and X.Z.; Investigation, H.Z. and X.L.; Resources, H.Z. and X.L.; Data curation, H.Z. and X.L.; Writing—original draft, H.Z. and X.L.; Writing—review & editing, H.Z., X.L. and X.Z.; Visualization, H.Z., X.L. and X.Z.; Supervision, H.Z. and X.L.; Project administration, X.L.; Funding acquisition, X.L. All authors have read and agreed to the published version of the manuscript.

Funding: This research was funded by the National Social Science Foundation of China, grant number 22BJY156.

Data Availability Statement: The data presented in this study are available on request from the corresponding author. The data are not publicly available due to the extensive manual processing required for dataset consolidation and institutional intellectual property policies.

Conflicts of Interest: The authors declare no conflict of interest.

References

1. García-Hernández, M.; De la Calle-Vaquero, M.; Yubero, C. Cultural Heritage and Urban Tourism: Historic City Centres under Pressure. *Sustainability* **2017**, *9*, 1346. [CrossRef]
2. Zhao, Y.; Liu, J.; Zheng, Y. Preservation and Renewal: A Study on Visual Evaluation of Urban Historical and Cultural Street Landscape in Quanzhou. *Sustainability* **2022**, *14*, 8775. [CrossRef]
3. UNESCO Recommendation on the Historic Urban Landscape. Available online: https://whc.unesco.org/en/hul/ (accessed on 29 December 2025).
4. Fan, J.; Maliki, N.Z.; Abidin, N.A.Z. Spatial Patterns of Historical and Cultural Blocks Based on Multisource Data and Protection and Development Strategies within the Context of Urban Renewal: A Case Study of Xi'an, China. *Landsc. Res.* **2024**, *50*, 4–22. [CrossRef]
5. Jiao, L.; Wu, Y.; Fang, K.; Liu, X. Typo-Morphological Approaches for Maintaining the Sustainability of Local Traditional Culture: A Case Study of the Damazhan and Xiaomazhan Historical Area in Guangzhou. *Buildings* **2023**, *13*, 2351. [CrossRef]
6. Chen, S.; Meng, B.; Liu, N.; Qi, Z.; Liu, J.; Wang, J. Cultural Perception of the Historical and Cultural Blocks of Beijing Based on Weibo Photos. *Land* **2022**, *11*, 495. [CrossRef]
7. Bai, X.; Jiao, X.; Sakai, T.; Xu, H. Mapping the Past with Historical Geographic Information Systems: Layered Characteristics of the Historic Urban Landscape of Nanjing, China, Since the Ming Dynasty (1368–2024). *Herit. Sci.* **2024**, *12*, 283. [CrossRef]
8. Gale, D.E. The Impacts of Historic District Designation Planning and Policy Implications. *J. Am. Plan. Assoc.* **1991**, *57*, 325–340. [CrossRef]
9. National Park Service. *National Register Bulletin 24: Guidelines for Local Surveys: A Basis for Preservation Planning*; U.S. Department of the Interior, National Park Service, Interagency Resources Division: Washington, DC, USA, 1985.
10. Zhang, F.; Liu, Q.; Zhou, X. Vitality Evaluation of Public Spaces in Historical and Cultural Blocks Based on Multi-Source Data, a Case Study of Suzhou Changmen. *Sustainability* **2022**, *14*, 14040. [CrossRef]
11. Ding, W.; Wei, Q.; Jin, J.; Nie, J.; Zhang, F.; Zhou, X.; Ma, Y. Research on Public Space Micro-Renewal Strategy of Historical and Cultural Blocks in Sanhe Ancient Town under Perception Quantification. *Sustainability* **2023**, *15*, 2790. [CrossRef]
12. Liang, F.; Pan, Y.; Gu, M.; Liu, Y.; Lei, L. Research on the Paths and Strategies of the Integrated Development of Culture and Tourism Industry in Urban Historical Blocks. *Front. Public Health* **2022**, *10*, 1016801. [CrossRef]
13. Chen, F.; Romice, O. Preserving the Cultural Identity of Chinese Cities in Urban Design through a Typomorphological Approach. *Urban Des. Int.* **2009**, *14*, 36–54. [CrossRef]
14. Kuo, S.-L.; Chen, C.-L. Operation Analysis of Cultural Heritage Service Ecosystems: Empirical Study Based on Dihua Street and Guansi Shihdianzih Old Street in Taiwan. *Appl. Syst. Innov.* **2022**, *5*, 42. [CrossRef]
15. Zhou, J.; Zhou, Y.; Li, D. Exploring the Complex Mechanism of the Influence of a Multisensory Tourism Experience on Tourists' Mental Restoration in the Research Context of Historical and Cultural Blocks. *Curr. Issues Tour.* **2024**, *27*, 2682–2702. [CrossRef]
16. Chen, S.; Dong, Y.; Chen, X.; Xu, X.; Gong, J. Analysis of the Characteristics and Ideas of Ancient Urban Land-Use Based on GIS and an Algorithm: A Case Study of Chang'an City in the Sui and Tang Dynasties. *Remote Sens.* **2023**, *15*, 2962. [CrossRef]
17. Mérai, D.; Veldpaus, L.; Pendlebury, J.; Kip, M. The Governance Context for Adaptive Heritage Reuse: A Review and Typology of Fifteen European Countries. *Hist. Environ. Policy Pract.* **2022**, *13*, 526–546. [CrossRef]

https://doi.org/10.3390/land15020239

18. Zhang, Y.; Han, Y. Vitality Evaluation of Historical and Cultural Districts Based on the Values Dimension: Districts in Beijing City, China. *Herit. Sci.* **2022**, *10*, 137. [CrossRef]
19. Cheng, Z.; Zhou, S.; Zhang, B. The Spatial Factors of Cultural Identity: A Case Study of the Courtyards in a Historical Residential Area in Beijing. *Sustainability* **2018**, *10*, 2587. [CrossRef]
20. Isa, M.M.; Ismail, N.A.; Rasam, A.R.A.; Teh, M.Z. Landscape Characterization Using GIS-Based Landscape Character Assessment in Hulu Langat District, Malaysia. *Geogr.-Malays. J. Soc. Space* **2023**, *19*, 17–33. [CrossRef]
21. Fotheringham, A.S.; Wong, D.W.S. The Modifiable Areal Unit Problem in Multivariate Statistical Analysis. *Environ. Plan. A* **1991**, *23*, 1025–1044. [CrossRef]
22. Wong, D.W. Modifiable Areal Unit Problem. *Int. Encycl. Hum. Geogr.* **2009**, *7*, 169–174. [CrossRef]
23. Anselin, L. *Spatial Econometrics: Methods and Models*; Springer Science & Business Media: Dordrecht, The Netherlands, 1988; ISBN 978-90-247-3735-2.
24. Cliff, A.D.; Ord, J.K. *Spatial Processes: Models and Applications*, 1st ed.; Pion Ltd.: London, UK, 1981; ISBN 978-0-85086-081-8.
25. Wand, M.P.; Jones, M.C. *Kernel Smoothing*; Chapman and Hall/CRC: New York, NY, USA, 1994; ISBN 978-0-429-17059-1.
26. Silverman, B.W. *Density Estimation for Statistics and Data Analysis*; Routledge: New York, NY, USA, 2018; ISBN 978-1-315-14091-9.
27. Haining, R. *Spatial Data Analysis: Theory and Practice*; Cambridge University Press: Cambridge, UK, 2003; ISBN 978-0-521-77437-6.
28. Guo, L.; Rong, X. Taking the High Ground: Construction of the Regional Spatial Order of Chang'an Area in Tang Dynasty. *Plan. Perspect.* **2020**, *35*, 115–141. [CrossRef]
29. Tong, Y. "Share the Pleasures with the Public": Evolution of Urban Landscapes in Nanjing in Song Dynasty China (976–1279). *Cult. Soc. Hist.* **2024**, *21*, 1–22. [CrossRef]
30. The People's Republic of China. *The People's Republic of China Fourth National Communication on Climate Change*; Ministry of Ecology and Environment of the People's Republic of China: Beijing, China, 2023.
31. World Bank. World Development Indicators: GDP (Current US$) and Population, Total for China (2023). Available online: https://data.worldbank.org/ (accessed on 28 January 2026).
32. Wright, A.F. The Study of Chinese Civilization. *J. Hist. Ideas* **1960**, *21*, 233–255. [CrossRef]
33. Kuhn, D. *The Age of Confucian Rule: The Song Transformation of China*, 1st Harvard University Press Paperback ed.; History of Imperial China; Belknap Press of Harvard University Press: Cambridge, MA, USA; London, UK, 2011; ISBN 978-0-674-06202-3.
34. Ezimand, K.; Aghighi, H.; Ashourloo, D.; Shakiba, A. The Analysis of the Spatio-Temporal Changes and Prediction of Built-up Lands and Urban Heat Islands Using Multi-Temporal Satellite Imagery. *Sustain. Cities Soc.* **2024**, *103*, 105231. [CrossRef]
35. Wang, J.; Chen, M.; Zhang, H.; Ye, F. Intangible Cultural Heritage in the Yangtze River Basin: Its Spatial Distribution Characteristics and Influencing Factors. *Sustainability* **2023**, *15*, 7960. [CrossRef]
36. Xiong, Y.; Zhang, F. Effect of Human Settlements on Urban Thermal Environment and Factor Analysis Based on Multi-Source Data: A Case Study of Changsha City. *J. Geogr. Sci.* **2021**, *31*, 819–838. [CrossRef]
37. Fei, X. The Formation and Development of the Chinese Nation with Multi-Ethnic Groups. *Int. J. Anthropol. Ethnol.* **2017**, *1*, 1. [CrossRef]
38. Brook, T. *The Confusions of Pleasure: Commerce and Culture in Ming China*; First Paperback Printing; University of California Press: Oakland, CA, USA; London, UK, 2011; ISBN 978-0-520-22154-3.
39. Chen, S. Demography of Qing China. In *Oxford Research Encyclopedia of Asian History*; Oxford University Press: Oxford, UK, 2022; ISBN 978-0-19-027772-7.
40. Kong, X.; Fu, M.; Zhao, X.; Wang, J.; Jiang, P. Ecological Effects of Land-Use Change on Two Sides of the Hu Huanyong Line in China. *Land Use Policy* **2022**, *113*, 105895. [CrossRef]
41. Chen, T.; Kung, J.K. War Shocks, Migration, and Historical Spatial Development in China. *Reg. Sci. Urban Econ.* **2022**, *94*, 103718. [CrossRef]
42. Feng, L.R. *Senses of the City: Perceptions of Hangzhou and Southern Song China, 1127–1279*; The Chinese University of Hong Kong Press: Hong Kong, China, 2017; ISBN 978-962-996-786-4.
43. Li, H. The Song Dynasty's Fiscal and Economic Policy and Its Social Economy. *Soc. Sci. China* **2023**, *44*, 90–114. [CrossRef]
44. Herman, J.E. Empire in the Southwest: Early Qing Reforms to the Native Chieftain System. *J. Asian Stud.* **1997**, *56*, 47–74. [CrossRef]
45. Dai, Y. Qing Military Institutions and Their Effects on Government, Economy, and Society, 1640–1800. *J. Chin. Hist.* **2017**, *1*, 329–352. [CrossRef]
46. Chen, Y. Regional Decline and Structural Changes in Northeast China: An Exploratory Space–Time Approach. *Asia-Pac. J. Reg. Sci.* **2024**, *8*, 397–427. [CrossRef]
47. Yin, S.; Gao, G.; Huang, A.; Li, D.; Ran, L.; Nawaz, M.; Xu, Y.J.; Fu, B. Streamflow and Sediment Load Changes from China's Large Rivers: Quantitative Contributions of Climate and Human Activity Factors. *Sci. Total Environ.* **2023**, *876*, 162758. [CrossRef]
48. Zhu, M.; Shao, L. An Analysis on the Economic Cooperation and the Industrial Synergy of the Main River Region: From the Perspective of the Yangtze River Economic Zone. *J. Ambient Intell. Humaniz. Comput.* **2020**, *11*, 1055–1064. [CrossRef]

49. Liu, Y.; Chen, M.; Tian, Y. Temporal and Spatial Patterns and Influencing Factors of Intangible Cultural Heritage: Ancient Qin-Shu Roads, Western China. *Herit. Sci.* **2022**, *10*, 201. [CrossRef]
50. Huang, B.-X.; Chiou, S.-C.; Li, W.-Y. Study on Courtyard Residence and Cultural Sustainability: Reading Chinese Traditional Siheyuan through Space Syntax. *Sustainability* **2019**, *11*, 1582. [CrossRef]
51. Bi, S.; Du, J.; Tian, Z.; Zhang, Y. Investigating the Spatial Distribution Mechanisms of Traditional Villages from the Human Geography Region: A Case Study of Jiangnan, China. *Ecol. Inform.* **2024**, *81*, 102649. [CrossRef]
52. Reardon-Anderson, J. *Reluctant Pioneers: China's Expansion Northward, 1644–1937 (Studies of the Weatherhead East Asian Institute, Columbia University)*; Stanford University Press: Redwood City, CA, USA, 2005; ISBN 978-0-8047-5167-4.
53. Gibbs, L.S. Going Beyond the Western Pass: Chinese Folk Models of Danger and Abandonment in Songs of Separation. *Mod. China* **2021**, *47*, 178–207. [CrossRef]

https://doi.org/10.3390/land15020239

land

MDPI

Article

Landscape Transformations and Heritage Management in Galicia: Insights from the Ulló Saltworks Complex

Gilberto Duarte Carlos [1,2,*], Ana Lima [1,2] and Javier Piñeiro [3]

[1] CIAUD—Research Centre for Architecture, Urbanism and Design, Lisbon School of Architecture, Universidade de Lisboa, Rua Sá Nogueira, Polo Universitário do Alto da Ajuda, 1349-063 Lisboa, Portugal; analima@upt.pt

[2] CIAUD-UPT—Branch of CIAUD Research Center, Department of Architecture and Multimedia Gallaecia, Portucalense University, Rua Dr. António Bernardino de Almeida, 541, 4200-072 Porto, Portugal

[3] DAMG—Department of Architecture and Multimedia Gallaecia, Portucalense University, 4200-072 Porto, Portugal; j.palvarez.1992@gmail.com

* Correspondence: gilbertocarlos@upt.pt

Abstract

The Galician Rias represent a unique and distinctive geographical entity within the context of the Iberian territory. Characterised by their estuarine geomorphology, resulting from the submersion of ancient fluvial valleys, they hold significant ecological and cultural value. Currently perceived as natural heritage, their utilisation for productive purposes has been a key driver of structured human settlement and economic development in the surrounding region. This article aims to systematise the main historical transformations of the Galician Rias through an interdisciplinary methodology, combining the overlaying of historical and contemporary cartographic sources with the documentary analysis of primary and secondary records. This approach enables the reconstruction and interpretation of landscape transformations, producing a narrative of the evolutionary processes that have shaped these areas. The research seeks to contribute to the formulation, extension, and diversification of protective measures for Galicia's riparian zones, while simultaneously raising awareness of the importance of safeguarding archaeological remains from classical, medieval, and pre-industrial periods. These vestiges are increasingly under threat from profound alterations to their natural and anthropogenic contexts. Additionally, the study highlights the value of preserving this heritage to foster public awareness of the need to protect these fragile landscapes.

Keywords: cultural landscape; cultural heritage; urban landscape; Roman built heritage; pre-industrial heritage; protected natural landscapes; Galician Rias

1. Introduction

1.1. Subject and Objectives of the Research

The region of *Rías Baixas* constitute one of the most distinctive coastal cultural landscapes of Atlantic Europe. Drowned river valleys, terraced slopes and dispersed settlement patterns form a tightly interwoven socio-ecological system shaped by long-term interactions between human activity and a complex maritime environment [1,2]. This hybrid condition, simultaneously rural, littoral and increasingly urbanised, has produced a landscape characterised by a high degree of permeability between land and sea, where vernacular practices and spatial structures reveal enduring cultural continuity [3,4]. These attributes make the region a paradigmatic setting for examining the relationships between heritage, landscape and territorial planning [5].

Within this framework, the research addresses the growing tension between landscape heritage and contemporary spatial-planning instruments. Current European planning policies frequently undervalue areas of archaeological potential, treating them as secondary constraints rather than as primary cultural resources. Broad land-use classifications often obscure local archaeological sensitivities, while environmental assessments tend to privilege ecological indicators over cultural stratigraphy [6,7]. The fragmentation of heritage governance further limits early intervention and weakens predictive and preventive approaches [8]. As a result, many archaeological landscapes remain exposed to irreversible transformation [9].

The objective of this research is to assess how culturally stratified coastal landscapes can be more effectively identified, interpreted and safeguarded within planning and heritage-management frameworks. By focusing on a historically layered site, the study seeks to demonstrate the relevance of cultural landscape approaches in integrating archaeological potential, environmental dynamics and spatial planning.

1.2. The Study Area

The Ulló Saltworks (*Salinas de Ulló*), located on the northern margin of the Vigo Ria, within the municipality of Vilaboa (Pontevedra), constitute a significant case study within the *Rías Baixas* context (Figure 1). They represent a dynamic yet highly vulnerable cultural landscape, shaped by centuries of interaction between tidal processes, productive infrastructures and human settlement [10].

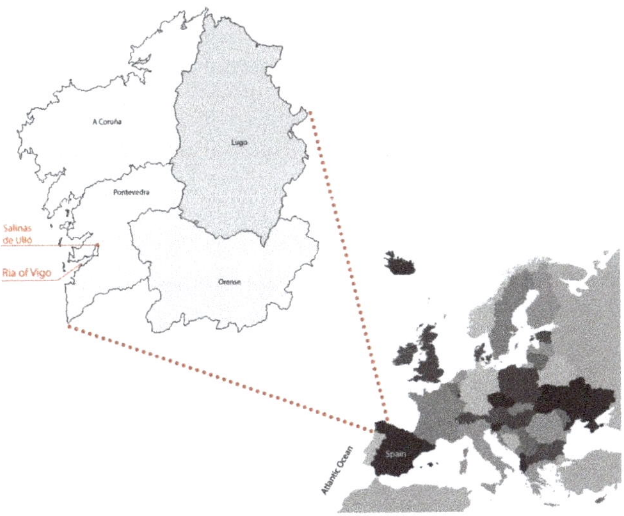

Figure 1. Geographical location of the Ulló saltworks.

Although most documentary references date the development of the saltworks to the seventeenth century, territorial appropriation can be traced back to the Roman period. The density of archaeological remains in the surrounding area, together with the strategic importance of salt production within the Roman economic system, suggests a high probability of earlier occupation [11,12].

The Ulló saltworks exploited tidal forces to extract salt, a vital commodity for food preservation. The complex consists of two contiguous rectangular ponds constructed within a natural inlet at the end of the estuary. In the larger basin, seawater introduced

https://doi.org/10.3390/land15030358

by tidal action underwent partial evaporation through solar exposure and ventilation, remaining stagnant for approximately thirty days until reaching around 18 °C, enabling pre-crystallisation [13]. The brine was then transferred to the adjacent basin, where crystallisation occurred at higher temperatures within orthogonal compartments, facilitating salt harvesting [14]. Recent documentation initiatives have highlighted both the heritage value and the vulnerability of this coastal industrial landscape, underlining the need for integrated conservation and monitoring strategies [15].

Conceptually, the Ulló Saltworks exemplify a cultural landscape understood as the combined work of nature and humankind, reflecting the evolution of human settlement under specific environmental, social and economic conditions [16,17]. Their tangible elements include the salt ponds, tidal-mill remains and associated built structures, while intangible dimensions encompass historical production practices and collective cultural memory. Effective preservation requires coordinated legal, technical and community-based instruments capable of addressing both physical integrity and cultural significance [18,19].

2. Materials and Methods

This study investigates how vulnerable coastal cultural landscapes, such as the Ulló saltworks, can be more coherently integrated into spatial-planning and heritage-management frameworks. It addresses the research question of whether current planning instruments adequately recognise archaeological and cultural stratification in dynamic coastal environments. The analysis supports the hypothesis that cultural landscape-based approaches offer a more effective framework for anticipating risk, strengthening protection and reconciling heritage values with territorial development.

The present research was raised during an exploratory academic essay, focussing the Ulló Saltworks heritage value, which provided the conceptual and analytical basis for its prerogatives [13]. The cited study, following a descriptive approach, provided a detailed characterisation of the principal stages in the territorial development of the site. The present article highlights one of its key long-term ambitions: To enhance the development of the technical and conceptual basis required to critically reassess and reformulate the legislative framework governing the site's heritage protection. Accordingly, the article is organised around the integration of two core components:

1. A systematic exposition of the territorial evolution of the Ulló saltpans, with particular emphasis on their historical and natural specificities, thereby highlighting the attributes that underpin their heritage significance.
2. A comparative analysis of the current legal framework, aimed at identifying potential inconsistencies, limitations, and opportunities for its critical reformulation. The identification of landscape alterations involves analysing physical, ecological, and cultural changes over time. This process often employs a combination of historical cartographic analysis, remote sensing, and field surveys to trace shifts in land use, geomorphological features, and vegetation cover [20]. Key factors include natural processes such as erosion, sediment deposition, and sea-level fluctuations, as well as anthropogenic influences like urbanisation, industrialisation, and infrastructure development. The goal is to understand how these changes reflect broader socio-environmental dynamics and their implications for regional ecosystems, heritage, and human activity [17].

Systematising the evolutionary process requires integrating historical, ecological, and archaeological data into a coherent narrative that traces the transformation of landscapes through time [21]. This involves identifying major phases of change, linking them to driving factors, natural or human-induced, and situating them within a broader context of historical development. Such systematisation provides critical insights into the interaction

 https://doi.org/10.3390/land15030358

between human societies and their environments, highlighting patterns of adaptation, resilience, and impact. It is a foundational step for informed management and conservation strategies, as well as for fostering public awareness of landscape heritage [22].

In order to pursue these aims, and considering the nature of the collected data, the research team resorted to methos of documental analysis, dividing the process in two major segments.

For the first segmented, oriented to geographical characterization, the research team resort to the analysis of historical cartography, cadastral register, aerial orthophotos and actual satellite images. Using digital superimposing techniques, the research team have created speculative mappings of the major territorial manifestations over the corresponding time span [23]. The correlation between them is corroborated with regional archaeological reports and historic description. The attempted georeferenced accuracy enables to configurate a visual landscape simulation, providing valid insights to deeper and extensive territorial relations, wither natural or artificial. The usage of historical cartography reveal itself full of challenges (Figure 2). If, by one side the medieval techniques did not permit a precise spatial configuration, the symbolic depiction presented very significant information regarding the cultural relevance of the represented events [24]. The scale and nature of the surveyed object, the Ria of Vigo, admitted a significative error tolerance, in order to achieve an overall interpretation of the region and of the legislative figure.

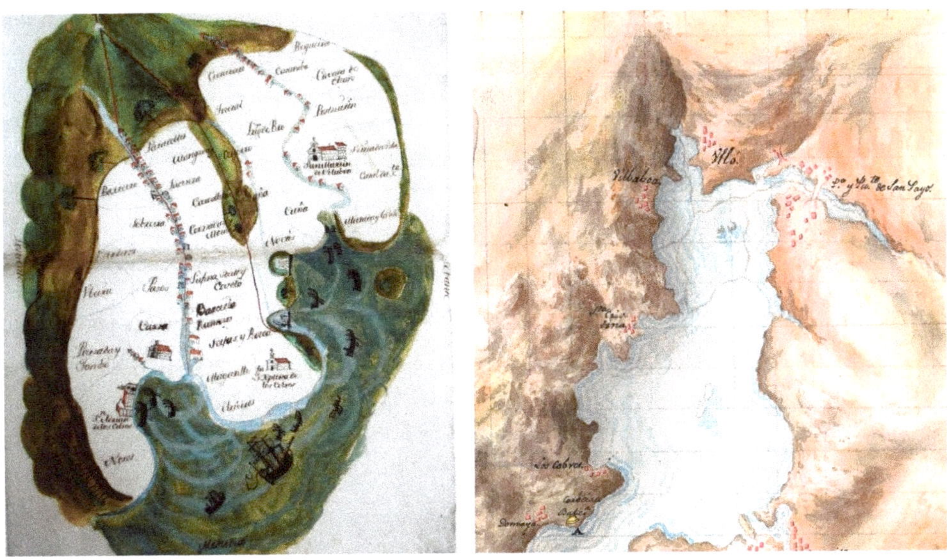

Figure 2. Examples of early historical cartographic sources depicting the study area, based on eighteenth-century maps dated 1752 and 1788.

In the second research segment, documental analysis was applied to all active Spanish Heritage laws applicable to such scale. Textual analysis was conducted to compare and differentiate legal frameworks on the protection of cultural landscapes [25]. From the Spanish heritage legislation, 8 laws were considered, from which 3 are national and 5 are regional, covering from 1985 to 2021. Except 1, all of them relate to historic, cultural and natural value. The exception addresses the Galician territorial management, and its transversal to all type of urban intervention. The legislative framework was systematised according to the primary values addressed by each regulation, distinguishing between

https://doi.org/10.3390/land15030358

cultural heritage protection, natural heritage conservation, landscape protection, and territorial planning instruments (Table 1).

Table 1. Primary legislative framework, 1985–2021 (historical and currently applicable).

Law Mentioned in the Text (UK English)	Scope	Main Protection Focus
Law 16/1985 on Spanish Historical Heritage	National	Historical and cultural heritage
Law 8/1995 on the Cultural Heritage of Galicia	Regional	Cultural heritage
Law 3/1996 on the Pilgrims' Routes to Santiago (Caminos de Santiago)	Regional	Cultural landscape/cultural routes
Law 42/2007 on Natural Heritage and Biodiversity	National	Natural heritage and biodiversity
Law 7/2008 on the Protection of the Landscape of Galicia	Regional	Landscape protection
Law 33/2015 amending Law 42/2007 on Natural Heritage and Biodiversity	National	Natural heritage (amendment)
Law 5/2019 on the Natural Heritage and Biodiversity of Galicia	Regional	Natural heritage and biodiversity
Law 1/2021 on Territorial Planning in Galicia	Regional	Land-use and territorial planning

3. Identification of Structural Transformations and Systematization of the Evolution of the Ulló Saltworks

Within the framework of the Galician Rias, the *Ría de Vigo* stands out as one of the most extensively transformed landscapes due to human intervention. Urbanisation has progressively extended along the entire coastal margin, from the mouth of the Verdugo River at the head of the estuary to the Cíes Islands, which enclose and protect the estuary from the Atlantic Ocean [26].

From a geomorphological perspective, the origins of this landscape date back to the end of the last glacial period, when the subsequent flooding of the fluvial valley generated a landscape enclosed between multiple mountain ranges to the north and south [27].

Anthropogenic alterations have occurred continuously from prehistoric times to the present, driven by both productive activities and the expansion of settlements and built infrastructure [28]. Archaeological studies indicate that early human activity shaped the local environment, with specific sites documented in the PXOM of Pontevedra plans. These findings offer insights into the origins of settlement patterns and the long-term relationship between humans and the coastal landscape [29].

The main categories of human-induced landscape transformation are outlined below:

(a) Coastal urbanisation: Archaeological evidence indicates early human presence along the Ría de Vigo, with evidences of occupation dating back to 5000 BC [30]. These have gradually contributed to the structuring of the coastal environment.

(b) Infrastructure development: The earliest major infrastructures were communication routes, established from the Roman period onwards [31]. Over time, maritime infill and port constructions were added to facilitate access to the sea.

(c) Productive and extractive activities: The exploitation of natural resources, including the Ulló saltworks, has been a key driver of landscape modification.

(d) Deforestation and agricultural expansion: Cycles of forest clearance, agricultural use, and subsequent abandonment have altered the local terrain, increasing sediment runoff into the sea and contributing to the formation of estuaries and saltmarshes.

The evolutionary trajectory of the *Ría de Vigo* can thus be structured into distinct phases, reflecting the interplay between natural processes and human interventions (Table 2). This

https://doi.org/10.3390/land15030358

systematisation not only documents the transformative processes but also identifies critical moments in which human-environment interactions reshaped the landscape [32]. Understanding these phases provides essential guidance for policymakers and conservationists seeking to balance the preservation of the Vigo Ria's natural and cultural heritage with ongoing socio-economic development [30].

Table 2. Chronology of Landscape Transformations at the Ulló Saltworks.

Stage	Period	Key Intervention/Event	Landscape Impact
1	before 1694	Pre-Anthropogenic State	Untouched estuarine landscape, baseline morphology
2	1694–1695	Initial earth embankment	Hydraulic manipulation, brine retention begins
3	1695–1727	Northern and western embankments; active salt exploitation	Establishment of functional saltworks, intensive landscape modification
4	1727–1870	Abandonment; conversion to As Casas farm	Partial integration into agricultural landscape; reduced hydraulic use
5	1870–1990	Southern dyke, tidal mill and territorial reconfiguration	Hybridisation of hydraulic and mechanical systems; renewed landscape activity
6	1990–2025	Contemporary infrastructure and interventions	Stabilisation of structures; heritage valorisation
7	2025–2100	Climate change scenario	Projected sea-level rise and tidal impacts; potential vulnerability of saltworks

At a local scale, the Ulló Saltworks have undergone a series of landscape transformations throughout their history, forming an additive process, yet whose ultimate trajectory appears to be a persistent reversion towards the natural state. The dynamic interplay between natural tidal regimes, human engineering, and socio-economic utilisation has produced a cultural landscape of remarkable complexity, where structural interventions have intermittently modified the estuarine environment while leaving traces of the original morphology. These successive modifications reflect both adaptive strategies to environmental conditions and evolving technological and economic imperatives in salt production.

The earliest identifiable stage corresponds to the natural state of the site during the early Middle Ages, prior to significant anthropogenic interventions, which persisted until the sixteenth century. The first major alteration occurred with the construction of the initial earth embankment in 1694–1695, signalling a deliberate manipulation of hydrological flows to facilitate controlled brine retention [33]. Subsequently, the construction of the northern and western embankments, alongside active salt exploitation between 1695 and 1727, established the functional core of the saltworks, marking a period of intensive landscape transformation and economic activity [34].

Following this productive phase, the site experienced abandonment and partial repurposing as the As Casas farm from 1727 to 1880, during which the previous hydraulic structures were gradually integrated into agricultural uses. Later interventions included the construction of the southern embankment and the installation of a tidal mill (1880–1980), which exemplified the continued hybridisation of natural and engineered elements within the estuarine landscape. Aerial surveys conducted in 1956 and 1980 provided valuable documentation of these changes, supporting subsequent conservation and reinterpretation efforts.

From 1980 onwards, contemporary infrastructures and restorative interventions sought to stabilise the saltworks while preserving both its historical and environmental values. Looking forward, the projected impacts of climate change on sea level and tidal

https://doi.org/10.3390/land15030358

dynamics (2025–2100) present new challenges for the preservation of this highly vulnerable coastal cultural landscape. Understanding these layered transformations provides a critical framework for managing heritage landscapes that are simultaneously ecological, economic, and cultural in nature.

3.1. Pre-Anthropogenic State (Before 1694)

The natural configuration of the landscape corresponds to the estuarine convergence of the *Villil* and *Tuimil* waterstreams, directly associated with the coastal dynamics of the area and the formation of potential sandbanks [35]. The extent of saltmarsh vegetation was likely considerably more restricted than at present, confined to the initial metres of the coastal fringe and shaped by tidal fluctuations and sedimentary processes typical of low-energy estuarine environments [36]. During the Roman period, *Vía Romana XIX* is documented nearby, running north of the area under study [37]. Recently several Roman settlements have been identified in the vicinity, reinforcing the economic relevance of the local connectivity between *Tude* (Tui) and *Turoqua* (Pontevedra), thereby situating the Ria landscape of Galicia within broader territorial networks of mobility and exchange [38].

However, the earliest written reference to the Ulló Saltworks derives from the transfer of land to the Jesuit Order of Pontevedra, offered as compensation for the construction of a college in the city. This institutional initiative followed the 1655 testament of Licenciado Jorge de Andrade, a priest residing in *Ciudad de Los Reyes* (Peru). In response to this bequest, the family of Don Antonio Mosquera Villar y Pimentel, Knight of the Order of Alcántara and General Administrator, on behalf of the Crown, of the Saltworks of the Kingdom of Galicia, and his wife, Doña Antonia Pimentel de Sotomayor, contributed the lands of Ulló, identified as the *zonas del Ulló* within the parish (*Feligresía*) of *San Martín de Vilaboa*. These lands were ceded explicitly for the establishment and operation of salt-production facilities [39].

Documentary records provide the only known evidence of human occupation prior to the seventeenth century, citing a defensive watchtower located on the northern elevation, *Monte Raxado*, and a fortification to the southeast on *A Cerca*, which controlled access to the bay. The presence of these structures is further supported by toponymic traces, indicating a strategic use of elevated points to oversee maritime circulation and safeguard the inlet [40].

On 14 July 1694, the rector of the Jesuit College of Pontevedra formally assumed possession of the property, delineating its limits by means of a sequence of access points extending from *Porta a Muíños* to *Punta de Agüeiros* [39]. This act of territorial definition constitutes the first explicit demarcation of the site's administrative boundaries.

The characterisation of the pre-seventeenth-century landscape reveals a territory shaped primarily by natural estuarine dynamics, with limited and strategically oriented human occupation. The convergence of hydrological systems, early mobility infrastructures and incipient defensive structures establishes a foundational geography upon which subsequent phases of landscape transformation, particularly those associated with salt production, would later be superimposed.

3.2. Initial Earth Embankment (1694–1695)

Existing documentation indicates that the first constructed elements in the area comprised the western access dyke on land, together with a set of small, isolated buildings located on the site where the farmstead would later be established [41]. These structures were complemented by a perimeter enclosure built in stone, marking an initial and deliberate act of territorial consolidation. Such interventions represent the earliest known attempts to stabilise the terrain, regulate water flows and create the basic infrastructural

https://doi.org/10.3390/land15030358

framework upon which subsequent productive activities, particularly those associated with salt extraction, would be developed.

By this period, the southern fortification at *A Cerca* had already fallen into disuse. Although the precise history of the structure remains uncertain, several accounts record its occupation by groups engaged in acts of pillage, who took advantage of its strategic position near *Vía XIX* and the surrounding maritime routes to conduct raids and other forms of illicit activity. This pattern of opportunistic appropriation reflects the broader vulnerabilities of sparsely controlled coastal landscapes during the transition from the late medieval to the early modern period [42].

In synthesis, this stage marks the initial shift from a predominantly natural landscape to one undergoing intentional human modification. The construction of hydraulic and architectural elements, combined with the progressive abandonment of earlier defensive structures, illustrates a reorientation of territorial functions: from surveillance and protection towards productive exploitation and managed occupation. This juncture therefore establishes the structural foundations for the organised anthropogenic transformation that would define the subsequent evolution of Ulló.

3.3. *Northern and Western Embankments and Salt Exploitation (1695–1727)*

The period between 1695 and 1727 corresponds to the most intensive phase of anthropogenic reconfiguration of the landscape. In 1695, the principal stone dyke was constructed, forming the northern boundary of the saltworks known as *San Ignacio de la Cruz*. This structure was anchored at both ends to the adjacent hills, thereby creating a stable and continuous barrier capable of controlling tidal inflow and regulating saline water levels across the newly delimited production area.

A second stone dyke, built between 1694 and 1695, subsequently enclosed the saltworks of *San José*. The completion of this hydraulic infrastructure marked a significant technical milestone, enabling systematic management of saline waters and providing the necessary conditions for large-scale salt production, following the full operational consolidation of the dyke system, until 1727 [13].

At the extremities of the hydraulic complex, two *alfolíes* (salt warehouses) were erected. The northern warehouse, considered the earliest, was constructed in stone, while its southern counterpart was built in timber. These buildings functioned as essential components of the production chain, supporting storage, processing and the organisation of labour associated with salt extraction [42].

This phase represents the culmination of deliberate landscape engineering, characterised by the construction of substantial hydraulic and architectural systems that redefined the ecological and spatial dynamics of Ulló. The establishment of dykes, the formal enclosure of productive basins and the erection of specialised storage facilities collectively signify the transition to a fully anthropised landscape, where environmental processes were subordinated to the operational demands of a strategically planned salt-production complex.

3.4. *Abandonment and Conversion to as Casas Farm (1727–1870)*

This stage extends from 1727 until the construction of a tidal mill on the southern dyke at the end of the nineteenth century. During this period, the ownership of both the saltworks and the *granja* underwent significant change, passing into the hands of the Arana family, who were responsible for erecting the principal structures that remain today. This reorganisation of property holdings resulted in the fragmentation of the former productive unit, thereby undermining the integrated operational capacity that had previously defined the site. As a consequence, the landscape's productive system experienced a progressive and irreversible decline [39].

https://doi.org/10.3390/land15030358

With the loss of unified management, the saltworks ceased to function as a mining–industrial landscape and were gradually repurposed for agricultural and pastoral uses. This shift entailed a fundamental reorientation of land use, replacing the precise hydrological and infrastructural regulation required for salt extraction with activities that imposed significantly lower technical demands on the territory. The residual hydraulic structures, increasingly neglected, deteriorated over time, accelerating the functional obsolescence of the former salt-production complex.

In summary, this period marks the transition from a highly specialised productive landscape to a predominantly agrarian one, characterised by the dilution of its former industrial coherence and the gradual degradation of its hydraulic infrastructure. The change in ownership, coupled with the abandonment of salt extraction, initiated a long-term process of landscape decay that would redefine the spatial and functional identity of *Ulló* throughout the eighteenth and nineteenth centuries.

3.5. Southern Dyke, Tidal Mill and Territorial Reconfiguration (1870–1990)

The final productive transformation of the Ulló landscape is associated with the southern dyke and the tidal mill, both attributed to the French engineer Felipe Auguste Cazaux. His name is reflected in the current toponym of the site, *bancas de Casó*. Cazaux was responsible for the construction of the Madrid viaduct in Redondela in 1876 and the international bridge at Tui in 1885, and he owned the house at *Punta do Carregal*, adjacent to the tidal mill [43].

The mill operated for approximately half a century, from around 1870 until 1930. According to Reinoso Maset [42], its mechanism also supplied power to a *fábrica de ropa de aguas* (water-based cloth factory) and, at times, even generated electricity.

Despite the evidences of the dyke or the mill impact, it is important to note the scarcity of information regarding these elements [13]. In particular, it remains unknown whether the two waterwheels located at the eastern end of the dyke were covered, or whether the cloth factory possessed additional infrastructural elements to support its operations [44].

In the 1956 American aerial survey, which is the first record of this type for the region [45], a parcelled landscape can be distinguished, corresponding to agricultural use, within a territory showing limited industrial transformation and minimal infrastructure development. The most prominent feature is the railway line between Arcade and Pontevedra, located to the east.

The sparse vegetation also allows the location of the *As Casas* Farm to be inferred, appearing as a clearing within the woodland; even the shadows reveal the differences in elevation between this area and the saltmarsh to the north. The aerial photograph from the 1980s, the Interministerial Survey, further illustrates the extent of abandonment, with visible signs of vegetation encroachment, the dissolution of the original parcel structure, and the gradual increase in dispersed built-up areas.

Moreover, the same aerial survey captures the construction works of the Autopista del Atlántico (AP-9), situated 600 m from the site, reflecting the significant economic activity taking place in Galicia during that decade.

Taken together, these developments illustrate the final stage in the long process of landscape transformation at Ulló, in which the remnants of nineteenth-century hydraulic and industrial ingenuity coexist with clear signs of twentieth-century decline and territorial reconfiguration. The limited documentation on the tidal mill and southern dyke, combined with the progressive agricultural repurposing, infrastructural encroachment and subsequent abandonment visible in mid- and late-century aerial imagery, reveals a landscape in transition: one where former productive logic has been gradually eroded, giving way to new spatial dynamics shaped by infrastructural modernisation, vegetation succession and diffuse patterns of settlement [30].

https://doi.org/10.3390/land15030358

In summary, this phase represents the last significant productive reconfiguration of the *Ulló* landscape, combining hydraulic engineering and industrial function within a singular, technologically advanced intervention. The integration of the tidal mill and the southern dyke demonstrates a shift toward mechanised exploitation, while the lack of detailed documentation highlights the need for further research to fully understand the operational and architectural characteristics of these late-nineteenth and early-twentieth-century infrastructures.

3.6. Contemporary Infrastructure and Interventions (1990–2025)

The most recent intervention documented within the study area concerns the creation of a new seafront promenade to the east of the largest saltmarsh, executed by the National Service of Coasts [46]. The 2002 aerial imagery reveals the opening of a compacted-earth track extending from the existing road located on the northern dike towards the *Casó* estate, continuing along the coastal edge of *A Cerca*.

Between 2014 and 2017, the trees growing on the central masonry dike were removed. Their root systems had begun to compromise the structural integrity of the dike, accelerating its deterioration and posing a risk to its long-term stability.

In the final decades of the twentieth century, the most significant landscape transformations were driven both by the introduction of major infrastructures—such as the Atlantic Motorway (AP-9)—and by the consolidation of *O Toural* as the central settlement of the municipality of Vilaboa, accompanied by the dispersed construction of new housing throughout the surrounding area.

The productive sector also shaped the evolving landscape, exemplified by the canning industry, for which the *Muelle Dos Caralletes* was built, clearly visible from the tidal mill.

In the twenty-first century, landscape transformation has continued apace with the implementation of major transport infrastructures, including the high-speed rail line connecting Vigo and Pontevedra, as well as the construction and opening of the new A-57.

Overall, the study area has undergone sustained and multifaceted transformation, driven by the expansion of transport infrastructure, the consolidation of residential nuclei, and sector-specific industrial developments. These interventions, both protective and developmental, have collectively reshaped the coastal and peri-urban landscape, reinforcing the constant interplay between environmental conservation and territorial modernisation.

3.7. Climate-Change Scenario (2025–2100)

Although the principal anthropogenic transformations of the twenty-first century have been the development of major infrastructures and the consolidation of settlement nuclei such as *O Toural* and Arcade, one of the most profound and already tangible drivers of change in the area is climate change.

A clear example of its effects occurred during the spring tides of 19 September 2024. At their peak, water levels exceeded 4 m at multiple points along the Galician coast, including the Vigo–Pontevedra estuarine sector ($\approx$4.2 m), overtopping the Ulló dikes and recreating a landscape reminiscent of the pre-anthropogenic configuration of the site.

Looking ahead, projected sea-level rise and increasing average temperatures suggest that existing heritage assets may experience accelerated deterioration, surpassing current rates of degradation.

In this context, climate change has emerged as a transformative force of equal or greater significance than infrastructural development in the study area. Recent extreme tidal events highlight the increasing exposure of coastal systems, while future climatic trends are expected to intensify the vulnerability of both natural and cultural heritage [7].

https://doi.org/10.3390/land15030358

The third frame of Figure 3 presents a cartographic simulation for the year 2100, conceived as a scenario-based interpretative map rather than a deterministic forecast. The simulation integrates official municipal and regional cartographic datasets, in DWG and GIS format, with projected sea-level thresholds derived from MeteoGalicia reports and land-use dynamics defined in the PXOM, establishing a coherent long-term spatial framework. The resulting representation combines climatic drivers with social and territorial trends, following a methodological approach consistent with previous historical and prospective landscape reconstructions.

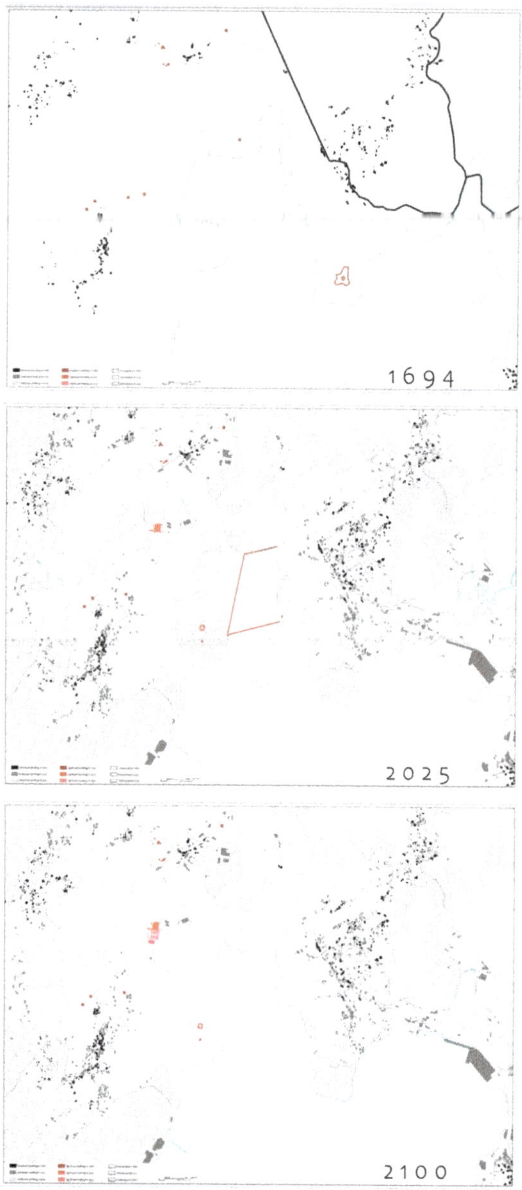

Figure 3. Cartography simulations interpreting the territorial transformation operated between major chronologic parameters (1694–2025–2100).

https://doi.org/10.3390/land15030358

Within this framework, several anticipated transformations are depicted. A coastal path currently under planning to the south of A Granja, along Monte Rañado, is represented as a potential new linear infrastructure. In parallel, the progressive abandonment of agricultural land is assumed, leading to the expansion of eucalyptus woodland around the perimeter of the salt pans, with associated ecological degradation. Finally, in accordance with the PXOM [29], the settlement of O Toural is shown expanding both northwards and southwards, reinforcing existing patterns of peri-urban growth.

4. Cultural Landscape Protection in Galicia

The protection of the cultural landscape in Galicia is supported by a comprehensive legal framework, encompassing international agreements, as well as national and regional legislation. At the international level, the European Landscape Convention [47] out as a key instrument promoting the protection, management, and integrated planning of landscapes. This perspective is further reinforced by UNESCO's approach to cultural and natural landscapes, particularly through the concept of Biosphere Reserves as learning sites for sustainable development, where the interaction between cultural practices, natural systems, and local communities is explicitly recognised [48].

At the national level, the Spanish Historical Heritage Act [49] already foresaw the protection of culturally significant sites, defining categories such as historic sites associated with historical events, traditions, or past cultural creations. This cultural approach is complemented by environmental legislation, notably the Spanish Natural Heritage and Biodiversity Act [50], later amended by Law 33/2015 [51], which introduces an integrated vision of natural and cultural heritage, emphasizing landscape as a key element of territorial cohesion, sustainability, and environmental quality.

In Galicia, this legal framework is reinforced by the Galician Cultural Heritage Act, which serves as the cornerstone of regional legislation on this matter [52]. This law is complemented by specific regulations, such as Law 3/1996 on the Camino de Santiago, which regulates the cultural and landscape protection of this historic route [53], and the Galician Landscape Protection Act, which introduces a cross-cutting approach to landscape management [54]. These regulatory provisions are further supplemented by environmental and urban planning laws at the regional level, including nature conservation and territorial planning regulations that integrate landscape-related criteria [55].

In recent decades, significant legislative changes have taken place. Until the 1990s, cultural landscape protection was primarily addressed through traditional heritage categories—monuments, historic ensembles and historic sites—and through specific measures targeting particular environments, including early national protection decrees affecting the Camino de Santiago and the subsequent Galician Act of 1996 safeguarding this historic route [53]. Since the 2000s, international influence, especially UNESCO's recognition of cultural landscapes and the implementation of the European Landscape Convention, has encouraged a broader approach to territorial protection [56].

In response to this evolution, Galicia enacted a specific landscape law in 2008, integrating cultural, environmental, and urban dimensions into territorial planning [54]. Simultaneously, heritage laws in various Spanish regions, including Galicia, began incorporating new protection categories—such as cultural landscapes, historic routes, and cultural parks, recognising environments shaped by the interaction between nature and human history [57]. This legislative shift reflects the transition from the isolated protection of monuments to a more holistic approach to landscape as heritage.

https://doi.org/10.3390/land15030358

4.1. Integration of Green Infrastructure into Territorial Planning

Alongside this transformation in landscape management, territorial and environmental planning in Spain has increasingly incorporated approaches aimed at ensuring ecological connectivity and ecosystem sustainability. The approval of the Natural Heritage and Biodiversity Act (Law 42/2007) marked a milestone in integrating the concept of green infrastructure into national legislation [50]. The 2015 amendment further strengthened this principle [51] by establishing the obligation to develop a National Green Infrastructure, Connectivity, and Ecological Restoration Strategy (ENIVCRE).

This strategy, aimed at consolidating a network of functional natural and semi-natural spaces by 2050, is coordinated with regional governments through the State Commission for Natural Heritage and Biodiversity, ensuring a structured planning approach that integrates cultural landscape protection with ecological restoration and territorial connectivity [58].

Within this framework, Galicia has developed its own regulatory approach through the Natural Heritage and Biodiversity Act [55] and the Territorial Planning Act [59], aligning regional policies with national and European guidelines on conservation and ecological restoration. Law 5/2019 establishes, in Chapter VIII, the obligation to develop a Galician Green Infrastructure, Connectivity, and Ecological Restoration Strategy, ensuring coordination across administrative levels and coherence in the management of ecological corridors and key landscape elements.

The Territorial Planning Act reinforces this regulatory framework by defining green infrastructure as a fundamental territorial system, ensuring its integration into spatial-planning instruments as presented in Figure 4 [59]. This legislation highlights the importance of preserving ecological connectivity and ecosystem services in both urban and rural environments, promoting sustainable territorial planning that prevents fragmentation and environmental degradation.

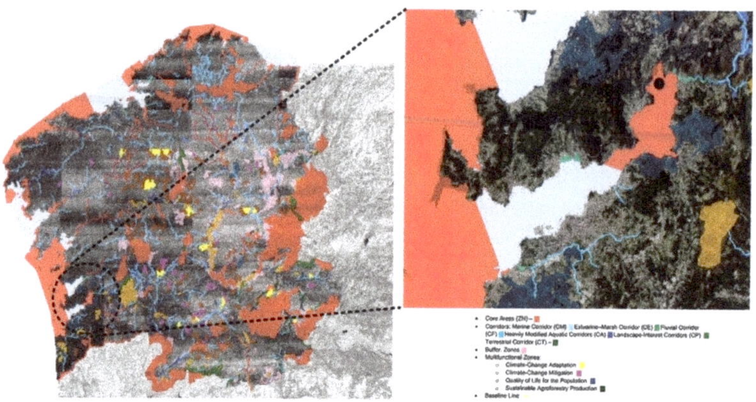

Figure 4. Territorial Elements and Configuration of the Green Infrastructure of Galicia (Regional Level).

4.2. Implementation and Prioritisation of Green Infrastructure in Galicia

The Galician Green Infrastructure Strategy is designed to be implemented at multiple scales, prioritising planning in major urban and peri-urban systems, where enhancing connectivity and sustainability objectives is essential. To support this process, the Scientific-Technical Bases for the Establishment of Green Infrastructure in Galicia [58] provide key information for configuring local-scale green infrastructure.

At a subregional level, the strategy focuses on multi-municipal areas and identifies Biosphere Reserves as an appropriate model for implementation, as they already incorporate zoning systems and participatory governance mechanisms [48]. In addition, the

 https://doi.org/10.3390/land15030358

inclusion of neighbouring municipalities with similar environmental and socio-economic characteristics is proposed in order to ensure a more integrated and effective restoration and connectivity strategy across Galicia.

4.3. Priority Areas

As part of the Galician Green Infrastructure Strategy, specific Priority Areas for Ecological Restoration (PARE) have been identified, focusing on the recovery of degraded habitats and the mitigation of invasive species. According to the Scientific-Technical Bases [58], the conservation status of protected species and habitats of Community interest in the Atlantic biogeographical region is predominantly unfavourable, highlighting the urgency of restoration measures.

At the regional level, 293 locations have been identified as priority restoration areas, grouped into three main categories:

1. Coastal areas
2. Fluvial corridors
3. Interior valleys and mountain systems

These areas were selected through scientific studies and public-participation processes and require detailed intervention plans, including habitat restoration, hydro-ecological recovery, public-use impact management and control of invasive species.

Within this ecological restoration framework, the Ulló Saltworks (*Salinas de Ulló*) have been identified as a Priority Area for Ecological Restoration (PO-031) in the Galician Green Infrastructure Strategy. This classification entails the implementation of specific measures for restoring its structure, composition, and ecological functionality. Restoration interventions must focus on terrestrial habitat recovery, hydro-ecological restoration, public-use impact control, and invasive species management, ensuring improved ecological connectivity and the regeneration of degraded natural processes, in accordance with management instruments and existing sectoral regulations [58].

Table 3 presents the officially designated Priority Areas for Ecological Restoration (PARE) established within the Galician Green Infrastructure Strategy. Each site is assigned a restoration code (01–05), which identifies the specific type of ecological intervention required: 01—Terrestrial habitat restoration, 02—Hydro-ecological restoration, 03—Control of invasive species, 04—Public-use impact management, and 05—Measures to reinforce ecological connectivity. These categories guide the design of restoration plans according to the ecological condition and restoration needs of each area.

Table 3. Priority Areas for Ecological Restoration (PARE) in Galicia.

Ecological Restoration Priority Areas	Code	1	2	3	4	5
Tourón Mine	OU_015				•	
Carballeda de Avia Mine	OU_016				•	
Pena Corneira Mines	OU_017				•	
Río Limia	OU_018	•	•	•	•	•
Pena Trevinca SAC	OU_019	•				
Rubía Mine	OU_020				•	
Xagoaza Mine	OU_021				•	
O Barco Industrial Area	OU_022	•		•		•
San Vicente Mine	OU_023					•
Regueiro Teixeira Mine	OU_024					•

https://doi.org/10.3390/land15030358

Table 3. *Cont.*

Ecological Restoration Priority Areas	Code	1	2	3	4	5
Penouta Mine	OU_025					•
Praia do Bao	PO_001		•	•	•	•
Punta Carreirón	PO_002	•	•	•		•
Praia Area da Secada	PO_003			•		•
Illote Areoso	PO_004					•
Umia Estuary	PO_005		•		•	
Illa da Toxa	PO_006	•	•	•	•	•
A Lanzada	PO_007	•		•	•	•
O Grove	PO_008	•		•	•	•
Sanxenxo Coast	PO_009	•	•	•	•	•
Cabo Udra SAC	PO_010			•		•
Praia Areabrava	PO_011			•	•	
Costa da Vela SAC	PO_012	•		•	•	•
A Ramallosa SAC	PO_013	•	•	•	•	•
Cabo Silleiro	PO_014			•	•	•
Baixo Miño SAC	PO_015	•	•	•	•	•
Cortegada Island	PO_016	•			•	
Ons Archipelago	PO_017		•	•	•	
Cíes Archipelago	PO_018	•	•	•	•	•
AG-57 Highway	PO_019			•	•	
Monte Alba	PO_020	•		•	•	•
Monte Aloia Natural Park	PO_021	•	•	•	•	•
Ulla-Deza River System SAC	PO_022	•	•	•	•	•
Breañas de Xestoso SAC	PO_023	•		•		•
Xunqueira de Alba	PO_024	•	•	•	•	•
Río Lérez SAC	PO_025	•	•	•	•	
Gándaras de Budiño SAC	PO_026	•	•	•	•	
Monte Castrove	PO_027			•		•
Torres do Oeste	PO_028					•
Monte Xiabre	PO_029	•			•	•
Río Verdugo	PO_030	•			•	
Salinas de Ulló	**PO_031**		•		•	
Serra do Galiñeiro	PO_032	•	•	•	•	•
Serra da Groba	PO_033	•	•	•	•	•
Porriño Quarries	PO_034		•	•	•	
Salvaterra de Miño Industrial Area	PO_035	•	•	•	•	
Río Tea	PO_036			•	•	
Río Louro	PO_037	•		•	•	
Río Miño 06	PO_038	•			•	

https://doi.org/10.3390/land15030358

Table 3. *Cont.*

Ecological Restoration Priority Areas	Code	1	2	3	4	5
Serra do Suído	PO_039		•	•		
Outeiro da Porca Mine	PO_040				•	
Meis Mines	PO_041			•		
Montes de Xinzo Mines	PO_042				•	
Prado Mine	PO_043			•		

5. The Legal Framework of the Ulló Saltworks

The Ulló Saltworks complex, including the saltworks, the tide mil and the farm, as listed in the Catalogue of the Plan Básico Autonómico de Galicia (PBA), is subject to a dual system of environmental and cultural protection. The PBA, approved by Decree 83/2018, constitutes the mandatory baseline spatial-planning instrument for the entire territory of Galicia and establishes uniform territorial criteria, including a binding Catalogue of Cultural Heritage applicable to all municipal and sectoral planning instruments.

Environmentally, the site lies within the Special Area of Conservation (SAC) "Enseada de San Simón" (ES1140016), designated under the EU Habitats Directive and formally approved by Decree 37/2014 of the Xunta de Galicia [60], which also sets out the management provisions for the Galician Natura 2000 Network. As part of this SAC, the saltworks are embedded in a protected estuarine landscape composed of intertidal habitats, saltmarsh communities and significant birdlife, falling under the legal regime of the Galician Natural Heritage and Biodiversity Act [55]. This framework requires strict conservation of ecological values, restricts land-use transformations and mandates prior environmental authorisation for any intervention.

From a cultural-heritage perspective, both the tidal mill and the salt-production structures are included in the PBA Catalogue of Cultural Heritage. This catalogue operates within the framework of the Galician Cultural Heritage Act (Law 5/2016) [61], which establishes binding protection categories to be applied in spatial planning at all scales. Although the Ulló Saltworks ensemble has not been formally declared a Bien de Interés Cultural (BIC), its inclusion in the PBA Catalogue confers immediate legal protection [62]. The 2016 Act explicitly recognises traditional hydraulic infrastructures, including tidal mills, and ethnographic production systems as protected heritage assets, providing a solid legal basis for safeguarding the site.

6. Revising the Legal Framework of the Ulló Saltworks

The current legal framework governing the Ulló Saltworks reveals a pronounced imbalance between the breadth of environmental protection and the fragmentation of cultural-heritage safeguards. As outlined above, the site benefits from robust ecological protection through its designation as the Natura 2000 Special Area of Conservation "Enseada de San Simón" [60] and its identification as a Priority Area for Ecological Restoration (PO-031) within the Galician Green Infrastructure Strategy, which mandates habitat recovery, hydro-ecological restoration and connectivity enhancement. In contrast, cultural protection remains limited to the individual listing of the tide mil (*Muíño de Mareas das Salinas de Ulló*), the farm (*Granxa*) and the saltworks of Ulló (*Salinas de Ulló*) and associated structures in the Catalogue of the Plan Básico Autonómico [62], without recognising the saltworks as a unified industrial ensemble or cultural landscape.

This disjunction contrasts with the conceptual and legal instruments already available in the Galician Cultural Heritage Act [61], which provides broader protection categories—

https://doi.org/10.3390/land15030358

historic ensemble, historic site and cultural landscape—intended to safeguard coherent territorial systems shaped by sustained interactions between human activity and the environment. While individual listing may be adequate for discrete monuments, it is insufficient for a pre-industrial salt landscape whose material, functional and symbolic values emerge precisely from the interdependence of ponds, embankments, hydraulic infrastructures, agricultural outbuildings and the estuarine environment. Treating these components as isolated artefacts dilutes their industrial unity and weakens the basis for integrated management [63,64].

A comparative analysis of other Spanish saltworks with higher levels of protection reinforces this conclusion. The Valle Salado de Añana (Álava) has been designated as a Bien Cultural in the category of Conjunto Monumental—a heritage designation protecting an integrated architectural or industrial ensemble—and is explicitly recognised as a cultural landscape in the Inventory of Basque Cultural Heritage [65,66]. This designation covers both the historic core and associated archaeological zones under a single protection perimeter and regulatory framework. Similarly, the *Conjunto Las Salinas* in *Arcos de las Salinas* (Teruel) has been declared a Bien de Interés Cultural in the category of Lugar de Interés Etnográfico, acknowledging the indivisible character of its hydraulic, architectural and productive system [67]. The *Salinas de Arinaga* (Gran Canaria) is likewise protected as a Bien de Interés Cultural, Spain's highest heritage-protection category, within a broader coastal salting complex where geomorphological and climatic conditions are considered inherent to heritage value [68]. Across these cases, saltworks are not reduced to isolated elements but are legally framed as unitary industrial landscapes in which heritage, landscape structure and ecological functions are jointly regulated.

Spanish experience also demonstrates that ensemble-based heritage protection can coexist with, and indeed be reinforced by, environmental designations. The *Ses Salines d'Eivissa i Formentera* Natural Park, for example, is managed as a protected area of exceptional ecological, landscape, historical and cultural value, where conservation policies address biodiversity, traditional salt-production infrastructures and associated cultural practices in an integrated manner [69,70]. This approach is highly relevant to Ulló, where Natura 2000 status provides ecological protection but does not articulate a cultural-landscape strategy capable of guiding active conservation, adaptive reuse and public interpretation.

In the specific case of the Ulló Saltworks, the absence of ensemble or cultural-landscape designation has at least three critical implications. First, it prevents recognition of the site as a coherent industrial system, complicating the justification of interventions that must operate at the scale of the ensemble—such as rehabilitating the entire hydraulic circuit. Second, it restricts archaeological protection to small perimeters around individually listed items, despite documentary and cartographic evidence indicating significant subsurface and intertidal archaeological potential [71]. Third, it fails to address pressing climate-change risks, including sea-level rise and storm-surge impacts, as highlighted by the 2024 spring tides, which temporarily overtopped the dikes and recreated quasi-natural flooding conditions [59,72,73].

On this basis, legislative reform for the Ulló Saltworks should prioritise reclassifying the site from a fragmented set of protected elements to a single protected industrial cultural landscape as presented in Figure 5. Several lines of action can be proposed:

1. initiating a formal procedure under Lei 5/2016 to declare the Ulló Saltworks a historic ensemble or cultural landscape, encompassing the full hydraulic system, productive structures and estuarine context;
2. extending archaeological safeguards to a continuous terrestrial and intertidal reserve area aligned with documented historical and archaeological potential;

 https://doi.org/10.3390/land15030358

3. integrating cultural-heritage objectives explicitly into the management instruments of the SAC "Enseada de San Simón" and the PO-031 ecological-restoration area;
4. incorporating climate-adaptation requirements into binding regulations, including systematic tidal-impact monitoring, stability assessment of dikes under sea-level-rise scenarios and the development of reversible protective measures.

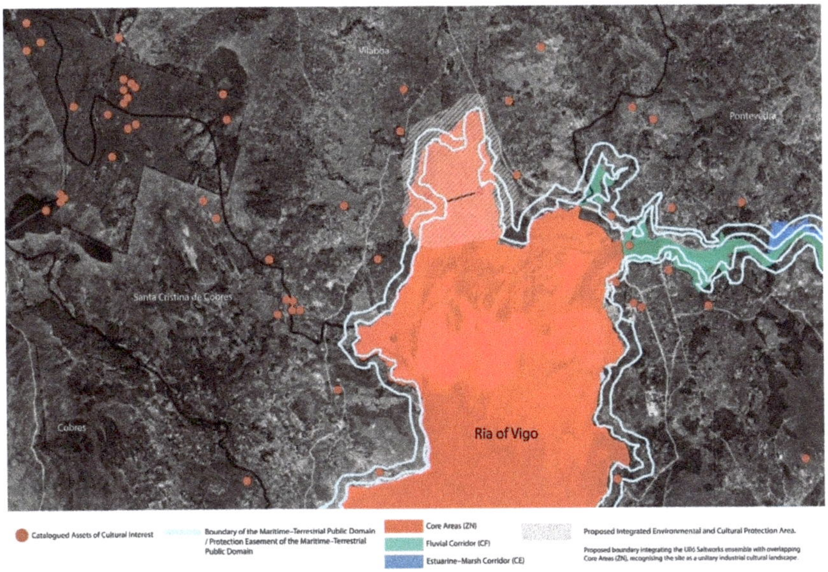

Figure 5. Representation of the proposed Integrated Environmental and Cultural Protection Area.

These proposals require no new legal categories; instead, they activate existing instruments in a more coherent and site-specific manner. Transitioning from isolated listings to a unified industrial cultural landscape would strengthen legal clarity, improve access to funding and enable more robust participatory governance, while laying the groundwork for broader discussions on regulatory innovation and landscape-scale heritage governance [74].

Within a broader European context, these challenges are embedded in current debates on cultural landscapes as living systems, in which heritage conservation, ecological sustainability, and social practices must be addressed in an integrated manner [75]. Recent research has highlighted the need to move beyond static protection frameworks and towards adaptive and participatory models capable of responding to environmental change and socio-economic transformations [76]. This perspective reinforces the relevance of landscape-governance approaches promoted at the European level [77,78].

7. Discussion

The Galician Rias, despite their well-recognised ecological and natural significance, remain insufficiently acknowledged as distinct cultural landscape assets [79]. This omission poses fundamental challenges for integrated heritage and environmental management, particularly in regions such as the Ulló Saltworks, where human intervention and natural dynamics intersect in complex ways. The lack of formal recognition as a cultural landscape limits the application of context-sensitive protective measures and risks reducing management to a reactive rather than proactive strategy.

One of the most pressing issues concerns the convergence of archaeological sensitivity and environmental vulnerability, increasingly exacerbated by climate change. Advances in digital documentation, both subaquatic and terrestrial, have enabled the timely record-

 https://doi.org/10.3390/land15030358

ing of information from features undergoing accelerated degradation [7]. Nevertheless, interventions aimed at artificially stabilising the coastline, primarily to safeguard built heritage and recently developed infrastructure, have, paradoxically, compromised older archaeological deposits, often in a fragile state [80]. This tension underscores a central paradox of heritage management in the region: the necessity of reconciling short-term preservation imperatives with the long-term conservation of irreplaceable historical and pre-industrial elements.

The regulatory framework further complicates this scenario. Instruments such as the Natura 2000 network, transposed into regional legislation, presuppose the existence of assets that fall under comparable categories across Europe. Yet, the mechanisms for safeguarding these assets vary substantially, depending on location and the administrative context [81]. Consequently, sites like the Ulló Saltworks are subject to legal inconsistencies that hinder the development of integrated and coherent management strategies.

Several interrelated constraints shape both the characterisation of the Ulló Saltworks and the effectiveness of the current regulatory framework. The site constitutes a highly heterogeneous landscape, where natural systems, historical infrastructures and pre-industrial productive elements coexist within tightly interwoven terrestrial, intertidal and subaquatic environments. This complexity complicates the definition of coherent spatial boundaries for research, protection and management, while accelerated environmental dynamics are progressively altering coastal morphology (Figure 4), ecological functioning and long-established landscape structures. At the same time, ongoing transformations in traditional agricultural systems, linked to soil salinisation and rising temperatures, are reshaping both land use and the ecological and cultural continuity of the area.

These spatial and environmental challenges are compounded by institutional and regulatory constraints. Fragmented property regimes, embedded in Galicia's cadastral system, hinder coordinated management and long-term stewardship, while existing heritage protection instruments tend to apply uniform categories to structurally diverse assets, reducing their capacity to address site-specific vulnerabilities. Moreover, insufficient coordination between heritage legislation, territorial planning tools and sectoral environmental policies limits the development of integrated conservation strategies. Together, these factors undermine proactive landscape governance and constrain the implementation of adaptive approaches capable of responding to the combined pressures of environmental change and cultural-heritage preservation.

While the present study does not claim to resolve these complex challenges fully, it systematically identifies critical gaps and inconsistencies in the regulatory and environmental frameworks affecting the Ulló saltworks. By highlighting these issues, it provides a foundation for context-sensitive management strategies that reconcile ecological, historical, and cultural imperatives [8]. Ultimately, the case of the Ulló Saltworks underscores the need for adaptive frameworks that are both legally coherent and ecologically informed, capable of supporting the sustainable evolution of culturally significant coastal landscapes in Galicia and, by extension, the wider Mediterranean region. Nevertheless, several limitations of the present study must be acknowledged. The analysis is based on a single case study, which limits the direct transferability of the results to other contexts. In addition, the legal and spatial assessment is not supported by detailed economic evaluations or quantitative modelling of future climate-change scenarios, which would be necessary to assess long-term management feasibility. Furthermore, while institutional and regulatory dimensions are examined in depth, social dynamics and stakeholder perspectives are only indirectly addressed. Future research should therefore integrate socio-economic analyses, participatory approaches and predictive environmental modelling to strengthen landscape-scale governance frameworks for coastal cultural heritage.

https://doi.org/10.3390/land15030358

8. Conclusions

This paper contributes to debates on coastal cultural landscape protection by demonstrating the limitations of fragmented legal frameworks when applied to complex pre-industrial systems. It integrates historical geography, coastal ecology, heritage cartography and heritage legislation, highlighting the structural disconnect between environmental protection instruments, such as Natura 2000, and cultural-heritage safeguarding. The study proposes a conceptual framework that recognises industrial cultural landscapes as functional territorial units aligned with European landscape-scale governance approaches.

The research emphasises the value of dynamic digital cartography based on multi-temporal geospatial data and interactive visualisation, which improves the interpretation of landscape change and its implications for heritage management. By combining historical sources with digital surveys, this approach supports more robust ecological and cultural assessments and facilitates adaptive management. The paper advances legally feasible measures through the coordinated use of existing instruments, offering a transferable model for the reclassification of historic saltworks and similar coastal systems.

However, as stated in the previous chapter, future research should expand comparative analyses, integrate hydro-sedimentary modelling and predictive cartography, and further examine multi-level governance, participation and fiscal tools for coastal cultural landscapes.

Natura 2000 is one of the legal protection instruments that has most significantly enhanced public awareness of the vulnerability of environmentally sensitive territories in the Iberian Peninsula, particularly coastal habitats and their associated river basins. However, its translation into regional legislation often proves limited, given the heterogeneity of the natural and cultural assets involved and the need to reconcile multiple territorial interests.

The Ulló Saltworks exemplify this complexity. The site, characterised by substantial archaeological potential and a distinctive pre-industrial technological heritage, reflects the asymmetric and discontinuous patterns of development typical of rural Galicia [1], where public infrastructural modernisation was both late and, at times, oversized [82]. In this context, the current regulatory categorisation, although designed to safeguard environmental values, may inadvertently constrain active management approaches and, consequently, accelerate the degradation of its cultural heritage.

At present, in terms of legal heritage protection, the Ulló Saltworks, or more specifically the part of the estuary in which they are located, is subject to the same regulatory framework as the historic centre of any town in the region or a religious built complex, regardless of their state of conservation.

The natural and urban specificity of the *Rías Baixas* demands a sensitive approach based on the integrated articulation of their occupation, exploitation and protection, as a number of authors have recently proposed [80]. The revision of the existing legal framework, in a progressive and structured manner, without compromising the general applicability of the underlying principles to analogous cultural assets, emerges as one of the most urgent challenges [83]. The rapid pace of social change and the technical implementation processes, highly dependent on political decision-making, have favoured a significant mismatch between inevitable landscape transformations and their regulatory mechanisms [25].

The results of this research shows that the Legislative reform for the Ulló Saltworks should prioritise their reclassification as a single protected industrial cultural landscape. Proposed actions include formal designation as a historic ensemble, the extension of continuous archaeological safeguards, and the explicit integration of heritage values into existing environmental management instruments. Climate-adaptation requirements, such as tidal monitoring and dike stability assessments, should also be incorporated. Without

https://doi.org/10.3390/land15030358

creating new legal frameworks, this approach enhances regulatory coherence, access to funding, and participatory governance.

The context of coastal heritage in the Rias (Iberian coastal lagoons), interweaving environmental valorisation with the built environment, and historical with pre-industrial heritage, constitutes an exceptional testing ground for a critical redefinition of landscape protection in Europe, as well as a means of supporting territorial evolution more attuned to its cultural identity [84].

Author Contributions: Conceptualization, G.D.C., A.L. and J.P.; methodology, G.D.C., A.L. and J.P.; software, J.P.; validation, G.D.C., A.L. and J.P.; formal analysis, G.D.C., A.L. and J.P.; investigation, G.D.C., A.L. and J.P.; resources, G.D.C., A.L. and J.P.; data curation, G.D.C., A.L. and J.P.; writing—original draft preparation, G.D.C., A.L. and J.P.; writing—review and editing, G.D.C., A.L. and J.P.; visualization, G.D.C. and A.L.; supervision, G.D.C.; All authors have read and agreed to the published version of the manuscript.

Funding: This work is financed by national funds through FCT—Fundação para a Ciência e a Tecnologia, I.P., under the Strategic Project with the references UID/04008: Centro de Investigação em Arquitetura, Urbanismo e Design.

Data Availability Statement: The original contributions presented in this study are included in the article. Further inquiries can be directed to the corresponding author.

Acknowledgments: The authors acknowledge the Directorate of the DAMG, UPT.

Conflicts of Interest: The authors declare no conflicts of interest.

References

1. Dalda Escudero, J.L.; García Docampo, M.; González Harguindey, J. *Cidade Difusa en Galicia*; Xunta de Galicia–Consellería de Política Territorial, Obras Públicas e Transportes, Dirección Xeral de Urbanismo: Galicia, Spain, 2005.
2. Bouhier, A. *Galicia–Ensaio Xeográfico de Análise e Interpretación dun Vello Complexo Agrário*; Tomo I; Xunta de Galicia; Caixa Nova: Santiago de Compostela, Spain, 2001.
3. Santos Solla, X. Galicia. In *La Investigación de la Geografía del Turismo en las Comunidades Autónomas Españolas: Orígenes, Desarrollo y Perspectivas de una Disciplina en el Horizonte de la Geografía*; Fernández Tabales, A., García Hernández, M., Ivars Baidal, J.A., Eds.; Asociación de Geógrafos Españoles, Grupo de Geografía del Turismo, Ocio y Recreación: Madrid, Spain, 2010; ISBN 978-84-936555-8-7.
4. Duarte Carlos, G. O Legado Morfológico da Arquitectura Vernácula. Ph.D. Thesis, ETSA | UdC, A Coruña, Spain, 2014.
5. Nogueira Román, C. *Camilo Nogueira: Pensar Galicia, Pensar o Mundo*; Fundación Otero Pedrayo: Galicia, Spain, 2016; p. 110, ISBN 8460890511, 9788460890515.
6. Hølleland, H.; Skrede, J.; Holm, S. Long-Term Monitoring of Protected Cultural Heritage Environments in Norway: Development of Methods and First-Time Application. *Land* **2019**, *8*, 75.
7. Bykowa, E.; Dyachkova, I. Modeling the Size of Protection Zones of Cultural Heritage Sites Based on Factors of the Historical and Cultural Assessment of Lands. *Land* **2021**, *10*, 1201. [CrossRef]
8. Calvano, R. Cultural Heritage Recognition through Protection of Historical Value and Urban Regeneration: CSOA Forte Prenestino. *Land* **2024**, *13*, 453. [CrossRef]
9. Liu, H.; Zhao, B.; Liu, J.; Shen, Y. Land-Use Evolution and Driving Forces in Urban Fringe Archaeological Sites: A Case Study of the Western Han Imperial Mausoleums. *Land* **2025**, *14*, 1554. [CrossRef]
10. De Vilaboa, C. As Paisaxes do Sal do Ulló; Concello de Vilaboa: Vilaboa, España. 2023. Available online: https://vilaboa.gal/wp-content/uploads/2023/12/As-Paisaxes-do-sal-do-Ullo.pdf (accessed on 21 December 2025).
11. Fernández, A.F.; Abad, P.V.; Nóvoa, A.A.R. 3D photogrammetry as a tool for studying erosive processes at a Roman coastal site: The case of the Roman fish salting plant at Sobreira (Vigo, Spain). *Archaeol. Anthropol. Sci.* **2022**, *14*, 32. [CrossRef]
12. Costa García, J.M.; Fonte, J.; Gago, M. Hallazgos arqueológicos recientes para el estudio de la presencia militar romana en el oriente gallego. *Gallaecia* **2016**, *35*, 39–70. [CrossRef]
13. Piñeiro Álvarez, J. *Las Salinas de Ulló: Intervención Para la Revalorización de la Arquitectura y Paisaje Cultural de la sal en Galicia*; Universidade Portucalense: Porto, Portugal, 2023. Available online: https://repositorio.upt.pt/entities/publication/a2f06a90-7b79-483f-a303-314626fad2e2 (accessed on 14 December 2025).

https://doi.org/10.3390/land15030358

14. Salinas do Ulló. Timeline; Salinas do Ulló: Vilaboa, España. 2023. Available online: https://salinasdoullo.gal/en/time-line/ (accessed on 12 December 2025).
15. Instituto de Historia (CSIC). Concluye la Valorización de las Salinas del s. XVIII de Vilaboa. Nota de prensa. CSIC: Madrid, España, 2024. Available online: https://ih.csic.es/es/press-release/article/concluye-valorizacion-salinas-s-xviii-vilaboa (accessed on 24 November 2025).
16. Gong, Z.; Zhang, Z.; Zhou, J.; Zhou, J.; Wang, W. The Evolutionary Process and Mechanism of Cultural Landscapes: An Integrated Perspective of Landscape Ecology and Evolutionary Economic Geography. *Land* **2022**, *11*, 2062. [CrossRef]
17. Gil-Docampo, M.; Peña-Villasenín, S.; Bettencourt, A.M.S.; Ortiz-Sanz, J.; Peraleda-Vázquez, S. 3D Geometric Survey of Cultural Heritage by UAV in Inaccessible Coastal or Shallow Aquatic Environments. *Archaeol. Prospect.* **2023**, *30*, 299–316. [CrossRef]
18. Maxim, C.; Chasovschi, C.E. Cultural landscape changes in the built environment at World Heritage Sites: Lessons from Bukovina, Romania. *J. Hosp. Mark. Manag.* **2021**, *20*, 100583. [CrossRef]
19. Gfeller, A.E. Negotiating the meaning of global heritage: 'Cultural landscapes' in the UNESCO World Heritage Convention, 1972–1992. *J. Glob. Hist.* **2013**, *8*, 483–503. [CrossRef]
20. Zohar, M. GIScience and Historical Visual Sources: A Promising Look at Past Scenarios and Sceneries. *ISPRS Int. J. Geo-Inf.* **2022**, *11*, 286. [CrossRef]
21. Carlos, G.D.; Sousa, G.; Correia, M.; Alcindor, M.; Bermudez, T. Vernacular Architecture in Chazhashi Settlement, Upper Svaneti, Georgia. In *From Vernacular to World Heritage*; Dipasquale, L., Mecca, S., Correia, M., Eds.; Firenze University Press: Florence, Italy, 2021; pp. 194–209, ISBN 978-88-5518-293-5. [CrossRef]
22. Sousa, G.; Carlos, G.; Florentino, R.; Bermudez, T. Multidisciplinary Analysis of the Vernacular Settlements in Ushguli (Upper Svaneti, Georgia). *Int. Arch. Photogramm. Remote Sens. Spat. Inf. Sci.* **2020**, *XLIV-M-1-2020*, 189–193. [CrossRef]
23. Gobbi, S.; Ciolli, M.; La Porta, N.; Rocchini, D.; Tattoni, C.; Zatelli, P. New Tools for the Classification and Filtering of Historical Maps. *ISPRS Int. J. Geo-Inf.* **2019**, *8*, 455. [CrossRef]
24. Armas, D. *Livro das Fortalezas; Facsimile of Codex 159 of the Arquivo Nacional da Torre do Tombo, with Introduction and Commentary by João José Alves Dias*; Caleidoscópio: Casal de Cambra, Portugal, 2015; ISBN 978-989-658-260-9.
25. Koumoulidis, D.; Varvaris, I.; Hadjimitsis, D.; Gabriele, M.; Brumana, R.; Gitas, I.; Georgopoulos, N.; Abdollahnejad, A.; Gkounti, E.; Stavrakoudis, D.; et al. Profiling Land Use Planning: Legislative Structures in Five European Nations. *Land* **2025**, *14*, 1261. [CrossRef]
26. Taboada Leal, N. Descripción topográfico-histórica de la ciudad de Vigo. Imprenta de la Viuda e Hijos de Compañel. 1840. Available online: https://es.wikipedia.org/w/index.php?title=Archivo:Descripci%C3%B3n_topogr%C3%A1fico-hist%C3%B3rica_de_la_ciudad_de_Vigo,_Nicol%C3%A1s_Taboada_Leal,_1840.pdf&page=5 (accessed on 21 December 2025).
27. Abril Gómez, H.; Abril Hurtado, J.; Priego Dones, D.; Corretge, L.G. *Mapa geológico y Memoria de la Hoja n°0185 (Pontevedra). Mapa Geológico de España*; Instituto Geológico y Minero de España: Madrid, Spain, 1981.
28. López-Romero, E.; Bóveda-Fernández, M.J.; Güimil-Fariña, A.; Mañana-Borrazás, P.; Sanjurjo Sánchez, J.; Vázquez-Collazo, S.; Vilaseco Vázquez, X.I.; Arce Chamorro, C. The Potential of Analysing Prehistoric Human Occupation in the Western Rias of Galicia (Northwest Iberia): Methods and Prospects. In *Investigate the Shore, Sound out the Past: Methods and Practices of Maritime Prehistory/Explorer la côte, Sonder le Passé: Méthodes et Pratiques de la Préhistoire Maritime*; Marchand, G., Pailler, Y., Stéphan, P., Eds.; Société préhistorique française: Paris, France, 2023; pp. 91–106.
29. Concello de Pontevedra. Planeamiento Municipal. Plan Xeral de Ordenación Municipal (PXOM) de Vilaboa, Boletín Oficial de Pontevedra, 86, "5 de mayo de 2022", 2022. Available online: https://boppo.depo.gal/detalle/-/boppo/2022/05/05/2022029938 (accessed on 21 December 2025).
30. De Agüero, E.G.-G.; Fernández-Rodríguez, C.; Roselló-Izquierdo, E.; Llorente-Rodriguez, L.; Bejega-García, V.; Fuertes-Prieto, N.; Morales-Muñiz, A. Fish- and Shellmiddens from Galicia (Northwest Spain): Reflections upon a Neglected Coastal Cultural Heritage from the Iberian Peninsula. *Humanities* **2019**, *8*, 85. [CrossRef]
31. Casal Fernández, L. La villa romana de Toralla y la explotación de los recursos marinos en las Rías Baixas. In *Proceedings of the Actas de las IX Jornadas de Jóvenes en Investigación Arqueológica, Santander, Spain, 8–11 Jun 2016*; Agudo Pérez, L., Duarte Simöes, C., García-Escárzaga, A., Geiling, J.M., Higuero-Pliego, A., Núñez de la Fuente, S., Rodríguez Santos, F.J., Suárez Revilla, R., Eds.; Editorial Universidad Cantabria: Cantabria, Spain, 2018; pp. 299–306, ISBN 978-84-697-9482-1.
32. Criado Boado, F.; Parcero Oubiña, C.; Otero Vilariño, C.; Cabrejas Domínguez, E.; Rodríguez Paz, A. *Atlas Arqueolóxico da Paisaxe Galega*; Edicións Xerais de Galicia, S.A.: Vigo, Spain, 2016; ISBN 978-84-9121-048-1.
33. Lara Coira, M. *Pegadas da industrialización en Galicia; A Enseada de San Simón*; Ézaro e Buxa: A Coruña, Spain, 2016.
34. Román Losada, A. Sal, Sardiñas e Peirados. Achega ao Patrimonio Marítimo na Enseada de San Simón. Grafischapela, SL. 2008. Available online: https://portalcientifico.uvigo.gal/documentos/679bd123209eb455aa1fd096?lang=en (accessed on 14 December 2025).
35. Menéndez Pérez, E. La sal en Galicia: Salinas, curtidos y salazones. In *Las rutas de la sal*; Netbiblo: A Coruña, Spain, 2008; pp. 35–43.

36. Vila-Concejo, A.; Lastra, M.; de Castro, F.; Gómez-Pujol, L.; Menéndez, M. Estuarine Shoreline Processes in a Dynamic Low-Energy System, 2010. Available online: https://www.researchgate.net/publication/226695725_Estuarine_shoreline_processes_in_a_dynamic_low-energy_system (accessed on 21 December 2025).
37. González Crespán, C.M. A Estrada Miscelánea Histórica e Cultural, 2018; Volumen 21. pp. 95–125. Available online: http://www.dspace.aestrada.gal/jspui/bitstream/123456789/967/1/06%20-%20Vias%20romanas%20XX%20y%20XIX.pdf (accessed on 21 December 2025).
38. Currás Refojos, B.X.; Rodríguez-Fernández, A.; Cortegoso, M.; Almeida, A.; Costa, M.; Brochado, T.; Machado, J.; Brochado, P.; Rodríguez-Martínez, R.; Brochado de Almeida, C.A. The Roman Saltworks of the Atlantic Coast of Gallaecia: Traces and Evidence of a Large Sea Salt Production Complex. *Quat. Sci. Rev.* **2024**, *339*, 108832. [CrossRef]
39. Castiñeiras Guerra, M. *Vilaboa do Morrazo*; Artes Gráficas Portela, S.L.: Pontevedra, Spain, 1979.
40. *A Cerca*. Galicia Máxica. 2021. Available online: https://www.galiciamaxica.eu/galicia/pontevedra/comarca-de-pontevedra/vilaboa/acerca/ (accessed on 21 December 2025).
41. Reboreda, A. Salinas do Ulló. 2022. Available online: https://salinasdoullo.gal/historia-das-salinas/as-salinas-e-o-sal/ (accessed on 21 December 2025).
42. Reinoso Maset, U. Vigo Industrial. 2022. Available online: https://www.vigoindustrial.com/copia-de-salinas-de-ullo-f-008 (accessed on 21 December 2025).
43. Bas López, B. *Muíños de mares e de Vento en Galicia. Catalogación Arqueolóxica e Artística de Galicia do Museo de Pontevedra*; Fundación Pedro Barrié de la Maza: A Coruña, Spain, 1991; ISBN 978-84-87819-13-1.
44. Fernández-Rodríguez, E.; Iglesias-Portas, A.; Pazos-López, A.; González, T.F.J.; Cabaleiro-Amil, M.C. *Estudio de la Ensenada de San Simón*; Servizo de Publicacións da Excma, Deputación de Pontevedra: Pontevedra, Spain, 1997.
45. Sanz Nieto, T. El "Vuelo Americano", el Primer Registro Completo y Preciso de lo Que Fueron Nuestros Pueblos, 2014. Available online: https://www.asajacyl.com/vida-rural/el-vuelo-americano-el-primer-registro-completo-y-preciso-de-lo-que-fueron-nuestros-pueblos (accessed on 21 December 2025).
46. Novoa Rodríguez, X. *Proyecto de: Recuperación Ambiental de las Salinas y Marisma de Vilaboa. Término Municipal de Vilaboa (Pontevedra)*; Servicio Provincial de Costas en Pontevedra: Pontevedra, Spain, 2002.
47. Council of Europe. *European Landscape Convention*; Council of Europe: Florence, Italy, 2000.
48. UNESCO. *Biosphere Reserves: Learning sites for Sustainable Development*; UNESCO: Paris, France, 2017.
49. Gobierno de España. *Ley 16/1985, de 25 de Junio, del Patrimonio Histórico Español*; Boletín Oficial del Estado: Madrid, Spain, 1985.
50. Gobierno de España. *Ley 42/2007, de 13 de Diciembre, del Patrimonio Natural y de la Biodiversidad*; Boletín Oficial del Estado: Madrid, Spain, 2007.
51. Gobierno de España. *Ley 33/2015, de 21 de Septiembre, por la que se Modifica la Ley 42/2007 del Patrimonio Natural y de la Biodiversidad*; Boletín Oficial del Estado: Madrid, Spain, 2015.
52. Xunta de Galicia. *Lei 8/1995, do Patrimonio Cultural de Galicia*; Diario Oficial de Galicia: Santiago de Compostela, Spain, 1995.
53. Xunta de Galicia. *Lei 3/1996 de Protección dos Camiños de Santiago*; Diario Oficial de Galicia: Santiago de Compostela, Spain, 1996.
54. Xunta de Galicia. *Lei 7/2008, de Protección da Paisaxe de Galicia*; Diario Oficial de Galicia: Santiago de Compostela, Spain, 2008.
55. Xunta de Galicia. *Lei 5/2019, do Patrimonio Natural e da Biodiversidade de Galicia*; Diario Oficial de Galicia: Santiago de Compostela, Spain, 2019.
56. Fairclough, G.; Herring, P. *Lens of Landscape: Integrating Cultural and Natural Heritage*; Routledge: Milton Park, UK, 2016.
57. Roma, F. Paisaje cultural y gestión patrimonial: Nuevas perspectivas legislativas en España. *Rev. Patrim. Cult. España* **2010**, *3*, 45–60.
58. Xunta de Galicia. *Bases Científico-Técnicas Para o Establecemento da Infraestrutura Verde de Galicia*; Dirección Xeral de Patrimonio Natural: Santiago de Compostela, Spain, 2020.
59. Xunta de Galicia. *Lei 1/2021, de Ordenación do Territorio de Galicia*; Diario Oficial de Galicia: Santiago de Compostela, Spain, 2021.
60. Xunta de Galicia. *Decreto 37/2014, do 27 de Marzo, polo que se Declaran Zonas Especiais de Conservación os Lugares de importancia Comunitaria de Galicia e se Aproba o Plan Director da Rede Natura 2000 de Galicia*; Diario Oficial de Galicia: Santiago de Compostela, Spain, 2014.
61. Xunta de Galicia. *Lei 5/2016, do 4 de maio, do Patrimonio Cultural de Galicia*; Diario Oficial de Galicia: Santiago de Compostela, Spain, 2016.
62. Xunta de Galicia. *Decreto 83/2018, do 26 de Xullo, Polo que se Aproba o Plan Básico Autonómico de Galicia (PBA)*; Diario Oficial de Galicia: Santiago de Compostela, Spain, 2018.
63. González-Álvarez, D. Cultural landscapes and industrial heritage: Legal challenges in Spain. *Int. J. Herit. Stud.* **2020**, *26*, 923–940.
64. López, J.; Pérez-Ruiz, C.; Suárez, A. Managing coastal industrial heritage landscapes: A comparative analysis of Spanish saltworks. *Herit. Soc.* **2019**, *12*, 87–106.
65. Gobierno Vasco. *Inventario General del Patrimonio Cultural Vasco: Valle Salado de Añana*; Gobierno Vasco: Vitoria-Gasteiz, Spain, 2010.

66. Fernández, M.; Uriarte, A. Industrial salt landscapes and heritage management in the Basque Country. *J. Cult. Herit.* **2017**, *28*, 115–126.
67. Gobierno de Aragón. *Decreto por el que se Declara Bien de Interés Cultural el Conjunto Las Salinas de Arcos de las Salinas*; Boletín Oficial de Aragón: Zaragoza, Spain, 2006.
68. Cabildo de Gran Canaria. *Declaración de Bien de Interés Cultural: Salinas de Arinaga*; Boletín Oficial de Canarias: Canary Islands, Spain, 2011.
69. Govern Balear. *Pla de Gestió del Parc Natural de ses Salines d'Eivissa i Formentera*; Conselleria de Medi Ambient: Valencia, Spain, 2015.
70. Rodríguez-Becerra, M.; Muntaner, J. Managing salt-making cultural landscapes in the Mediterranean: The case of Ses Salines. *Landsc. Res.* **2018**, *43*, 633–650.
71. Lamas, R. Arqueología del paisaje salinero en Galicia: Evolución histórica y potencial patrimonial. *Estud. Do Patrim.* **2022**, *14*, 45–67.
72. IPCC. *Climate Change 2021: The Physical Science Basis*; Cambridge University Press: Cambridge, UK, 2021.
73. Xunta de Galicia. *Estratexia Galega de Infraestrutura Verde e da Conectividade e Restauración Ecolóxica*; Xunta de Galicia: Santiago de Compostela, Spain, 2019.
74. Fairclough, G. Cultural landscape, sustainability and living heritage. *Landsc. Res.* **2018**, *43*, 618–630.
75. Bandarin, F.; van Oers, R. *Reconnecting the City: The Historic Urban Landscape Approach and the Future of Urban Heritage*; Wiley-Blackwell: Hoboken, NJ, USA, 2015.
76. Stephenson, J. The cultural values model: An integrated approach to values in landscapes. *Landsc. Urban Plan.* **2008**, *84*, 127–139. [CrossRef]
77. Council of Europe. *European Landscape Convention: Landscape and Sustainable Development*; Council of Europe: Strasbourg, France, 2019.
78. Jones, M.; Stenseke, M. *The European Landscape Convention: Challenges of Participation*; Springer: Berlin/Heidelberg, Germany, 2011.
79. Castro Carrera, J.C. Architecture, construction system and functioning of the Roman saltworks of O Areal (Vigo, Galicia, Spain). A reference in the investigation of salt production in the Roman world. *Quat. Sci. Rev.* **2024**, *339*, 108797. [CrossRef]
80. Nieto, C.E.; Martínez-Graña, A.M.; Merchán, L. Landscape Analysis and Coastal Planning: Ría de Arosa (Pontevedra, Spain). *Land* **2024**, *13*, 645. [CrossRef]
81. Wu, J. Landscape of culture and culture of landscape: Does landscape ecology need culture? *Landsc. Ecol.* **2010**, *25*, 1147–1150. [CrossRef]
82. Domingues, Á. *Vida no Campo*, 1st ed.; Dafne: Porto, Portugal, 2011.
83. Pardellas de Blas, X.; Santos Solla, X. Turismo e lecer. In *Unha Estratexia Marítima Para Galicia*; Varela Lafuente, M.M., Ed.; Galaxia: Galicia, Spain, 2010; pp. 79–94, ISBN 978-84-9865-341-0.
84. Gómez-Sal, A.; Bonet, A.; Saldaña-López, A.; Muñoz-Rojas, J. Landscape ecology and landscape approaches in the Iberian context: Challenges, opportunities, and future prospects. *Landsc. Ecol.* **2024**, *39*, 153. [CrossRef]

https://doi.org/10.3390/land15030358

MDPI AG
Grosspeteranlage 5
4052 Basel
Switzerland
Tel.: +41 61 683 77 34

Land Editorial Office
E-mail: land@mdpi.com
www.mdpi.com/journal/land

www.ingramcontent.com/pod-product-compliance
Lightning Source LLC
Chambersburg PA
CBHW061617220926
51850CB00005B/529
* 9 7 8 3 7 2 5 8 7 3 6 6 1 *